MS324 Waves, diffusion and variational principles

Block I

Waves

This publication forms part of an Open University course. Details of this and other Open University courses can be obtained from the Student Registration and Enquiry Service, The Open University, PO Box 197, Milton Keynes MK7 6BJ, United Kingdom (tel. +44 (0)845 300 6090, email general-enquiries@open.ac.uk).

Alternatively, you may visit the Open University website at www.open.ac.uk where you can learn more about the wide range of courses and packs offered at all levels by The Open University.

To purchase a selection of Open University course materials visit www.ouw.co.uk, or contact Open University Worldwide, Walton Hall, Milton Keynes MK7 6AA, United Kingdom, for a brochure (tel. +44 (0)1908 858793, fax +44 (0)1908 858787, email ouw-customer-services@open.ac.uk).

The Open University, Walton Hall, Milton Keynes, MK7 6AA.

First published 2005. Second edition 2006. Third edition 2010.

Copyright © 2005, 2006, 2010 The Open University

All rights reserved. No part of this publication may be reproduced, stored in a retrieval system, transmitted or utilised in any form or by any means, electronic, mechanical, photocopying, recording or otherwise, without written permission from the publisher or a licence from the Copyright Licensing Agency Ltd. Details of such licences (for reprographic reproduction) may be obtained from the Copyright Licensing Agency Ltd, Saffron House, 6–10 Kirby Street, London EC1N 8TS (website www.cla.co.uk).

Open University course materials may also be made available in electronic formats for use by students of the University. All rights, including copyright and related rights and database rights, in electronic course materials and their contents are owned by or licensed to The Open University, or otherwise used by The Open University as permitted by applicable law.

In using electronic course materials and their contents you agree that your use will be solely for the purposes of following an Open University course of study or otherwise as licensed by The Open University or its assigns.

Except as permitted above you undertake not to copy, store in any medium (including electronic storage or use in a website), distribute, transmit or retransmit, broadcast, modify or show in public such electronic materials in whole or in part without the prior written consent of The Open University or in accordance with the Copyright, Designs and Patents Act 1988.

Edited, designed and typeset by The Open University, using the Open University TeX System.

Printed in the United Kingdom by Cambrian Printer Limited, Aberystwyth.

The paper used in this publication is procured from forests independently certified to the level of Forest Stewardship Council (FSC) principles and criteria. Chain of custody certification allows the tracing of this paper back to specific forest-management units (see www.fsc.org).

ISBN 978 0 7492 5167 3

Contents

CHAPTER 1   THE WAVE EQUATION 7

1.1   Introduction to wave phenomena 7
    1.1.1   Some properties of wave motion 9
    1.1.2   Content and organisation of this block 11
1.2   Springs, rods and elastic strings 12
    1.2.1   Elasticity 12
    1.2.2   Hooke's law 15
    1.2.3   Tension 16
    1.2.4   Density 18
1.3   Transverse waves on a stretched string 19
    1.3.1   Derivation of the wave equation 19
    1.3.2   Some solutions of the wave equation 22
1.4   Non-uniformly deformed springs 24
    1.4.1   Tension 26
    1.4.2   Linear density 26
1.5   Longitudinal waves 28
    1.5.1   Derivation of the wave equation 29
    1.5.2   Longitudinal waves in a stretched spring 31
    1.5.3   Transverse and longitudinal waves in a stretched string 34
1.6   The wave equation 35
    1.6.1   Scope of the wave equation 35
    1.6.2   A few remarks on the history of the wave equation 36
1.7   Summary and Outcomes 36
1.8   Further Exercises 37
1.9   Appendix: Sound waves in a tube 40
    1.9.1   Volume and pressure 40
    1.9.2   Derivation of the wave equation 42
Solutions to Exercises in Chapter 1 45

CHAPTER 2   SOLUTIONS OF THE WAVE EQUATION 59

2.1   Introduction 59
    2.1.1   Boundary and initial conditions 59
    2.1.2   Content of this chapter 60
2.2   Harmonic wave motion 60
2.3   General solution for the infinite string 63
    2.3.1   D'Alembert's solution 64
    2.3.2   Initial conditions 66
2.4   General solution for the semi-infinite string 68
    2.4.1   A left-moving pulse 69
    2.4.2   A right-moving pulse 71
    2.4.3   General solution 72

2.5	The finite string: solution by multiple reflection		73
	2.5.1 Multiple reflections of a left-moving pulse		73
	2.5.2 Multiple reflections of a right-moving pulse		75
	2.5.3 General solution		75
2.6	Separation of variables		76
	2.6.1 Solution of the position-dependent part		78
	2.6.2 Solution of the time-dependent part		80
	2.6.3 Normal mode solutions of the boundary-value problem		80
	2.6.4 Superposition of normal modes and Fourier series		81
	2.6.5 Semi-infinite strings and Fourier sine integrals		84
	2.6.6 Infinite strings and Fourier integrals		85
2.7	Summary and Outcomes		86
2.8	Further Exercises		87
	Solutions to Exercises in Chapter 2		89

CHAPTER 3 FOURIER SERIES AND FOURIER TRANSFORMS 103

3.1	Introduction	103
3.2	Fourier series	104
	3.2.1 The trigonometric Fourier series	104
	3.2.2 The complex exponential Fourier series	106
	3.2.3 Calculating Fourier coefficients	107
3.3	Fourier transforms	112
	3.3.1 Introduction and definition	112
	3.3.2 Some examples of Fourier transforms	114
	3.3.3 Derivation of the Fourier transform	115
3.4	Some properties of Fourier transforms	117
3.5	Convolutions and the convolution theorem	122
	3.5.1 Blurred images	122
	3.5.2 Time-delay processes	123
	3.5.3 The convolution theorem	124
3.6	Fourier transforms of derivatives	127
3.7	Summary of Fourier transforms	129
3.8	Outcomes	130
3.9	Further Exercises	130
3.10	Appendix: Multi-dimensional Fourier transforms	131
	Solutions to Exercises in Chapter 3	133

CHAPTER 4   FOURIER METHODS IN ONE DIMENSION   143

4.1 Introduction   143
4.2 Initial-value problem for strings with fixed ends   144
4.3 Strings with moving ends   149
    4.3.1 A free end   149
    4.3.2 Damped and sprung ends   150
    4.3.3 Normal modes for strings with moving ends   152
    4.3.4 Inhomogeneous boundary conditions   157
4.4 Damped waves   160
    4.4.1 Separation of variables   161
    4.4.2 Solution of the damped wave equation   163
4.5 Inhomogeneous damped wave equation   165
    4.5.1 Harmonic driving   166
4.6 Summary and Outcomes   170
4.7 Further Exercises   171
Solutions to Exercises in Chapter 4   173

CHAPTER 5   WAVES IN TWO AND THREE DIMENSIONS   185

5.1 Introduction   185
5.2 Vibrating membranes   185
    5.2.1 Free transverse vibrations of a rectangular membrane   186
    5.2.2 A triangular membrane   194
    5.2.3 Non-separable boundaries   195
    5.2.4 Circular membranes   196
    5.2.5 Bessel functions   198
    5.2.6 Normal mode solutions for circular membranes   203
    5.2.7 Can one 'hear' the shape of a drum?   205
5.3 Isotropic waves in two and three dimensions   206
    5.3.1 Generalisation of d'Alembert's solution   207
    5.3.2 The Fourier transform of the wave equation   209
    5.3.3 Isotropic waves in two dimensions   213
    5.3.4 Isotropic waves in three dimensions   215
5.4 Summary and Outcomes   218
5.5 Further Exercises   219
5.6 Appendix: Power series for Bessel functions   220
Solutions to Exercises in Chapter 5   221

CHAPTER 6   WAVE PROPAGATION — 231

6.1	Introduction	231
6.2	The Doppler effect	232
6.3	Reflection and transmission	234
	6.3.1 Reflection and transmission at interfaces	235
	6.3.2 Reflection and transmission in two dimensions	238
	6.3.3 Total reflection	244
6.4	Dispersion	245
	6.4.1 Dispersion relation	246
	6.4.2 Phase and group velocity	247
	6.4.3 A discrete system of coupled oscillators	248
6.5	Summary and Outcomes	251
6.6	Further Exercises	251
Solutions to Exercises in Chapter 6		255

INDEX — 263

CHAPTER 1
The wave equation

1.1 Introduction to wave phenomena

Wave phenomena are abundant in the physical world. There is at least one instance of wave motion with which everyone is familiar – *water waves*. For example, when throwing a pebble into a pond, you can observe waves moving away from the point of impact in a circular pattern (see Figure 1.1).

Figure 1.1   A stone thrown into a pond creates a circular ripple pattern

Photograph courtesy of Tini Garske and Kit Logan

The wind blowing over a puddle or small pond creates patterns of ripples. On a different scale, the wind may cause huge *ocean waves* that reveal their power when hitting the coastline. Besides the wind, underwater earthquakes, landslides or meteorite impacts may cause particularly destructive ocean waves which are known as tsunamis; fortunately, you will probably have encountered such an event only in one of the numerous motion pictures about natural disasters.

Water waves shape our perception of the notion of waves, but there are many other instances where waves are encountered in everyday life. For instance, a plucked guitar string can be seen to vibrate: this is an example of *mechanical wave motion*. You may also have heard about *sound waves*, *radio waves* or *light waves*: for these, you cannot observe the wave motion directly. However, you may sense indirectly the vibration of the membrane in a loudspeaker that produces low bass tones, or feel the tremor that they cause in your stomach. The beautiful iridescent colours seen on a thin film of oil on water, or on the wings of a butterfly, are consequences of the wave character of visible light.

You may be astonished to realise to what extent our daily lives are governed by waves. Our interaction with the environment and communication is almost entirely based on waves. This applies to the sound that our ears can hear, and to the light that our eyes can see. Light waves are a particular kind of *electromagnetic waves*, an exceedingly important class of waves. These form the basis of the electronic transmission of signals, used for radio and television programmes or telephone calls. Electromagnetic waves also encompass the harmful ultraviolet light emitted by the sun, X-rays used in medical examinations, and part of the radiation emitted by radioactive material. Finally, a microwave oven does not carry its name by chance; once more, electromagnetic waves do the job.

However diverse the physical systems and the wave phenomena may appear, on a more abstract level, the basic features of wave motion are the same in all these systems. Essentially, a wave can be imagined as a propagating displacement in a medium. This displacement at position x in space and at time t is described by a function $u(x,t)$ (see Figure 1.2).

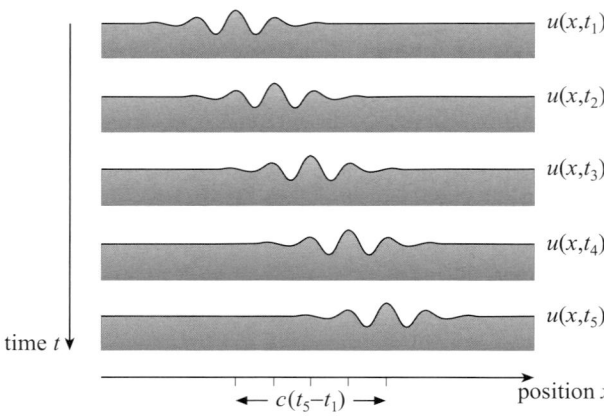

Figure 1.2 Schematic view of a typical wave pattern moving in the x direction. The function $u(x,t)$ describes the displacement at position x and at time t. The figure shows snapshots $u(x,t_j)$ of the displacement, taken at regular time intervals Δt for five consecutive equally-spaced instants t_j, with $j = 1, 2, 3, 4, 5$. The marks along the x-axis denote the positions of the maximum displacement at time t_j. In this example, the wave pattern moves to the right with a constant speed c, moving by a distance $c\Delta t$ between each snapshot.

The mathematical formula describing basic wave motion is the wave equation,

One-dimensional linear wave equation
$$u_{tt}(x,t) = c^2 \, u_{xx}(x,t). \tag{1.1}$$

Here, $u_{tt} = \partial^2 u/\partial t^2$ and $u_{xx} = \partial^2 u/\partial x^2$ denote the second partial derivatives of the function u with respect to t and x, respectively. Thus the wave equation is a second-order partial differential equation, which can alternatively be written as

$$\frac{\partial^2 u}{\partial t^2}(x,t) = c^2 \, \frac{\partial^2 u}{\partial x^2}(x,t). \tag{1.2}$$

Both notations for partial derivatives are conventional and will be used in what follows.

Some equations (like this one) are distinguished by a title, either because they are of particular importance, or because they are frequently referred to by this title.

The wave equation contains a constant c. Distinctions between different types of waves arise due to different values of this constant and the physical meaning of the quantities involved. The positive constant c is the *speed* of the wave motion.

We shall see in the next chapter that this name is justified.

The wave equation governs the behaviour of many physical systems, which is one of the reasons why wave motion comprises a major part of this course. However, the wave equation also marks a very important step in the historical development of mathematics. Together with the diffusion equation, which is considered later in this course, it was among the first partial differential equations studied in detail, and its investigation stimulated the development of the foundation of modern analysis.

Before considering the physics and mathematics of wave motion in detail, we discuss some typical properties and characteristics of wave motion on the basis of our everyday experience with waves.

1.1.1 Some properties of wave motion

What are waves? A precise definition of a wave is difficult to formulate, though we may have an intuitive perception of waves. For this reason we tend to characterise waves by their properties, so our discussion starts by considering some examples.

Throwing a pebble into a pond produces a circular pattern of surface waves moving outwards from the point of impact. What is happening here?

The pebble, when hitting the surface of the water, transfers part of its *kinetic energy* to the water. This initiates, among other effects, the observed surface ripples. The ripple pattern stores some of the energy released by the pebble, and in fact transports it away on its outward motion. Thus, very much like a moving particle, waves carry *energy* through space. The warmth of sunlight on your skin is another instance where you can immediately sense the energy transported by waves.

For surface waves on water generated by a pebble, although it may look as if water is moving outwards with the ripples, this is not, in fact, the case. There is no net flow of water away from the point of impact – otherwise there would be a permanent dip left in the surface, which clearly does not happen. The motion of the ripples is just a fluctuation of the water surface: it is this fluctuation that travels, not the water itself. A 'Mexican wave' or 'La Ola' created by the spectators at a football match is a good illustration of this: the wave appears to move around the stadium, but the individuals just move up and down.

The ripples caused by a pebble in a pond move outwards, and they appear to move at constant speed. The speed of different types of waves can vary enormously. Light waves, for example, travel so swiftly that our senses cannot perceive their finite speed, about 300 000 kilometres per second. Sound waves travel very much slower than light: in air, sound takes about 3 seconds to travel 1 kilometre. You can appreciate this from the time difference between lightning and thunder in a distant thunderstorm. Knowing the speed of sound in air (which is about $343 \, \text{m s}^{-1}$, roughly a thousand feet per second, in normal conditions), the time difference tells you your distance from the lightning. Water waves are again much slower, so you can observe the motion directly.

What happens if two or more waves overlap? If you throw two pebbles into a pond, both will generate circular wave patterns (see Figure 1.3).

Figure 1.3 Throwing two stones into a pond results in intersecting ripple patterns Photograph courtesy of Tini Garske and Kit Logan

When the two ripple patterns meet, the resulting pattern may appear rather complicated, as in Figure 1.3. But, in fact, the two wave patterns pass through each other, and they continue to move as independent ripple patterns after they leave the intersection zone. The same happens with light and sound waves. For instance, when several people in a room speak simultaneously, their voices may be mixed, but you can still recognise the voice of a person you know, provided that you can hear it. It does not sound different to how it would sound without the other voices present. Many signals, such as radio and television broadcasts or mobile telephone calls, are transmitted simultaneously without affecting each other.

Let us consider what this means for the mathematical description of wave motion, assuming for the moment that we do not know the wave equation. Let $u_1(x,t)$ and $u_2(x,t)$ describe two waves moving in space, for instance the sound waves (which are distortions in air pressure) produced by two people standing at opposite ends of a long tunnel, shouting at each other. Here t denotes time and x denotes position; for simplicity we consider only one spatial dimension. After the few moments that it takes the sound to traverse the tunnel, the people hear each other. The fact that the sound waves necessarily overlap in the middle of the tunnel, and that there is no distortion of the sound, means that the presence of one sound wave does not influence the other. The combined wave motion $u(x,t)$, the *superposition* of the two waves, is therefore the sum $u(x,t) = u_1(x,t) + u_2(x,t)$, so the two contributions just add up. It also should not matter how loudly the people shout, so multiplying a wave motion $u(x,t)$ by a constant factor, which corresponds to changing the loudness, still describes a possible motion.

Phrased more generally, we want our wave equation to have the property that *linear combinations*,

Superposition of waves

$$u(x,t) = a_1 \, u_1(x,t) + a_2 \, u_2(x,t), \qquad (1.3)$$

of any two solutions $u_1(x,t)$ and $u_2(x,t)$, with arbitrary constants a_1 and a_2, are again solutions. This implies that the wave equation must be *linear*, which means that the function $u(x,t)$ and its derivatives enter only linearly, not as higher powers such as $(u(x,t))^3$ or as arguments of functions such as in $\sin[u(x,t)]$. Moreover, the wave equation must also be *homogeneous*, which means that it does not contain a term that is independent of the function $u(x,t)$. In other words, all terms in the equation vanish for $u(x,t) = 0$, which thus is a solution. The one-dimensional wave equation (1.1) on page 8 is indeed linear and homogeneous. Waves described by this equation are called *linear waves*, and they are characterised by the superposition property. In reality, linear behaviour need not be precisely obeyed, though it is often a very good approximation. Thus we shall concentrate on linear waves in this course.

Exercise 1.1

If $u_1(x,t)$ and $u_2(x,t)$ satisfy the wave equation (1.1), show that the function $u(x,t)$ given by equation (1.3) is also a solution of the wave equation.

Waves may travel in certain media, such as air, solids and liquids. However, not all kinds of waves require a medium. Electromagnetic waves, such as light emitted from distant stars in the universe, reach Earth through the vast expanse of empty space. Astronauts in a spaceship on route to the Moon can communicate with Earth by radio, except when their vessel is behind the Moon, which is not transparent to radio waves. (But sound waves cannot travel in empty space, although this fact is ignored in many science fiction movies featuring spaceships.)

There are more properties that we could list. For instance, waves can be reflected by objects, just as a mirror reflects light and a wall reflects sound (which you may perceive as an echo). Wave motion is influenced by the surrounding medium, which affects the propagation speed. This becomes particularly apparent at interfaces between different media. A familiar example is the bending of light at a water surface: when you look at a straight stick partially submerged in water, the stick appears to be bent at the surface. The same process, known as *refraction*, is utilised in optical instruments containing lenses, like magnifying glasses, telescopes and cameras.

Waves are everywhere

Where do we encounter waves in nature? We have already mentioned water waves, sound waves and light waves, as well as mechanical waves on a string which, for instance, produce the sound of a guitar. Other examples of mechanical waves are *seismic waves* produced by earthquakes, and waves in quartz crystals used in modern watches.

According to the modern view of the physical world, in particular *quantum mechanics*, the seemingly distinct concepts of particles and waves are intimately related. This means that particles may also show wave-like behaviour, and vice versa, although this will be directly apparent only in the microscopic world of atoms and elementary particles, not in our everyday life experience. But, depending on the type of experiment performed, light may show 'particle-like' behaviour or 'wave-like' behaviour, and the same is true of atoms and elementary particles. Viewed from this perspective, wave motion is in fact a fundamental aspect of our description of nature.

1.1.2 Content and organisation of this block

In this part of the course, we shall develop the mathematical description of wave motion. However, the motivation to study wave motion originates in physics. Electromagnetic waves are arguably most relevant to everyday life, but the physics involved, the theory of electromagnetism, is rather complicated, so we concentrate on systems governed by *Newton's laws of mechanics*.

Isaac Newton
(1642–1727)

The remaining part of this chapter deals with deriving the wave equation and the values of the wave speed c for different physical systems. As our main examples, we consider transverse waves on a stretched string and longitudinal waves in springs, rods and stretched strings. In the transverse case, the distortion takes place in a direction perpendicular (transverse) to the equilibrium position of the string, whereas in the longitudinal case the distortion is along the axis of the spring, rod or string. We start by recapitulating some basic notions of elasticity and Newtonian mechanics, and then discuss in detail how the wave equation arises in these systems. In an optional appendix, we give the derivation for sound waves in a gas confined to a narrow tube.

In the next five chapters, we turn to mathematics, and discuss the solutions of the wave equation and their mathematical properties. Whereas we shall often allude to the underlying physical situation, the main emphasis lies on the mathematical aspects.

This starts in Chapter 2, which considers solutions of the one-dimensional wave equation. Considering the general form of solutions naturally leads to the mathematical tools of Fourier series and Fourier transforms, which are discussed in detail in the subsequent chapter. This is followed by two chapters where these tools are employed to derive solutions of the wave equation, first in one dimension, then in two and three dimensions. The final chapter of this block is devoted to some important aspects of wave propagation, and you will see how some familiar physical properties of waves can be understood from the mathematical analysis.

This block is self-contained in the sense that it does not assume that you have seen the wave equation before. If you have previously studied the second-level course MST209 (or MST207) or a similar course, you will already be familiar with some of the material contained in the first two chapters of this block, such as the transverse vibrations of a stretched string. While this will be the first example discussed in this course, wave motion in one and more dimensions will be considered in depth, with a detailed discussion of boundary and initial conditions. We present several methods of solving the wave equation, introducing and exemplifying mathematical techniques that have important applications in other areas of mathematics.

1.2 Springs, rods and elastic strings

Elasticity plays a vital role in the wave motion in the systems that we shall discuss. Therefore we start by recalling a few basic facts about the physics of elasticity, in particular with respect to *springs*, *rods* and *elastic strings*. Wave motion takes place in a similar fashion in all these systems.

In what follows, we shall concentrate on elastic forces, assuming that other forces, such as gravity, can be neglected.

1.2.1 Elasticity

Knowingly or not, you continually experience elasticity in everyday life. Whether walking on concrete, on wooden planks or on soft soil, your weight is compensated by elastic forces resulting from small deformations. When driving on a suspension bridge, you rely on the elastic force of the stretched cables. Every time you touch something solid, elastic forces are involved.

1.2 Springs, rods and elastic strings

The concept of elasticity links the deformation of the solid to the forces that act on it. As an example, consider a cylindrical rod of cross-sectional area A, which without an external force has a natural length l_0. Pulling at one end with a force $\boldsymbol{F}_2$, while keeping the other end fixed, extends the rod to a length $l > l_0$. To keep the other end fixed, the fixture needs to exert a force $\boldsymbol{F}_1 = -\boldsymbol{F}_2$; see Figure 1.4.

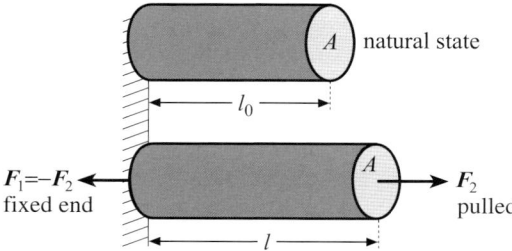

Figure 1.4 A rod in its natural state and extended by forces $\boldsymbol{F}_1$ and $\boldsymbol{F}_2 = -\boldsymbol{F}_1$

The relative extension $\varepsilon = (l - l_0)/l_0$ is the (longitudinal) *strain* in a solid, and the force per area $\sigma = |\boldsymbol{F}_1|/A = |\boldsymbol{F}_2|/A$ is known as the (longitudinal) *stress*. Experimental measurements are usually displayed by showing the stress as a function of the strain, in a stress–strain diagram. Not all solids behave in the same way, but a typical outcome of such a measurement is sketched in Figure 1.5.

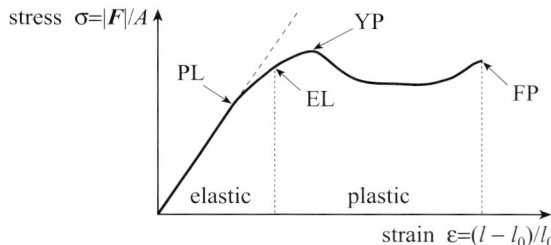

Figure 1.5 Sketch of a typical stress–strain diagram of a solid. The points PL, EL, YP and FP are the proportional limit, the elastic limit, the yield point and the failure point, respectively.

There are two qualitatively different behaviours that can be observed when deforming a solid. The deformation is called *elastic* if it is reversible. If the force is removed, the material regains its original shape. In our case, this means that the rod contracts back to its natural length l_0. If this is not the case, we have a *plastic* deformation, and the solid is permanently deformed. The largest deformation for which the material still behaves elastically is called the *elastic limit*. The corresponding point in the stress–strain diagram of Figure 1.5 is denoted by EL. The elastic limit can vary considerably between materials: while a rubber band may well be extended elastically to, say, twice its natural length, the elastic limit for a piano wire is of the order of 0.01, which corresponds to an extension of 1%.

For small deformations within the elastic limit, most materials obey *Hooke's law*, which states that the stress and the strain are proportional to each other. It is worth noting that a material can be elastic without obeying this linear relationship. In Figure 1.5, the *proportional limit* PL denotes a point up to which the behaviour is well described by a linear relationship. Note that PL is not a precisely defined point on the curve, but rather depends on how much deviation from linearity you are prepared to tolerate.

Robert Hooke (1635–1703)

Within the linear regime (i.e. where strain is smaller than proportional limit), the graph of stress against strain has constant slope Y, with

$$\sigma = \text{stress} = Y \times \text{strain} = Y\varepsilon. \tag{1.4}$$

Y is called *Young's modulus*; it is a characteristic feature of the material. Some typical values are given in Table 1.1, together with typical values of the *volume density* $\rho^{(V)} = m/V$, which is the mass per unit volume of the material.

Thomas Young (1773–1829)

Volume density is explained further in Subsection 1.2.4.

Table 1.1 Typical values of the volume density $\rho^{(V)}$, in kilograms per cubic metre, and of Young's modulus Y, in units of gigapascals ($1\,\text{GPa} = 10^9\,\text{Pa} = 10^9\,\text{N m}^{-2}$), for various solids

	$\rho^{(V)}$ (kg m^{-3})	Y (GPa)
Steel	7860	200
Aluminium	2710	70
Concrete	2320	30
Glass	2190	65
Polystyrene	1050	3
Wood	525	13

For a rod of cross-sectional area A, we define its *modulus of elasticity* or *elastic modulus* as

$$K = AY. \tag{1.5}$$

This is constant if we can neglect any changes in the cross-sectional area of the rod caused by the extension. Using equation (1.4), the magnitude F of the force is related to the strain by

$$F = A\sigma = AY\varepsilon = K\varepsilon. \tag{1.6}$$

There are two further points marked in Figure 1.5. The first is the *yield point* YP, which is defined as the strain where the stress starts to decrease with increasing strain, i.e. the material 'gives in'. In a rod, this may, for instance, happen due to pinching of the rod: the rod effectively becomes thinner and longer, and the stress decreases. Finally, if the strain reaches the *failure point* FP, the material breaks and the stress–strain curve terminates.

Exercise 1.2

(a) Calculate the elastic modulus K for uniform steel wires of cross-sectional areas $A = 1\,\text{mm}^2$ and $A = 2\,\text{mm}^2$.

(b) Consider two uniform wires of cross-sectional area $A = 1\,\text{mm}^2$ made of steel and aluminium, respectively. What force magnitude F is needed to extend each wire by 1%, such that $\varepsilon = (l - l_0)/l_0 = 1/100$?

Suppose that the extension is produced by the gravitational force of magnitude $F = mg$ of a mass supported by the wire, where $g \simeq 9.81\,\text{m s}^{-2}$ is the magnitude of the acceleration due to gravity. What are the corresponding masses m that have to be attached to the two wires to extend them by 1% (where the masses of the wires can be ignored)?

(c) A uniform aluminium wire has a natural length $l_0 = 1\,\text{m}$ and a cross-sectional area $A = 10\,\text{mm}^2$. What is the mass of the wire?

The wire is fixed at one end, and stretched to a length of $1.02\,\text{m}$. Calculate the magnitude of the force that the wire exerts on the fixture.

1.2.2 Hooke's law

You may already be familiar with helical (coil) springs and their properties from previous courses. Usually, the oscillatory motion of a particle attached to a spring is considered. In that case, the role of the spring is to provide a force acting on the particle, and it is the motion of the particle that is of interest. Here, it is the motion of the spring itself that will be of interest. This means that we have to investigate the forces and motion *within* the spring. Unless stated otherwise, we shall assume that the spring is *uniform*, which means that its material properties do not change along its length.

Most coil springs consist of metal rods, and their physical properties can, in principle, be derived from those of the metal rod. However, we shall not go into detail here, but instead consider idealised springs (model springs) which have a particularly simple behaviour. In what follows, the word 'spring' always refers to this idealisation, unless we explicitly talk about real springs.

For our purposes, a spring is completely characterised by three parameters: the *natural length* l_0, the *stiffness* k and the *mass* m. We assume that the spring obeys Hooke's law, which states that a uniformly stretched (or compressed) spring of natural length l_0 and stiffness k exerts a force

$$\boldsymbol{F} = -k(l - l_0)\hat{\boldsymbol{s}} \qquad (1.7)$$

on any object attached to an end, with the other end fixed. Here l is the *length* of the spring, and $\hat{\boldsymbol{s}}$ is a unit vector along the axis of the spring, directed *from the centre of the spring towards the end* where the force acts. The direction of the force $\boldsymbol{F}$ is thus along the axis of the spring. If the spring is stretched, $l > l_0$, the force is directed towards the centre of the spring; if it is compressed, $l < l_0$, the force is directed outwards, away from the centre of the spring. This force is equal in magnitude and opposite in direction to the force needed to keep the spring in the stretched or compressed state of length l.

If the spring is fixed at the ends $x = 0$ and $x = l$ along the $\mathbf{i}$ direction, as depicted in Figure 1.6, it exerts a force $\boldsymbol{F}_1 = k(l - l_0)\mathbf{i}$ at its left end ($x = 0$) and a force $\boldsymbol{F}_2 = -k(l - l_0)\mathbf{i} = -\boldsymbol{F}_1$ of equal magnitude and opposite direction at its right end ($x = l$). In equation (1.7), the unit vector $\hat{\boldsymbol{s}}$ is given by $-\mathbf{i}$ at the left end of the string, and by $\mathbf{i}$ at the right end of the spring.

Note that this differs from the notion used in MST209/MST207, where model springs are defined to be massless.

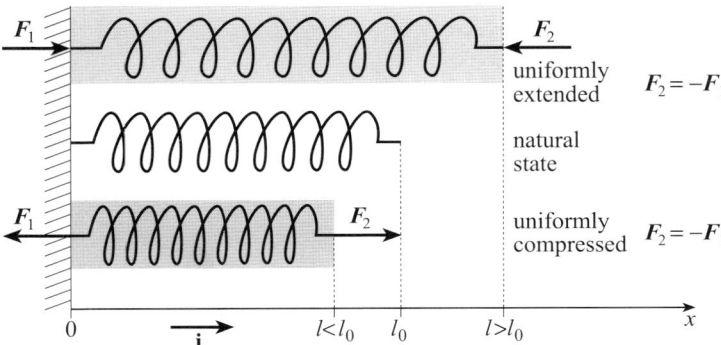

Figure 1.6 Elastic forces *exerted by* a uniformly extended or compressed spring

The larger the stiffness k of the spring, the larger the force required to extend or compress it by a given distance $|l - l_0|$. Instead of specifying the spring stiffness k, it is often preferable to use the *elastic modulus* K, which in this case is given by $K = kl_0$. With this substitution, Hooke's law becomes

Hooke's law
$$\boldsymbol{F} = -K\frac{l - l_0}{l_0}\hat{\boldsymbol{s}}, \qquad (1.8)$$

which is the form that we shall use mainly.

So K measures the magnitude of the force in terms of the *relative* change of length, the strain $\varepsilon = (l - l_0)/l_0$, of the spring (compare equation (1.6) on page 14). The units of K are those of force, so values are given in newtons, with $1\,\text{N} = 1\,\text{kg}\,\text{m}\,\text{s}^{-2}$. In contrast to the stiffness k, the elastic modulus is characteristic of the material and construction of the spring, and does not depend on its length. This is most easily seen in the example of equal springs of elastic modulus K and length l_0, joined together end to end, and considered as a single spring of length $2l_0$. Applying a force of magnitude F, the double spring extends twice as much as each of the single springs, so its stiffness that enters equation (1.7) is half that of its constituents. However, the elastic modulus of the double spring *equals* that of its constituents, as $(k/2)(2l_0) = kl_0 = K$.

> As an aside, we note that real springs are often 'pre-tensioned', where the coils are not separated in the natural state of the spring. This means that while these springs obey Hooke's law (1.8) for a certain range of deformations $l - l_0$, the value of l_0 that enters Hooke's law differs from the free length of the spring in its natural state. In that case, we need to distinguish between the natural length l_0, which is defined as the constant that appears in Hooke's law (1.8), and the actual length in equilibrium. For simplicity, we do not make this distinction in what follows, and assume that the lengths coincide.

Elastic strings

Comparing equation (1.6) for the rod in the linear regime with equation (1.8) for the spring, one finds that the relationship between force and deformation is the same in both systems. For our purposes, a rod behaves, for small deformations, just like a very stiff spring, so we shall not make a distinction between these two systems. Provided that we are in the linear elastic regime, Hooke's law applies.

A similar comment applies to *elastic strings*. For instance, a piece of rubber band behaves like a thin rod or a long spring: you need to pull it, which means to apply a force, in order to extend it. We assume that elastic strings are ideally *flexible*, so they can be bent without an effort. This means that there is an elastic force associated with extending, but *not* with bending the string. Clearly, this is an idealisation, but it is a reasonable approximation for flexible strings provided that the bending is not too sharp.

In contrast to the spring and the rod, a flexible string *cannot* be compressed by applying a force, because it would bend instead. But we may assume that the stretching of flexible strings is once more described by Hooke's law, with an appropriate value of the stiffness or the elastic modulus.

1.2.3 Tension

Let us return to springs for a moment. Hooke's law tells us about the forces that a spring exerts on objects attached to its ends. However, we are now interested in what happens *within* a spring. If you touch a stretched spring or a taut string, for instance a guitar string, and slightly deform it from its straight form, you can feel that it is in *tension*: the spring or string resists deformation, and if you pluck the guitar string, it will start to vibrate. A slack flexible string, on the other hand, can easily be bent and will not straighten out again. Let us make this more precise and define tension.

1.2 Springs, rods and elastic strings

Consider again a uniform spring in a static state of uniform compression ($l < l_0$) or extension ($l > l_0$), with no other forces acting. Imagine cutting the spring at position x ($0 < x < l$), holding the two cut ends together (see Figure 1.7).

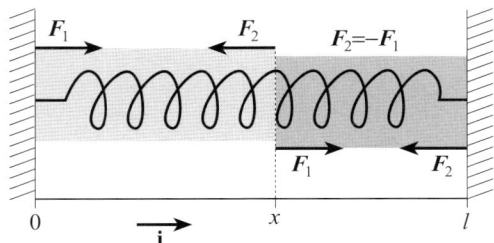

Figure 1.7 The tension $T(x)$ in an extended spring corresponds to the magnitude of the elastic forces $|\boldsymbol{F}_1| = |\boldsymbol{F}_2|$ that are needed to keep the ends at an imagined cut at position x in place

The forces needed to keep the two new ends together correspond to the forces that the two newly-formed springs exert at their ends. Because the spring was in a static state before we cut it, the two forces exerted by the two new ends must be of opposite direction and of equal magnitude. This holds true for any x along our original spring, so the magnitude of the force is constant along the spring, and the forces at the imagined cut are thus given by Hooke's law (1.8).

We define the tension T such that equation (1.8) becomes $\boldsymbol{F} = -T\hat{\boldsymbol{s}}$. So, in a spring that obeys Hooke's law (1.8) and has been stretched uniformly from its natural length l_0 to length l, the tension is constant and given by

Tension in a uniformly deformed spring
$$T = K\frac{l - l_0}{l_0} = K\frac{l}{l_0} - K. \tag{1.9}$$

The *magnitude* of the tension $|T|$ is the magnitude of the forces that keep the cut ends of the spring in place. The *sign* of T tells us whether the spring is in a state of extension or compression: if the spring is extended, $l - l_0 > 0$ and T is positive; if the spring is compressed, $l - l_0 < 0$ and T is negative.

The definition of tension raises an obvious question. As tensions always correspond to forces, why do we not just consider the forces rather than the signed scalar quantity T? The reason is the following. Consider a uniformly stretched spring with length $l > l_0$. The tension is constant and positive along the spring. If we imagine cutting the spring at a position x, we need forces of magnitude T to keep the two ends in position, but the two forces have opposite signs – the new (right) end of the left part of the spring at smaller values of x wants to move to the left, so it exerts a force $\boldsymbol{F}_2 = -T\mathbf{i}$, whereas the new (left) end of the right part pulls to the right, exerting a force $\boldsymbol{F}_1 = T\mathbf{i}$. So tension does not have a 'natural' direction, but it has a sign which tells us whether our two new ends are pulled apart or driven together by the forces exerted by the two parts of the spring.

In a flexible string, the tension cannot be negative, because the string cannot be in a state of compression. Later, we shall also be interested in the behaviour of flexible strings, which in general will not be straight. In that case, at any position along the string, the elastic forces are *tangential* to the string, and the (scalar) tension again corresponds to the magnitude of those forces.

The tension in non-uniformly compressed or extended springs will be discussed later.

1.2.4 Density

Usually, one defines the density of a material as the mass per unit volume. In general, the density can fluctuate in a material, and is defined as a limit:

Volume density
$$\rho^{(V)}(\boldsymbol{r}) = \lim_{\delta V \to 0} \frac{\delta m(\boldsymbol{r})}{\delta V}, \qquad (1.10)$$

where $\delta m(\boldsymbol{r})$ denotes the mass contained in the volume δV at position $\boldsymbol{r}$. However, for a rod, spring or stretched string, it is natural to use the mass per unit length, i.e. the linear density

Linear density
$$\rho(x) = \lim_{\delta l \to 0} \frac{\delta m(x)}{\delta l}, \qquad (1.11)$$

rather than the mass per unit volume.

In the uniform case, m and ρ are independent of position, and these definitions simplify to $\rho^{(V)} = m/V$ and $\rho = m/l$, respectively. For a uniform wire or rod with constant cross-sectional area A, the volume is related to the length by $V = Al$, so the linear and volume densities are related by

$$\rho = A\rho^{(V)}. \qquad (1.12)$$

In a uniformly stretched or compressed spring of length l, the linear density is constant, $\rho = m/l$. If l_0 denotes the natural length of the spring, the density in the natural state is $\rho_0 = m/l_0$. The mass m does not change, so

$$\rho = \frac{l_0}{l} \rho_0. \qquad (1.13)$$

Exercise 1.3

Consider a spring of natural length l_0, elastic modulus K and natural linear density ρ_0 that obeys Hooke's law.

(a) The spring is uniformly extended to a length $l > l_0$. Calculate the linear density ρ and the tension T in the uniformly stretched spring. Express ρ in terms of ρ_0, K and the constant tension T.

(b) Compare the results with those for a uniformly compressed spring of length $l < l_0$.

Exercise 1.4

Consider a spring of elastic modulus $K = 100\,\text{N}$ that obeys Hooke's law. The spring is uniformly extended or compressed to a length $l = 30\,\text{cm}$.

(a) If the tension in the spring is $T = 25\,\text{N}$, what is the natural length l_0 of the spring?

(b) Alternatively, if the tension in the spring is $T = -25\,\text{N}$, what is the natural length l_0 of the spring?

(c) What are the relative length changes $(l - l_0)/l_0$ in the two cases?

Exercise 1.5

Consider a spring of natural length $l_0 = 1\,\text{m}$, elastic modulus $K = 100\,\text{N}$ and linear density $\rho_0 = 0.1\,\text{kg}\,\text{m}^{-1}$. Assume that the spring obeys Hooke's law.

(a) What is the mass of the spring?

(b) What force is needed to extend the spring to length $l = 1.5\,\text{m}$? What is the tension in the extended spring? What is its linear density?

(c) What force is needed to compress the spring to length $l = 0.8\,\text{m}$? Calculate the tension and the linear density in this case.

1.3 Transverse waves on a stretched string

We are now going to analyse our first instance of wave motion, namely transverse waves on a stretched string.

As an example of transverse wave motion, you may think of a guitar string that is plucked and then released, vibrating freely. It is this vibration that makes the sound. We assume that the only forces that are relevant are elastic forces; any other forces, such as those due to gravity and friction, both within the string and due to the presence of air around the string, are neglected. In fact, frictional forces are present in the example of the guitar string. We could not hear the sound of a vibrating string if there were no energy carried away by the sound wave, and the energy losses due to the sound wave and due to friction are responsible for the damping of the vibration and the resulting fading of the sound. However, for a sufficiently short time, we can ignore the effects of friction on the motion.

1.3.1 Derivation of the wave equation

We consider a string of natural length l_0 and constant linear density ρ_0, which is uniformly stretched to a length $l > l_0$. In equilibrium, the string is along the $\mathbf{i}$ direction and is fixed at its end points at $x = 0$ and $x = l$. In equilibrium, the tension $T > 0$ is constant.

Now we consider the situation where the string is slightly disturbed *perpendicular* to the string, in the $\mathbf{j}$ direction (see Figure 1.8). We define a function $u(x, t)$ as the displacement of the string in the $\mathbf{j}$ direction at time t, and we assume that the deformation is sufficiently smooth that the function $u(x, t)$ is at least twice differentiable with respect to both arguments.

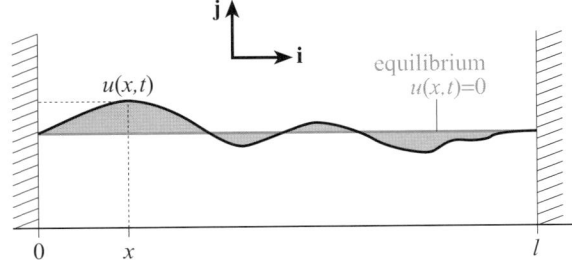

Figure 1.8 Transverse displacement $u(x, t)$ of a stretched string in the $\mathbf{j}$ direction at a fixed instant of time

We are interested in the *motion* of the string, described by the space and time dependence of the deformation $u(x, t)$.

More simplifying assumptions

We assume that the motion takes place solely in the $\mathbf{j}$ direction. This means that a marked point at a position $x\mathbf{i}$ along the undeformed string moves *only* in the $\mathbf{j}$ direction, not along the equilibrium string direction. At time t, the position of the marked point is thus $x\mathbf{i} + u(x, t)\mathbf{j}$; its component in the $\mathbf{i}$ direction stays fixed, and its transverse component, in the $\mathbf{j}$ direction, corresponds to the deformation $u(x, t)$.

We also assume that we may neglect the change in length of the string due to the displacement in the **j** direction. This can be justified if the deformation of the string is sufficiently small. More precisely, we require that, for all $x \in [0, l]$, the partial derivative u_x satisfies $|u_x| \leq \epsilon$ for some small positive $\epsilon > 0$. The changes in the length, tension and linear density of the string are then of order ϵ^2, so in a first-order approximation, we can neglect these changes.

These statements are verified in optional Exercise 1.23 at the end of this chapter.

In this approximation, the tension T remains constant throughout the motion. This also means that we do *not* need to assume that the string obeys Hooke's law. With the assumptions made above, the only information that will enter the argument is the tension $T > 0$ in the string.

A string element

We start by considering the situation at a fixed instant of time t. The deformation of a small element $[x, x + \delta x]$ of length δx is shown in Figure 1.9.

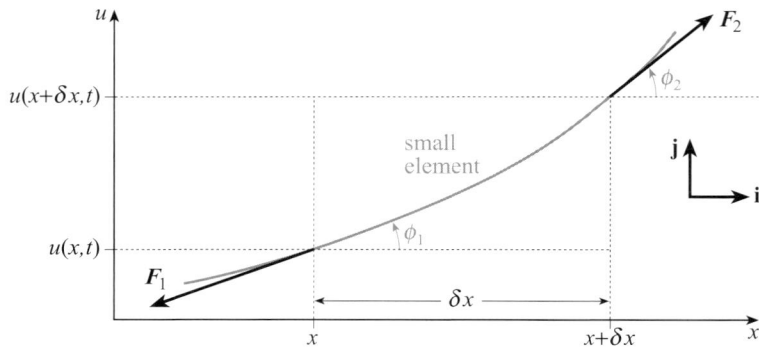

Figure 1.9 The forces acting on a small string element. For clarity, the angles ϕ_1 and ϕ_2, which are supposed to be small, have been exaggerated.

At time t, the displacements at the ends of the interval are $u(x, t)$ and $u(x + \delta x, t)$, and the angles between the string and the **i** direction are ϕ_1 and ϕ_2, respectively. The slope of the string is given by the tangents of the angles,

$$u_x(x, t) = \tan \phi_1 \simeq \phi_1, \quad u_x(x + \delta x, t) = \tan \phi_2 \simeq \phi_2, \quad (1.14)$$

where $u_x(x + \delta x, t)$ denotes the partial derivative of the function $u(x, t)$ with respect to the first variable, evaluated at $(x + \delta x, t)$. According to our assumptions, the slope is of order ϵ, so the angles ϕ_1 and ϕ_2 are themselves of order ϵ. From the Taylor expansion $\tan \phi = \phi + \frac{1}{3}\phi^3 + \cdots$, we find that we can approximate the tangents by the angles, because the difference is of order ϵ^3.

Motion of a string element

If we consider a small string element, we can neglect any changes within it, and think of it as a particle whose motion is described by Newton's second law,

Newton's second law
$$\text{force} = \text{mass} \times \text{acceleration.} \quad (1.15)$$

Applying this to the string element, we obtain a relation between the net force acting on the string element, its mass and its acceleration. The strategy is to use the equation of motion for the string element and then take the limit as $\delta x \to 0$.

Force on a string element

Due to the tension in the string, the string exerts forces $\boldsymbol{F}_1$ and $\boldsymbol{F}_2$ on our string element shown in Figure 1.9.

1.3 Transverse waves on a stretched string

These forces are tangential to the string, so

$$\boldsymbol{F}_1 = -T[\cos\phi_1\,\mathbf{i} + \sin\phi_1\,\mathbf{j}] \simeq -T(\mathbf{i} + \phi_1\,\mathbf{j}), \tag{1.16}$$
$$\boldsymbol{F}_2 = T[\cos\phi_2\,\mathbf{i} + \sin\phi_2\,\mathbf{j}] \simeq T(\mathbf{i} + \phi_2\,\mathbf{j}). \tag{1.17}$$

We have used again the fact that the angles ϕ_1 and ϕ_2 are small (of order ϵ), so $\sin\phi_i = \phi_i + O(\epsilon^3)$ and $\cos\phi_i = 1 + O(\epsilon^2)$, for $i = 1, 2$.

Considering the small string element as a particle, we can add the forces $\boldsymbol{F}_1$ and $\boldsymbol{F}_2$ to obtain the resulting net force $\delta\boldsymbol{F}$ on the string element as

$$\delta\boldsymbol{F} = \boldsymbol{F}_1 + \boldsymbol{F}_2 \simeq T(\phi_2 - \phi_1)\,\mathbf{j} \simeq T[u_x(x + \delta x, t) - u_x(x, t)]\,\mathbf{j}, \tag{1.18}$$

which acts in the $\mathbf{j}$ direction, perpendicular to the unperturbed string.

Exercise 1.6

Consider a transversely deformed string, fixed at its end points at $x = 0$ and $x = 1$, which is described by a function $u(x) = x(1-x)/10$. We assume that the string has constant tension T.

(a) Calculate the slope of the string.

(b) What is the maximal displacement of the string, and at which position x does it occur?

(c) Find the extremal values of the slope.

(d) Calculate the force $\delta\boldsymbol{F}(x)$ on a string element $[x, x+\delta x]$, and compare the magnitude of $\delta\boldsymbol{F}(x)$ with the second derivative $u''(x)$.

Mass of a string element

Because we have assumed that the string is uniform and its length remains constant, the mass of the string element is

$$\delta m = \rho\,\delta x, \tag{1.19}$$

where $\rho = \rho_0 l_0/l$ denotes the constant linear density of the stretched string.

Acceleration of a string element

The acceleration $\boldsymbol{a}$ of the string element in the $\mathbf{j}$ direction is

$$\boldsymbol{a} = a\,\mathbf{j} \simeq u_{tt}(x, t)\,\mathbf{j}, \tag{1.20}$$

where we ignore any variation of u_{tt} within the string element.

Applying Newton's second law

According to Newton's second law, the equation of motion for the string element is force equals mass times acceleration, $\delta\boldsymbol{F} = \delta m\,\boldsymbol{a}$. Using the expressions for the force $\delta\boldsymbol{F}$, mass δm and acceleration $\boldsymbol{a}$ in equations (1.18)–(1.20), we obtain

$$T[u_x(x + \delta x, t) - u_x(x, t)]\,\mathbf{j} \simeq \rho\,\delta x\,u_{tt}(x, t)\,\mathbf{j}. \tag{1.21}$$

Using the definition of the second partial derivative,

$$u_{xx}(x, t) = \lim_{\delta x \to 0} \frac{u_x(x + \delta x, t) - u_x(x, t)}{\delta x}, \tag{1.22}$$

we obtain the second-order partial differential equation

$$u_{tt}(x, t) = \frac{T}{\rho}\,u_{xx}(x, t) \tag{1.23}$$

in the limit as $\delta x \to 0$.

This is the wave equation (1.1) quoted on page 8,

$$u_{tt}(x,t) = c^2 \, u_{xx}(x,t), \tag{1.24}$$

with the wave speed c given by

Transverse wave speed

$$c = \sqrt{\frac{T}{\rho}}. \tag{1.25}$$

The wave equation is the crucial equation of this part of the course. It involves the second partial derivatives of the function $u(x,t)$ with respect to time t and position x. The second partial derivative with respect to time is a direct consequence of Newton's second law, which is used to describe the acceleration of a small element of the string. The second partial derivative with respect to position enters via the force on a small string element.

Exercise 1.7

(a) Show that the wave speed (1.25) can be expressed as

$$c = \sqrt{\frac{Tl}{\rho_0 l_0}}.$$

(b) Show that for a string that obeys Hooke's law with an elastic modulus K, the wave speed (1.25) can be expressed as

$$c = \sqrt{\frac{(K+T)T}{\rho_0 K}}.$$

1.3.2 Some solutions of the wave equation

The possible motions of the string are determined by the solutions of the wave equation. These will be discussed in detail in the subsequent chapters. However, some example solutions are considered in Exercises 1.8–1.11.

The wave equation was derived by considering the internal forces acting on, and the accelerations of, small elements of the string. It depends on the linear density ρ and the tension T in the string, but does not involve the length l of the string. Solutions of the wave equation do, however, depend on l, because it enters via additional conditions, known as *boundary conditions*, which prescribe the value of the function $u(x,t)$ at the ends of the string. Different boundary conditions lead to different solutions of the wave equation. If, for example, the string is fixed at its ends at $x=0$ and $x=l$, the deformation has to vanish at these points, so we require $u(0,t) = u(l,t) = 0$ for all values of t.

The role of boundary conditions will be discussed in detail in Chapters 2 and 4.

In the following exercise we show you a solution of the wave equation. It may be something of a disappointment, because once we apply the boundary conditions, it turns out to be a rather trivial solution.

Exercise 1.8

Consider the function $u(x,t) = A\sin(kx - \omega t)$ for $x \in [0,l]$, where A, k and ω are constants.

(a) Insert $u(x,t)$ into the wave equation (1.24). Find the conditions on k and ω such that $u(x,t)$ is a solution for a given wave speed c.

Note that the constant k in this case is distinct from the spring stiffness discussed earlier.

1.3 Transverse waves on a stretched string

(b) Suppose that $u(x,t)$ describes the transverse vibrations of a string that is fixed at its end points $x = 0$ and $x = l$. Show that the boundary condition $u(0,t) = 0$ for the above solution implies that $u(x,t)$ is time-independent.

(c) Show that requiring the wave speed c to be non-zero implies that $u(x,t)$ is the trivial solution $u(x,t) = 0$ for all x and t (which clearly satisfies the wave equation and the boundary conditions $u(0,t) = u(l,t) = 0$).

The next exercise introduces some more interesting solutions, which are important because we shall repeatedly return to them. For some of the solutions, there are points along the string other than the end points that do not move at all during a particular motion of the string. Such points are called *wave nodes* or simply *nodes*.

Exercise 1.9

Consider the function $u(x,t) = A\sin(kx)\sin(\omega t)$ for $x \in [0,l]$, where A, k and ω are constants.

(a) Find the relation between k and ω imposed by the wave equation (1.24) for $u(x,t)$ for a given value of the wave speed c.

(b) What are the conditions on the parameters A, k and ω imposed by the boundary conditions $u(0,t) = u(l,t) = 0$ for all values of t?

(c) Solve these conditions to derive functions $u(x,t)$ which are non-trivial solutions of the wave equation with these boundary conditions.

These solutions correspond to possible motions of the string.

(d) Which points x_s along the string, besides the boundary points $x_s = 0$ and $x_s = l$, are nodes? In other words, what are the values x_s for which $u(x_s, t) = 0$ for all t? Label the solutions by $n = 1, 2, \ldots$ according to the number of nodes, with $u_n(x,t)$ containing $n - 1$ nodes.

(e) Consider solutions $u_1(x,t)$ and $u_2(x,t)$ which have no nodes and just a single node, respectively. Sketch the functions $u_1(x,t_j)$ and $u_2(x,t_j)$ at several fixed instants of time t_j in order to visualise the corresponding motion.

The solutions discussed in this exercise correspond to *standing wave* motion. As for a plucked guitar string, there is no apparent motion of the wave pattern along the string. Later, we shall discuss *travelling waves*, an example of which are water waves produced by throwing a pebble into a pond.

Complex solutions of the wave equation

Although the deformation $u(x,t)$ clearly should be a real number for all values of x and t, it sometimes turns out to be useful to consider *complex solutions* of the wave equation. As will be shown in Exercise 1.10, the *real* and *imaginary* parts of a complex solution individually satisfy the wave equation. So a single complex solution encodes two real solutions which each correspond to a possible motion. However, there is also a more obvious reason why one might prefer to use complex solutions: real solutions, such as those of Exercises 1.8 and 1.9, often involve trigonometric functions, while complex solutions, like those considered in Exercise 1.11 below, involve the complex exponential function. Algebraic manipulation of the latter is often much easier than that of the former.

Exercise 1.10

Consider an arbitrary complex solution $u(x,t) = p(x,t) + iq(x,t)$ of the wave equation, where p, q, x and t are real.

(a) Show that the real and imaginary parts $p(x,t)$ and $q(x,t)$ of $u(x,t)$ separately satisfy the wave equation.

(b) Show that this means that the complex conjugate u^* of any complex solution $u(x,t)$ of the wave equation also satisfies the wave equation. (Again, x and t are assumed to be real.)

(c) Express $p(x,t)$ and $q(x,t)$ as linear combinations of the function u and its complex conjugate u^*.

Exercise 1.11

Consider the complex exponential functions

$$f(x,t) = A \exp[i(kx - \omega t)], \quad g(x,t) = B \exp[i(kx + \omega t)],$$

for $x \in [0, l]$, and their linear combination $u(x,t) = f(x,t) + g(x,t)$, where A and B are complex constants, and k and ω are real constants.

(a) Find the relation between k and ω imposed by the wave equation (1.24) for a solution of the form $f(x,t)$ at a given value of the wave speed c. Do the same for $g(x,t)$. What does this mean for linear combinations of the form $u(x,t)$ with arbitrary complex constants A and B?

(b) Show that the boundary condition $u(0,t) = 0$ cannot be fulfilled for such a linear combination for all times t unless $A = B = 0$ or $\omega = 0$.

(c) Show that for real $A = -B$, the real part $p(x,t)$ of the complex function $u(x,t) = p(x,t) + iq(x,t)$ is of the same form as the solutions of Exercise 1.9.

(d) Consider now the real and imaginary parts $p(x,t)$ and $q(x,t)$ of the function $u(x,t) = p(x,t) + iq(x,t)$ for real $A = -B$. Can you satisfy the boundary conditions $p(0,t) = p(l,t) = 0$ and $q(0,t) = q(l,t) = 0$? What follows from that for complex solutions $u(x,t)$ with boundary conditions $u(0,t) = u(l,t) = 0$?

You may be surprised by the apparent asymmetry between the real and imaginary parts of a complex solution with regard to the boundary conditions that occur in Exercise 1.11. This is caused by restricting the coefficient $A = -B$ to be real. For complex coefficients $A = a \exp(i\alpha) = -B$, we have $u(x,t) = 2a \sin(kx + \alpha) \sin(\omega t) - 2ia \cos(kx + \alpha) \sin(\omega t)$, and it *is*, in general, possible to find values for k where either the real or the imaginary part of $u(x,t)$ (but not both) vanishes at the boundaries.

1.4 Non-uniformly deformed springs

Let us now turn our attention to waves in springs. So far, we have considered springs that are uniformly stretched or compressed. Now we intend to examine springs that are stretched or compressed unevenly. Your intuition may lead you to suspect that such a spring cannot be in equilibrium unless some forces are applied so as to maintain such a deformation. At this moment, we are not yet interested in the motion of the spring, but in the tension in a non-uniformly deformed spring. Thus you may imagine that the non-uniformly deformed spring is being kept in static equilibrium by external forces, acting at various points along its length. An example of a spring that is non-uniformly stretched in equilibrium is given by a suspended spring

1.4 Non-uniformly deformed springs

that stretches due to its own weight. In this situation, upper parts of the spring have to support the weight of the lower parts, so the force that acts is non-uniform, and consequently the stretching will also be non-uniform.

In what follows, we shall often compare a spring in a general, non-uniformly stretched or compressed state with the spring in its natural state, or with a uniformly stretched or compressed spring. We use x as the variable for the distance along the uniform reference state. In order to describe the deformation in a general spring with respect to its reference state, we introduce a function $\tilde{x}(x,t)$ that relates the position $\tilde{x}$ at time t of a marked point along the spring to its position x in the reference state (see Figure 1.10). It is easy to keep in mind that the symbol $\tilde{x}$ refers to the deformed case, because the tilde is reminiscent of a wavy pattern.

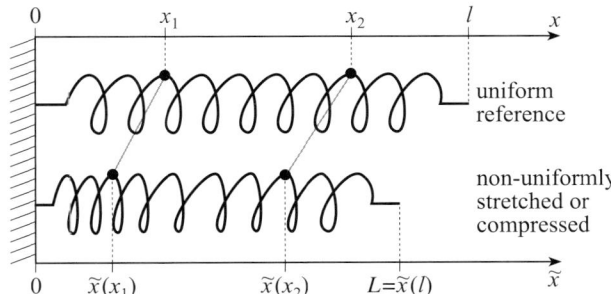

Figure 1.10 Positions of marked points along the axis of a spring in the reference state of length l and in a non-uniform general state of length L, at a fixed instant of time. Here the point x_1 is moved to the point $\tilde{x}(x_1)$ and the point x_2 to the point $\tilde{x}(x_2)$. The new length of the spring is $L = \tilde{x}(l)$.

In the general situation of a deformed spring, the tension T and the density ρ depend on the position $\tilde{x}$ along the spring. We are interested in local analogues of equations (1.9) on page 17 and (1.13) on page 18, which will give the tension and the density from the local deformation of the spring at a certain position. We assume that equations (1.9) and (1.13) can be applied to sufficiently small elements of the spring which may be considered as being uniformly stretched or compressed.

Consider a small element $[x, x + \delta x]$ of length δx in the natural state of the spring. Measured along the non-uniformly stretched spring, the same piece of the spring is now at $[\tilde{x}, \tilde{x} + \delta \tilde{x}]$, and has deformed length $\delta \tilde{x}$, as shown in Figure 1.11.

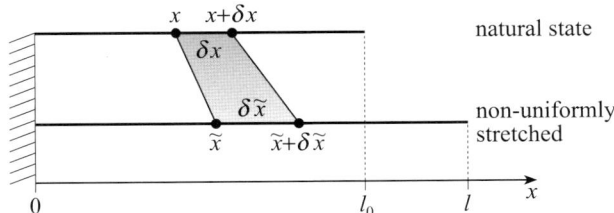

Figure 1.11 Schematic representation of a spring represented by a line along its axis. A small element $[x, x + \delta x]$ of the natural state of length l_0 is deformed to an element $[\tilde{x}, \tilde{x} + \delta \tilde{x}]$ in the deformed state of length l.

We consider the position $\tilde{x}$ as a function $\tilde{x}(x)$ of the original position x of a marked point along the spring in its natural state. The two end points of the deformed spring element are then $\tilde{x}(x)$ and $\tilde{x}(x + \delta x)$, so the deformed length becomes

$$\delta \tilde{x} = \tilde{x}(x + \delta x) - \tilde{x}(x). \tag{1.26}$$

1.4.1 Tension

In order to calculate the tension T at the deformed position $\tilde{x}$, we apply equation (1.9) on page 17 to the small element. This holds only approximately for finite δx because we need to assume that the small element is uniformly stretched or compressed, such that the tension T is constant on the small element. This gives

$$T \simeq K\frac{\delta\tilde{x} - \delta x}{\delta x} = K\frac{\delta\tilde{x}}{\delta x} - K. \tag{1.27}$$

Note that T now depends on $\tilde{x}(x)$, and thus on x. Later, the deformation of the spring will be allowed to vary with time, so it is advantageous to use x as an independent variable. Therefore we denote the tension in equation (1.27) by $T(x)$, keeping in mind that this is the tension at position $\tilde{x}(x)$ in the deformed spring.

In the limit as $\delta x \to 0$, the approximation in equation (1.27) becomes exact. Using equation (1.26), we find

$$\lim_{\delta x \to 0} \frac{\delta\tilde{x}}{\delta x} = \lim_{\delta x \to 0} \frac{\tilde{x}(x + \delta x) - \tilde{x}(x)}{\delta x} = \frac{\partial \tilde{x}}{\partial x}. \tag{1.28}$$

We use partial derivatives here because later, in the general case, the position $\tilde{x}$ will depend both on x and on time t. For the tension, we obtain

Tension in a non-uniformly deformed spring
$$T(x) = K\frac{\partial \tilde{x}}{\partial x} - K. \tag{1.29}$$

Solving equation (1.29) for the partial derivative $\partial\tilde{x}/\partial x$, we can express the local deformation in terms of the tension:

$$\frac{\partial \tilde{x}}{\partial x} = \frac{K + T(x)}{K}. \tag{1.30}$$

1.4.2 Linear density

Let us now consider the linear density. As the element of length $\delta\tilde{x}$ in the deformed state corresponds to an element of length δx in the natural state, both elements must have the same mass δm. Considering the densities ρ_0 and $\rho(x)$ in the natural and deformed states yields

$$\rho_0\,\delta x = \delta m \simeq \rho(x)\,\delta\tilde{x}, \tag{1.31}$$

where, as above, we use the approximation that the small element is uniformly stretched or compressed, and consequently the linear density ρ does not vary within the element. Solving for $\rho(x)$, this gives

$$\rho(x) \simeq \rho_0 \frac{\delta x}{\delta\tilde{x}} = \rho_0 \left(\frac{\delta\tilde{x}}{\delta x}\right)^{-1}. \tag{1.32}$$

Taking the limit as $\delta x \to 0$, we obtain

Density in a non-uniformly deformed spring
$$\rho(x) = \rho_0 \left(\frac{\partial \tilde{x}}{\partial x}\right)^{-1}, \tag{1.33}$$

which is an equality because our approximation becomes exact in the limit. Using equation (1.30), we can express the local density in terms of the tension as

$$\rho(x) = \frac{\rho_0 K}{K + T(x)}. \tag{1.34}$$

Locally, for positive tension (extended spring), the density is smaller than in the natural state; for negative tension (compressed spring), it is larger.

1.4 Non-uniformly deformed springs

Example 1.1

We consider a spring which is stretched by external forces from its natural length l_0 to a length $l > l_0$, such that $\tilde{x}(x) = xl/l_0$.

By equation (1.29), the tension is $T = K(l/l_0) - K = K(l - l_0)/l_0$, and the density is given by equation (1.32) as $\rho = \rho_0 l_0 / l$. This corresponds to a uniformly deformed spring with constant tension and density (compare equations (1.9) on page 17 and (1.13) on page 18, respectively). ∎

Example 1.2

We consider a spring stretched by external forces from its natural length l_0 to a length $l > l_0$, such that the equilibrium deformation is described by the non-uniform stretch

$$\frac{\tilde{x}(x)}{l} = \exp\left(\frac{x}{a}\right) - 1 \tag{1.35}$$

for $0 \leq x \leq l_0$, where $a = l_0 / \ln(2)$. With this choice, $\tilde{x}(l_0) = l$, because $\exp[\ln(2)] = 2$. If the stretching were uniform, the relation between $\tilde{x}$ and x would be linear, $\tilde{x}/l = x/l_0$, as in the previous example. Clearly, the stretching is non-uniform in this case.

According to equation (1.29), the tension $T(x)$ in the deformed spring is given by

$$\frac{T(x)}{K} = \frac{\partial \tilde{x}}{\partial x} - 1 = \frac{l}{a} \exp\left(\frac{x}{a}\right) - 1 = \frac{l-a}{a} + \frac{\tilde{x}(x)}{a}. \tag{1.36}$$

So, in this example, the tension increases linearly with $\tilde{x}$, in other words along the stretched spring.

The linear density is obtained from equation (1.34) as

$$\frac{\rho(x)}{\rho_0} = \frac{K}{K + T(x)} = \frac{a}{l + \tilde{x}(x)}. \tag{1.37}$$

To illustrate this result, Figure 1.12 shows the variation of the tension and the density in the deformed spring for the case $l = 2l_0$, with respect to $\tilde{x}/l$ along the stretched spring and x/l_0 along the unstretched spring.

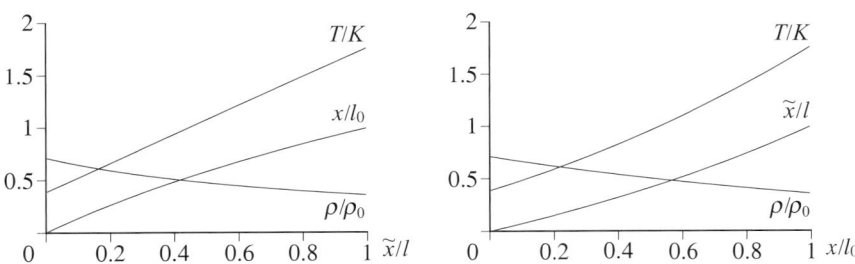

Figure 1.12 The tension T/K from equation (1.36) and the density ratio ρ/ρ_0 from equation (1.37) are shown as functions of $\tilde{x}/l$ (left) and x/l_0 (right) for the non-uniformly stretched spring with $l = 2l_0$. ∎

Exercise 1.12

Consider a spring of natural length $l_0 = 1\,\text{m}$, elastic modulus $K = 100\,\text{N}$ and linear density $\rho_0 = 0.1\,\text{kg}\,\text{m}^{-1}$. The spring obeys Hooke's law, and is kept in a non-uniform static state which is described by the function $\tilde{x}(x) = 2x^2$, where x and $\tilde{x}$ measure (in metres) the length along the spring in the natural state and in the non-uniform state, respectively.

(a) Where are the end points of the spring?

(b) What is therefore the length l of the spring?

(c) Draw a sketch of the function $\tilde{x}(x)$ and of the stretching $\partial\tilde{x}/\partial x$. Where is the spring compressed, and where is it extended?

(d) Calculate x as a function of $\tilde{x}$.

(e) What would the tension T and the mass density ρ be if the spring were uniformly stretched to the same length?

(f) Calculate the tension $T(x)$, and express the result in terms of the position $\tilde{x}$ along the non-uniformly stretched spring. Calculate the tension at the end points and in the middle of the non-uniformly stretched spring.

(g) Sketch the tension T as a function of $\tilde{x}$. Where does T vanish?

(h) In which part of the spring is the state described by the function $\tilde{x}(x) = 2x^2$ unrealistic, and why?

As an aside, we note that by considering small elements, we can also deal with a situation where the spring is non-uniform in its natural state, for instance if the elastic modulus K or the linear density ρ_0 varies with the position x along the spring in its natural state. In this case, K and ρ_0 are functions of x. However, at any position along the spring, the local relations (1.29) and (1.33) between the tension, the deformation and the linear density still apply, provided that the appropriate local values of the elastic modulus $K(x)$ and the linear density $\rho_0(x)$ are used.

1.5 Longitudinal waves

We now use the results that we have derived about springs to consider another example of wave motion, the time-dependent *longitudinal motion* of a spring. The word 'longitudinal' means that the deformation takes place only along the axis of the spring, such as for a compression pulse travelling along a spring as shown in Figure 1.13.

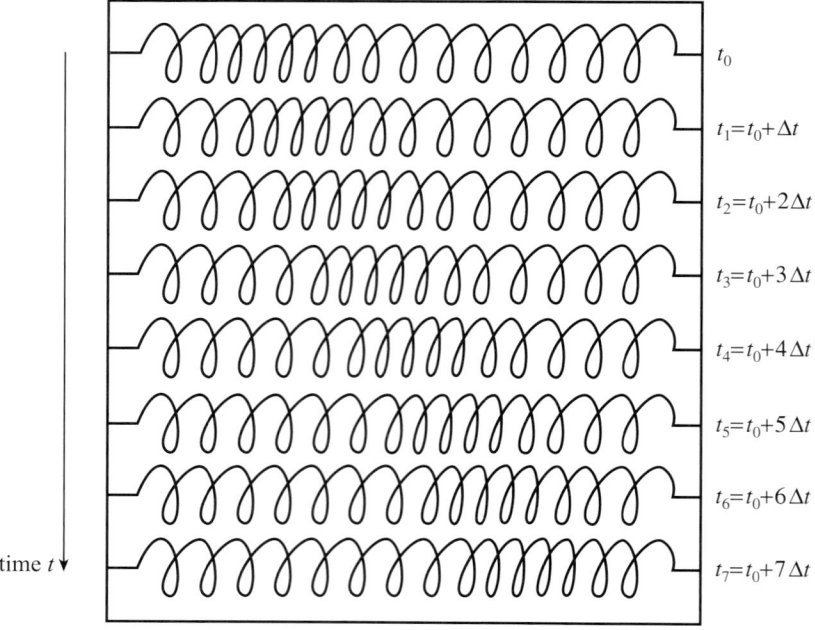

Figure 1.13 Snapshots of a longitudinal motion of a helical spring

1.5.1 Derivation of the wave equation

We consider a uniform spring, fixed at its end points, which is characterised by natural length l_0, linear density ρ_0 and elastic modulus K. We assume that we do not have to take into account gravitational forces, either because they do not affect the motion (you may think, for instance, of a spring that lies horizontally, supported on a smooth table on which it can move without friction), or because they are small and can be neglected. Note that the mass of the spring still affects the motion; it enters via Newton's second law.

We assume that the forces due to the tension in the spring are the only relevant forces for the motion; in particular, we neglect any frictional forces, for instance due to air resistance. In the static equilibrium situation, the spring is in its natural state, thus the tension is zero. We assume that the spring can be stretched and compressed, and that the deformations during the motion are such that Hooke's law applies.

Tension and density in a moving spring

If we consider a moving spring, the tension and density will, in general, depend on both the position along the spring and on the time t. We assume that we can derive the relations between the deformation and the tension by 'freezing' the motion at any instant of time, and thus apply the relations that we derived for the static case also in the dynamic case. This means that equations (1.29) and (1.33) hold at any instant of time t, and these equations give the local tension $T(x,t)$ and density $\rho(x,t)$ of the moving spring, provided that we know the function $\tilde{x}(x,t)$ that describes the deformation. We thus use the static equations to derive the forces in a moving spring, neglecting any influence that the accelerated motion might have on the tension.

Deformation of a spring

The function $\tilde{x}(x,t)$ describes the deformation of the spring at time t. We introduce a function

$$u(x,t) = \tilde{x}(x,t) - x, \tag{1.38}$$

which measures the distance that a marked point along the spring has moved with respect to its position x in the natural state. We deliberately use the same letter as for the transverse motion, because the function again describes the deviation from the static equilibrium case $u(x,t) = 0$ of the natural spring. We assume that the deformation is sufficiently smooth, such that the function $u(x,t)$ is at least twice partially differentiable with respect to both arguments. Differentiating equation (1.38) partially with respect to x, we have

$$u_x(x,t) = \frac{\partial \tilde{x}}{\partial x} - 1, \tag{1.39}$$

where we use partial derivatives because we consider the position x along the (undisturbed) spring and the time t as independent variables.

The tension along the deformed spring is then given by equation (1.29) as

$$T(x,t) = K\, u_x(x,t). \tag{1.40}$$

Motion of a spring element

We consider a small element $[x, x + \delta x]$ of the spring in its natural state, which at time t is positioned at $[\tilde{x}(x,t), \tilde{x}(x+\delta x, t)]$ and has length $\delta \tilde{x} = \tilde{x}(x+\delta x, t) - \tilde{x}(x,t)$ (see Figure 1.14). Applying Newton's second law to the element gives $\delta \boldsymbol{F} = \delta m \, \boldsymbol{a}$, which relates the force $\delta \boldsymbol{F}$ acting on the small spring element, its mass δm and its acceleration $\boldsymbol{a}$.

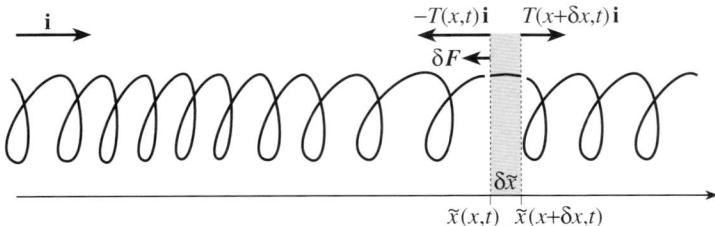

Figure 1.14 The force on a spring element

Force on a spring element

The force $\delta \boldsymbol{F}$ acting on the spring element is the sum of the forces acting at its two ends. These are the forces exerted by the spring on the element, compensating for the elastic forces in the spring element, so they act outwards for positive tension, as shown in Figure 1.14. The magnitude of the elastic force is the tension at the end, thus

$$\delta \boldsymbol{F} = T(x+\delta x, t)\,\mathbf{i} - T(x,t)\,\mathbf{i} = K[u_x(x+\delta x, t) - u_x(x,t)]\,\mathbf{i}. \quad (1.41)$$

Mass of a spring element

The mass of the spring element is related to the equilibrium density ρ_0 and element length δx by

$$\delta m = \rho_0 \, \delta x \quad (1.42)$$

(see equation (1.31)).

Acceleration of a spring element

The acceleration of the spring element is

$$\boldsymbol{a} \simeq \frac{\partial^2 \tilde{x}}{\partial t^2}(x,t)\,\mathbf{i} = u_{tt}(x,t)\,\mathbf{i}, \quad (1.43)$$

because x and t are independent variables. Again, we neglect any change of the acceleration within the small element.

Applying Newton's second law

According to Newton's second law, $\delta \boldsymbol{F} = \delta m \, \boldsymbol{a}$, and following from equations (1.41)–(1.43) we have

$$K[u_x(x+\delta x, t) - u_x(x,t)]\,\mathbf{i} \simeq \rho_0 \, \delta x \, u_{tt}(x,t)\,\mathbf{i}. \quad (1.44)$$

Dividing by δx and taking the limit as $\delta x \to 0$, we obtain

Linear wave equation for longitudinal waves in a spring

$$K \, u_{xx}(x,t) = \rho_0 \, u_{tt}(x,t). \quad (1.45)$$

Here, we used equation (1.22) on page 21, and our approximations for a finite interval δx become exact in this limit. So we once more arrive at the *wave equation* (1.24) for the function $u(x,t)$, but the wave speed is now

1.5 Longitudinal waves

$$c = \sqrt{\frac{K}{\rho_0}}.$$

Longitudinal wave speed for a spring (natural state) or rod (1.46)

This expression for the longitudinal wave speed c involves only ρ_0 and K, which characterise the spring.

You may wonder why equation (1.46) should also apply to rods. Suppose that we have a rod and strike it at one end, say with a hammer. A longitudinal compression wave will travel down the rod. If the elastic deformation is within the linear regime, the wave equation applies, and the wave speed is again given by equation (1.46) with the appropriate value of the elastic modulus K, so the rod just acts like a very stiff spring. These longitudinal waves correspond to sound waves travelling in the solid rod.

Exercise 1.13

Consider a rod of cross-sectional area A.

(a) What is the expression for the elastic modulus K in terms of Young's modulus Y? Express the linear density ρ_0 in terms of the volume density $\rho^{(V)}$.

(b) Use the results of part (a) to write the longitudinal wave speed c of equation (1.46) in terms of Y and $\rho^{(V)}$. Does the speed of sound in a rod depend on its cross-sectional area?

(c) Consider a rod made of steel, with $A = 10\,\text{cm}^2 = 10^{-3}\,\text{m}^2$. The values of the volume density $\rho^{(V)}$ and Young's modulus Y are given in Table 1.1 on page 14. Compute the corresponding values of K, ρ_0 and the speed of sound c in the steel rod. How does this speed compare to the speed of sound in air, $343\,\text{m\,s}^{-1}$?

Sound waves in gases are discussed in the optional Appendix at the end of this chapter (Section 1.9, page 40).

1.5.2 Longitudinal waves in a stretched spring

Above, we considered how small deviations from the natural state give rise to longitudinal waves in a spring. What happens if we consider a stretched or compressed spring and disturb it slightly? Clearly, we expect that this will also lead to longitudinal wave motion.

We consider a spring of natural length l_0 that has been uniformly stretched or compressed to length l, and fixed at its end points. What is the difference between the equilibrium situation in this case as compared to the case discussed above? In the uniformly stretched or compressed reference state, the linear density $\rho = m/l = \rho_0 l_0 / l$ and the equilibrium tension T, which was zero in the natural state, are constant. If we then further deform the spring slightly, the tension will change. We denote the tension in the general deformed case by $\widetilde{T}$. We assume that for a small change δl in the length l, the change in tension $\widetilde{T} - T$ is proportional to $\delta l / l$:

$$\widetilde{T} = T + K_{\text{eff}}\,\frac{\delta l}{l}, \tag{1.47}$$

where K_{eff} is an appropriate proportionality constant, an effective elastic modulus of the stretched or compressed spring.

Taking this into account, we can repeat the above calculations with minor modifications, with x now referring to the position of a marked point in the uniform reference state, and $u(x,t) = \tilde{x}(x,t) - x$ being its deformation at time t (see Figure 1.15).

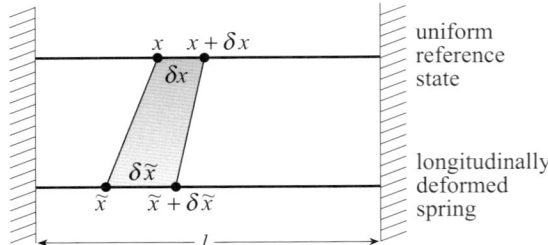

Figure 1.15 Schematic representation of a spring represented by a line along its axis. A small element $[x, x+\delta x]$ of the uniformly stretched or compressed spring of length l corresponds to an element $[\tilde{x}, \tilde{x}+\delta\tilde{x}]$ in the moving spring of length l, at a fixed instant of time.

As will be shown in Exercise 1.14, the tension $\widetilde{T}(x,t)$ is given by

$$\widetilde{T}(x,t) = T + K_{\text{eff}}\, u_x(x,t), \tag{1.48}$$

which involves the constant tension T; compare equation (1.40). The tension gives rise to a force

$$\delta \boldsymbol{F} = K_{\text{eff}}\left[u_x(x+\delta x, t) - u_x(x,t)\right]\mathbf{i} \tag{1.49}$$

acting on a string element $[\tilde{x}(x,t), \tilde{x}(x+\delta x, t)]$.

Following the same route as in equations (1.41)–(1.46) (see Exercise 1.14), we once more derive the wave equation (1.1),

$$u_{tt}(x,t) = c^2\, u_{xx}(x,t), \tag{1.50}$$

with the wave speed c now given by

Longitudinal wave speed for a stretched or compressed spring

$$c = \sqrt{\frac{K_{\text{eff}}}{\rho}}. \tag{1.51}$$

Comparing this result with equation (1.46), we find that the elastic modulus K and the density ρ_0 are replaced by the *effective* value K_{eff} and the density ρ of the uniformly deformed spring.

The following exercise goes through the derivation step by step.

Exercise 1.14

Consider a spring of natural length l_0 that has been uniformly stretched or compressed to an equilibrium length l. It has a constant tension T, and for small changes δl of its length it obeys the linear relation (1.47).

(a) Consider the spring in a general situation at a fixed time t. Calculate the tension $\widetilde{T}(x,t)$ by considering a small element $[\tilde{x}, \tilde{x}+\delta\tilde{x}]$ of length $\delta\tilde{x}$ which corresponds to an element of length δx in the uniformly stretched case, and derive equation (1.48).

(b) Derive the wave equation (1.50) and the wave speed (1.51) for this case, following the steps of equations (1.41)–(1.46).

1.5 Longitudinal waves

If throughout the motion we stay within the linear regime, such that any non-linear contributions can be neglected entirely, Exercise 1.15 below shows that the constant K_{eff} is given by $K_{\text{eff}} = K + T$, where K is the elastic modulus that describes the relationship between extension and tension over the whole range of deformation.

Exercise 1.15

Consider a spring of natural length l_0 and elastic modulus K. The spring is uniformly extended to a length $l > l_0$.

(a) Assuming that we stay within the linear regime, what is the constant tension T in the spring?

(b) Calculate the tension $\widetilde{T}$ corresponding to a change in length from l to L, assuming that we stay within the linear regime. Show that this yields a relation of the form (1.47) with $K_{\text{eff}} = K + T$.

In this situation, we obtain for the wave speed

$$c = \sqrt{\frac{K+T}{\rho}} = \sqrt{\frac{(K+T)\,l}{\rho_0 l_0}}, \tag{1.52}$$

where we use $\rho = \rho_0 l_0 / l$ for the density in the stretched spring. This equation for the wave speed c can be regarded as a generalisation of equation (1.46) to the case of a pre-stretched spring, and in the limit as $T \to 0$, $l \to l_0$ and $\rho \to \rho_0$, we recover equation (1.46). Furthermore, by Hooke's law, the equilibrium tension T is given by $T = K(l - l_0)/l_0$. Eliminating either T or l in the expression for the wave speed (1.52) gives $c = \sqrt{K/\rho_0}\,(l/l_0)$ or $c = (K+T)/\sqrt{\rho_0 K}$, respectively.

Exercise 1.16

Let us return to the solutions of the wave equation introduced in Exercise 1.9 (page 23). Because the wave equation for $u(x,t)$ and the boundary conditions $u(0,t) = u(l,t) = 0$ are the same for the longitudinal as for the transverse motion, the solutions $u_n(x,t)$ found in Exercise 1.9 also describe the longitudinal motion of a spring.

(a) Sketch the function $\tilde{x}(x, t_j)$ for the solution $u_2(x,t)$ of Exercise 1.9 at various time instances t_j.

(b) Use equation (1.40) to calculate the tension $T(x,t)$ for the function $u_n(x,t)$. Where is the tension minimal and maximal?

(c) Using equation (1.33), show that the relation $\rho(x,t) = \rho_0/[1 + u_x(x,t)]$ holds.

(d) Use this result to calculate $\rho(x,t)$ for $u_n(x,t)$.

In most real systems, non-linear terms cannot be neglected if large deformations are considered. However, some systems, such as a stretched helical spring, may be well described by a linear relationship for tensions of the order of the elastic modulus K, which means for extensions up to twice the natural length, or even more.

1.5.3 Transverse and longitudinal waves in a stretched string

So far, we have considered longitudinal waves only in springs. In principle, however, longitudinal wave motion also happens in a stretched *string*. Previously, we met the problem that a string cannot have negative tension, because the string will go slack. However, if we stretch the string first and then deform it only a little more, this will not happen. So the wave equation applies to small deformations of an originally stretched string, and the wave speed c is given by equation (1.51), with an appropriate value of the effective elastic modulus K_{eff}.

So both transverse and longitudinal waves can travel on a stretched string. Both motions are described, within our approximations, by the same linear wave equation (1.1), albeit with different values of the wave speed c.

For transverse waves, the wave speed is given by equation (1.25) as $\sqrt{T/\rho}$, whereas for longitudinal waves, provided the extension is within the linear regime, we obtain $\sqrt{(K+T)/\rho}$ in equation (1.52). Because K is positive, the wave speed for the longitudinal waves in a stretched string is always larger than the wave speed for transverse waves.

Where does this difference come from? For transverse waves, the restoring force given in equation (1.18) on page 21 is determined by the fixed uniform tension T of the stretched string and its transverse deformation. In this case, Hooke's law does not enter at all, at least within our approximation that changes in the length of the string and its tension during the motion are negligible. For longitudinal waves on a stretched string, however, the restoring force given in equation (1.49) is the elastic force due to the time-dependent stretching of the string. Here, we do consider the local extension of the string, and thus Hooke's law enters. A flexible string is easier to deform in the transverse direction than in the longitudinal direction; the string appears much 'stiffer' in the longitudinal case, and the wave speed is thus larger.

For transverse waves, the restoring force in equation (1.18) is proportional to the tension $T > 0$, and thus vanishes for $T = 0$, which corresponds to a slack string. This is reflected in a vanishing wave speed in equation (1.25) as $T \to 0$. In contrast, the longitudinal wave speed becomes $\sqrt{K/\rho_0}$ as $T \to 0$, which corresponds to the case of longitudinal waves in springs and rods considered above. As mentioned previously, a slack string will bend and thus cannot support longitudinal waves, but for a string with a small positive tension T, $\sqrt{K/\rho_0}$ is the approximate wave speed for longitudinal waves, whereas the wave speed for transverse waves is almost zero.

Exercise 1.17

Consider a string which has linear density $\rho_0 = 50\,\text{g}\,\text{m}^{-1}$ and elastic modulus $K = 2 \times 10^5\,\text{N}$. At equilibrium, the string is uniformly extended, within the linear regime, such that it has a tension $T = 10^3\,\text{N}$.

(a) Calculate the strain $\varepsilon = (l - l_0)/l_0$ for the uniformly stretched string.

(b) Calculate the wave speed for longitudinal waves in the stretched string.

(c) What is the wave speed for transverse waves on the stretched string? Compare the two wave speeds.

(d) What is the minimal wave speed for longitudinal waves in this string? More precisely, what is the longitudinal wave speed for the case $T \to 0$?

1.6 The wave equation

The same partial differential equation, the linear wave equation (1.1), describes both the transverse and the longitudinal motion of a stretched string. In fact, it describes a variety of physical systems. A further example, sound waves in a narrow tube, is discussed in an (optional) appendix at the end of this chapter.

Why is it the case that the same equation describes what appear to be completely different physical situations? And how general is this equation? In order to answer these questions, let us briefly consider the essential ingredients that were used in the derivations.

1.6.1 Scope of the wave equation

What were the main ideas that we used in order to derive the wave equation for transverse and longitudinal waves? First, we described the system by a function $u(x,t)$. This function, depending on space and time, quantifies the small deformation from the equilibrium situation. There are two basic ingredients that enter the derivations of the wave equation for $u(x,t)$.

The first ingredient is that we express the net force that acts locally on a small element of our system, at position x and at time t, in terms of partial derivatives of the displacement function $u(x,t)$. The main point is the assumption, or approximation, that a disturbance of the system causes a restoring force that increases *linearly* with the disturbance. This is a good approximation to many real systems; from a mathematical point of view, this might be considered as a consequence of Taylor's theorem. Provided that the function that describes the restoring force upon deformation is smooth, and that the amplitude of the deformation is small, we can use a Taylor expansion of that function and, as an approximation, truncate after the linear term.

In the first instance, this linear response approximation yields just one derivative with respect to x, as in equations (1.18) and (1.40) on pages 21 and 29, respectively. The second derivative comes from considering the sum of the forces when calculating the resulting net force on an element of length δx (see equations (1.18) and (1.41)) and taking the limit as $\delta x \to 0$.

The second important ingredient is that the system can be described by Newton's second law, which relates the acceleration $u_{tt}(x,t)$ to the net force acting on a small element of our system at position x.

Any system that shares these two basic ingredients can be described by the wave equation (1.1). Two further examples are the twisting motion of a thin rod or wire, where $u(x,t)$ is the angular displacement, and water waves in a narrow canal, where $u(x,t) = h(x,t) - h_0$ is the difference in the height of the water $h(x,t)$ with respect to the height h_0 at rest, and the restoring force is due to gravity.

One should keep in mind that waves are not restricted to systems that are described by Newtonian mechanics. For example, the relevant equation of motion for electromagnetic waves is not Newton's second law, but is given by the set of *Maxwell's equations* relating electric and magnetic fields to charges and currents, respectively. The wave equation still results in this case.

1.6.2 A few remarks on the history of the wave equation

The transverse vibrations of a string are not only a particularly instructive mechanical example; their investigation initiated the derivation and analysis of the wave equation. The motivation came primarily from the study of musical sounds, notably from a violin string. Already in 1638, Marin Mersenne (1588–1648) had stated a law that determined the frequency of the vibration of such a string in terms of the tension and the material constants. But for a long time there was no explanation for the motion of the string. A discrete version, a 'string of beads', was treated by Johann Bernoulli (1667–1748) in 1727. Then, in 1746, Jean le Rond d'Alembert (1717–1783) took the step to tackle the continuous string, and derived the partial differential equation (1.1). He also found a form of the general solution of the wave equation, now often referred to as d'Alembert's solution, which will be discussed in the next chapter.

D'Alembert's work was a major achievement at that time, because it was one of the first investigations and applications of partial differential equations. The resulting, often rather controversial, discussions involved an illustrious circle of mathematicians, notably Leonhard Euler (1707–1783), Johann Bernoulli's son Daniel Bernoulli (1700–1782), Joseph-Louis Lagrange (1736–1813) and Pierre-Simon Laplace (1749–1827). More complicated situations were soon attacked. In 1759, Euler and Lagrange independently studied two- and three-dimensional wave equations.

The investigation of the wave equation and the diffusion equation (which will be considered later in this course), and the controversies about solutions and their analytic properties, initiated the theory of differential equations, and led Jean Baptiste Joseph Fourier (1768–1830) to introduce an expansion of a function as a trigonometric series, now known as the *Fourier series*. The investigation of problems posed by physics proved very fruitful, and was vital in the development of analysis.

1.7 Summary and Outcomes

In this chapter we introduced the basic elastic properties of solids. In particular, we discussed Hooke's law, the limitations within which it is applicable, and the characterisation of materials by their elastic modulus or Young's modulus. The concept of tension in one-dimensional systems such as springs, rods or strings was introduced and discussed in detail. The wave equation was then derived for transverse and longitudinal waves, using Newton's second law. Some example solutions of the wave equation were investigated in the exercises.

After working through this chapter and the exercises, you should:
- know basic properties about the elastic behaviour of solids;
- know how to apply Hooke's law to a spring or stretched string;
- be able to calculate the tension in springs, rods and strings;
- be able to solve statics problems involving springs, rods and strings;
- know how to apply Newton's second law to continuous one-dimensional systems;
- understand the important steps in the derivation of the wave equation for transverse and longitudinal waves;
- understand the approximations that enter the derivation of the wave equation;
- know what the wave equation looks like;
- be able to relate the wave speed to the physical properties of the medium in both transverse and longitudinal cases;
- know why the wave equation arises in many systems in physics;
- be able to name some properties of wave motion;
- be able to test whether a function solves the wave equation.

1.8 Further Exercises

Exercise 1.18

Consider a spring that consists of two uniform parts that are joined together end to end. The first part has natural length $l_1 = 2\,\text{m}$ and elastic modulus $K_1 = 10\,\text{N}$; the second part has natural length $l_2 = 4\,\text{m}$ and elastic modulus $K_2 = 20\,\text{N}$. The combined spring is fixed at one end, and the other end is subject to an external force of magnitude F that extends the spring in equilibrium to a length $L = L_1 + L_2 = 8\,\text{m}$.

(a) What forces act at the place where the two parts are joined? What is the tension within each part of the spring, in terms of F?

(b) What is the extension of each part of the spring?

(c) What is the magnitude of the force applied to extend the combined string?

(d) What can you conclude for the tension in an extended spring in equilibrium, which has an elastic modulus $K(x)$ that varies along the spring, so it is non-uniform already in its natural state?

Exercise 1.19

Consider a transversely deformed string, fixed at its end points at $x = 0$ and $x = 1$, which is described by a function $u(x) = \sin(\pi x)/40$. We assume that the string has constant tension T.

(a) Calculate the slope of the string, $u'(x)$.

(b) What is the maximal displacement of the string, and at which position x does it occur?

(c) Find the extremal values of the slope.

(d) Calculate the force $\delta \boldsymbol{F}(x)$ on a string element $[x, x + \delta x]$, in first order of δx, and compare the result with the second derivative $u''(x)$.

Exercise 1.20

Consider an initially uniform spring of natural length l_0, total mass m, constant linear density $\rho_0 = m/l_0$ and elastic modulus K. The spring hangs vertically from the ceiling, stretched to a length $l > l_0$ by its own weight – see Figure 1.16. The gravitational force on a body of mass m is $\boldsymbol{W} = mg\,\mathbf{i}$, where $g \simeq 9.81\,\mathrm{m\,s^{-2}}$ is the acceleration due to gravity. We choose coordinates as shown in Figure 1.17.

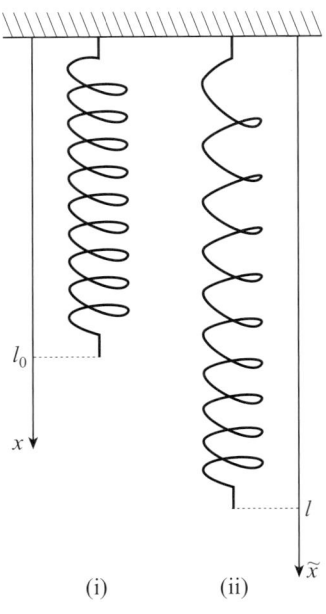

Figure 1.16 A suspended spring in (i) its natural state with gravity 'switched off' and (ii) a non-uniformly stretched state with gravity 'switched on'

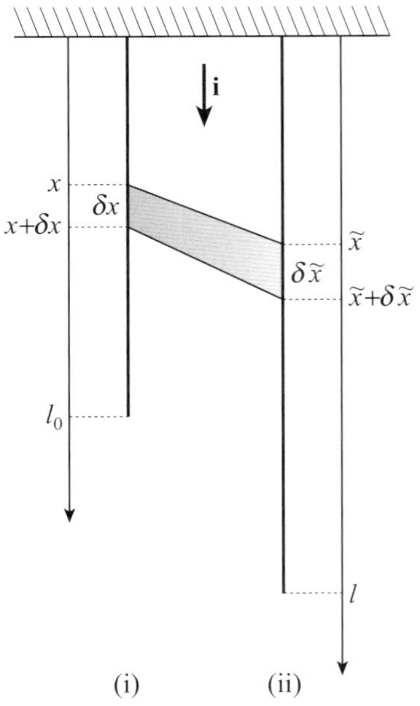

Figure 1.17 Coordinates x and $\tilde{x}$ along the spring of Figure 1.16

(a) Denote by $T(x)$ the tension at position $\tilde{x}(x)$ in the stretched spring. What is the tension $T(0)$ at the top $\tilde{x} = x = 0$ of the spring? What is the tension $T(l_0)$ at its lower end point $\tilde{x}(l_0) = l$?

(b) Show that the tension is given by
$$T(x) = mg\left(1 - \frac{x}{l_0}\right).$$

(c) Calculate the density ratio $\rho(x)/\rho_0$.

(d) Use equation (1.30) to calculate the deformation $\tilde{x}(x)$ from the tension $T(x)$.

(e) Calculate the length l of the stretched spring.

(f) Consider the case where $mg = 2K$ and $l_0 = 1\,\mathrm{m}$, and calculate the length l. Sketch the corresponding deformation $\tilde{x}(x)$, and calculate x as a function of $\tilde{x}$.

(g) Continuing with the values of part (f), calculate the tension $T(x)$ and the linear density $\rho(x)$, expressing the results in terms of the length $\tilde{x}$ along the stretched spring. Sketch the results, and compare these with the corresponding values for a *uniformly* stretched spring of length l.

Exercise 1.21

A flexible string of length L and constant linear density $\rho = m/L$ is suspended by its two ends, which are fixed at positions $(x, y) = (0, 0)$ and $(x, y) = (l, 0)$, where $l < L$ (see Figure 1.18).

1.8 Further Exercises

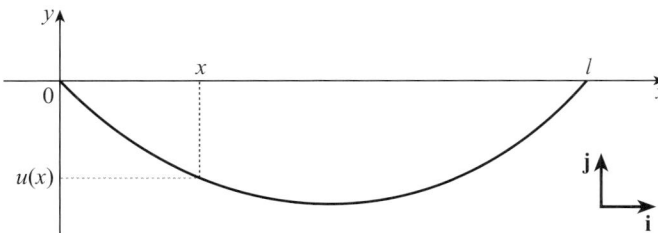

Figure 1.18 A suspended flexible string

We assume that we can neglect the extension of the string due to the tension $T(x)$. We shall calculate the equilibrium shape $y = u(x)$ of the string, which is known as a *catenary*.

(a) Consider a small element $[x, x + \delta x]$ of the string. Denote by $\phi(x)$ the angle between the tangent to the string and the $\mathbf{i}$ direction at position x. What are the forces acting on this string element? Resolve the net force $\delta \mathbf{F}$ on the string element in the $\mathbf{i}$ and $\mathbf{j}$ directions. (Note that because the element is in static equilibrium, $\delta \mathbf{F} = \mathbf{0}$.)

(b) By considering the component of the force $\delta \mathbf{F}$ in the $\mathbf{i}$ direction, show that $T_1 = T(x) \cos[\phi(x)]$ is constant.

(c) Consider now the component of $\delta \mathbf{F}$ in the $\mathbf{j}$ direction. Using the previous result, show that this component is given by $T_1(\tan[\phi(x + \delta x)] - \tan[\phi(x)])$.

(d) Denote by $u(x)$ the function that describes the displacement of the string at position x. Express $\tan[\phi(x)]$ in terms of $u(x)$, and derive the differential equation $u''(x) = k\sqrt{1 + [u'(x)]^2}$ for $u(x)$ from the $\mathbf{j}$ component of $\delta \mathbf{F}$, where $k = \rho g / T_1$ (k is not the spring stiffness in this case).

(e) Substitute $f(x) = du/dx$, and solve the differential equation for f by separation of variables. (You may find the integral $\int dy/\sqrt{1+y^2} = \operatorname{arcsinh}(y)$ useful.)

> Separable equations are discussed in Block 0, Subsection 1.3.2.

(f) Derive the solution for $u(x)$ in terms of two constants of integration.

(g) Insert the boundary conditions $u(0) = u(l) = 0$ to compute the values of the constants of integration, and interpret them.

(h) The constant k, which essentially involves T_1, is as yet undetermined. The only information that we have not yet exploited is the length of the string L. Use

$$L = \int_0^l dx \, \sqrt{1 + u'(x)^2} \qquad (1.53)$$

to derive the equation $kL = 2\sinh(kl/2)$ for k.

> This formula for the length of a curve is given in the Handbook.

(i) The resulting equation for k cannot be solved in terms of elementary functions. However, for small values of kl, we can use the Taylor expansion to third order, $2\sinh(kl/2) \simeq kl + (kl)^3/24$. Calculate k in this approximation. When is the approximation appropriate?

(j) Suppose that the string has length $L = 7\,\text{m}$, mass $m = 1\,\text{kg}$ and $l = 6\,\text{m}$. Calculate k in the approximation considered above. What is the maximum distance of the string from the horizontal?

Exercise 1.22

(a) Consider the function $u(x,t) = A/\cosh(x - ct)$. Show that this is a solution of the wave equation (1.1). What does this solution look like at time $t = 0$? How does it change with time?

(b) Show that the function $u(x,t) = A/\cosh(x + ct)$ also solves the wave equation (1.1). What does this solution look like?

Exercise 1.23

This exercise is optional.

Let us examine the simplifying assumptions made in the derivation of the wave equation for transverse motion of a stretched string (see page 19). If the string is deformed from its equilibrium position as described by the function $u(x,t)$, it has to be stretched to a length $L > l$ given by equation (1.53). The stretching means that the tension $\widetilde{T}$ and the linear density $\widetilde{\rho}$ vary along the string and hence differ from the equilibrium tension T and density ρ of the uniformly stretched string.

(a) We assume that $|u_x| \leq \epsilon$ for all $x \in [0, l]$. Show that for small ϵ, the relative length change $(L - l)/l$ is of order ϵ^2.

(b) Assume that the string satisfies Hooke's law (1.8). What are the changes $(\rho - \widetilde{\rho})/\rho$ in linear density and $(\widetilde{T} - T)/T$ in tension if the string is uniformly stretched to length $L > l$? Show that these changes are of order ϵ^2.

(c) Show that in the general case of non-uniform stretching, the change $[\rho - \widetilde{\rho}(x)]/\rho$ in the linear density is of order ϵ^2.

(d) Show that in the general case, the change $[\widetilde{T}(x) - T]/T$ in the tension is of order ϵ^2.

(e) Given the condition on the partial derivative $|u_x| \leq \epsilon$, and the condition that $u(0,t) = u(l,t) = 0$ at the ends of the string, argue that $|u(x,t)| \leq \epsilon l/2$, so $|u(x,t)|$ is of order ϵ.

1.9 Appendix: Sound waves in a tube (optional)

Here we consider a different example of wave motion: sound waves in a gas. We restrict ourselves to a particularly simple situation where the gas is confined to a narrow tube. In this case, we need to consider the variation of gas properties only along the length of the tube, neglecting any changes over its cross-section. This means that we can effectively treat it as a one-dimensional system. Again, we assume that we may neglect the effect of gravity or any frictional forces.

1.9.1 Volume and pressure

Consider a cylindrical tube of constant cross-sectional area A, filled with gas, which we can compress with a piston (see Figure 1.19). How do we characterise the gas inside the tube? First, it fills a volume $V = Al$, which depends on the position of the piston l. Secondly, if we move the piston in the $-\mathbf{i}$ direction, the gas becomes compressed, and it exerts a force on the piston. The force tries to restore a certain equilibrium position of the piston which depends on the surrounding environment. The net force on the piston is due to the *pressure* P of the gas inside the tube and the pressure P_{amb} of the ambient (surrounding) gas, for instance atmospheric pressure. Pressure measures the force per unit area that a gas exerts on a wall of a container, so the total force on the piston, when it is out of equilibrium, is

$$\boldsymbol{F} = \boldsymbol{F}_1 + \boldsymbol{F}_2 = AP\mathbf{i} + AP_{\text{amb}}(-\mathbf{i}) = A(P - P_{\text{amb}})\mathbf{i}, \qquad (1.54)$$

which is proportional to the difference in pressure $P - P_{\text{amb}}$. The pressure difference in equation (1.54) comes from adding the force $\boldsymbol{F}_1 = AP\mathbf{i}$ that the gas inside the tube exerts on the piston and the force $\boldsymbol{F}_2 = -AP_{\text{amb}}\mathbf{i}$ that the ambient air exerts on the piston (see Figure 1.19).

It does not matter if the cross-sectional area at the end of the piston differs from A, because the ambient pressure exerts forces of equal magnitude in all directions, thus forces on additional parts of the piston cancel.

1.9 Appendix: Sound waves in a tube

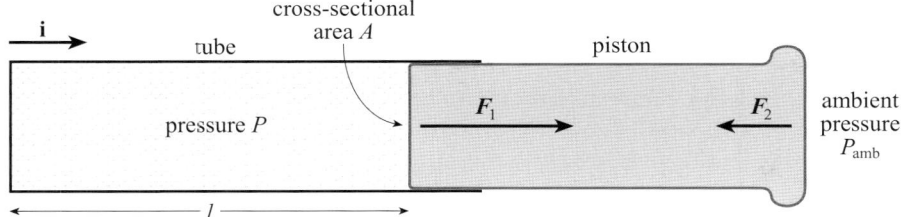

Figure 1.19 Forces $\boldsymbol{F}_1$ and $\boldsymbol{F}_2$ on a piston in a tube of cross-sectional area A, containing a volume $V = Al$ of gas at pressure P, in an environment with ambient pressure P_{amb}

In equation (1.54), the pressure difference can also become negative, when the pressure in the tube is lower than outside. In this case, the force acts in the opposite direction, trying to restore the equilibrium position by pushing the piston inwards.

How are the pressure and volume of the gas related? Clearly, the pressure increases when we compress the gas, i.e. when its volume is reduced. A small change $\delta V = A\,\delta l$ in volume results in a small change δP in pressure:

$$\frac{\delta P}{P} \simeq -\gamma \frac{\delta V}{V} = -\gamma \frac{\delta l}{l}, \qquad (1.55)$$

where γ is a constant. This relationship between pressure and volume holds for a large number of real gases, with proportionality constants γ roughly given by $\gamma = \frac{5}{3}$ for monatomic gases, such as the noble gases helium (He), neon (Ne) and argon (Ar), and $\gamma = \frac{7}{5}$ for gases which consist of diatomic molecules, such as oxygen (O_2) and nitrogen (N_2). The latter value is also appropriate for air, which consists of about 99% of a mixture of nitrogen and oxygen, and 1% other gases, mainly argon.

Note that the relation (1.55) is completely analogous to Hooke's law (1.8) for the spring, in the sense that the resulting force $\delta \boldsymbol{F}$ on the piston due to the pressure difference δP, which is

$$\delta \boldsymbol{F} \simeq A\,\delta P\,\mathbf{i} \simeq -AP\gamma \frac{\delta l}{l}\,\mathbf{i}, \qquad (1.56)$$

depends *linearly* on the change in length δl. The negative sign is due to the fact that the pressure of the gas *decreases* with increasing length l. For a positive change δl, the resulting change in force acts to reduce the volume, thus acts in the negative $\mathbf{i}$ direction.

Volume density

Another property of a gas is its volume density $\rho^{(V)}$, which is the mass per unit volume: $\rho^{(V)} = m/V$ in a uniform gas. If there is a fixed amount of gas, the mass m is constant when the volume V is varied, therefore

$$0 = \frac{\partial m}{\partial V} = \frac{\partial}{\partial V}\left(\rho^{(V)} V\right) = \frac{\partial \rho^{(V)}}{\partial V} V + \rho^{(V)}. \qquad (1.57)$$

Thus a small change δV in volume results in a change

$$\frac{\delta \rho^{(V)}}{\rho^{(V)}} \simeq -\frac{\delta V}{V} = -\frac{\delta l}{l} \simeq \frac{1}{\gamma}\frac{\delta P}{P} \qquad (1.58)$$

in density, where we have used equation (1.55) to relate this to the change in pressure.

1.9.2 Derivation of the wave equation

We now consider the gas in the cylindrical tube for a fixed position of the piston. This is analogous to a stretched string fixed at its end points. In equilibrium, at rest, the pressure and density of the gas are constant, say P_0 and ρ_0, respectively. Now we consider what happens if we suddenly disturb the system by applying an instantaneous small change in density or pressure. This disturbance can travel along the tube, and we eventually use Newton's second law to derive once more that the motion is governed by the wave equation. The waves can be regarded as *pressure waves* (sometimes called *compression waves*) or *density waves*, as both pressure and density, which are related by equation (1.58), vary with position and time. In the real world, our ears detect such changes in gas pressure as *sound*, which is why these waves are called *sound waves*.

Consider the small amount of gas of mass δm initially located within a slice of the tube of width δx at position x. At a fixed instant t, this gas has moved and now occupies a slice of width $\delta \tilde{x}$ located at position $\tilde{x}$ (see Figure 1.20).

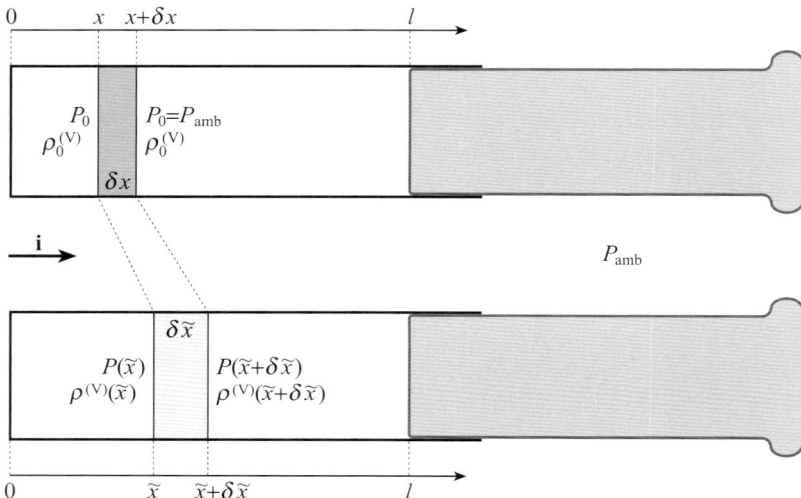

Figure 1.20 Snapshot of the gas in a slice of the tube in equilibrium (top), where $P_0 = P_{\text{amb}}$, and in a general state (bottom), at a fixed instant of time. The position of the piston is kept fixed.

We assume that the pressure and density of the gas depend only on the position $\tilde{x}$ along the tube and do not vary across its cross-section, so they become functions $P(x,t)$ and $\rho^{(V)}(x,t)$, respectively. Let $u(x,t)$ again be the function that describes the 'deformation' from the equilibrium position:

$$u(x,t) = \tilde{x}(x,t) - x. \tag{1.59}$$

Density fluctuations

The density $\rho^{(V)}(x,t)$ in the slice of width $\delta \tilde{x}$ can be related to the density $\rho_0^{(V)}$ at rest by noticing that the mass δm in the slice stays the same:

$$\delta m = A\,\rho_0^{(V)}\,\delta x \simeq A\,\rho^{(V)}(x,t)\,\delta \tilde{x}. \tag{1.60}$$

1.9 Appendix: Sound waves in a tube

The approximation arises because we neglect any variations of $\rho^{(V)}(x,t)$ over the interval of length $\delta\tilde{x}$. This yields

$$\rho_0^{(V)} \simeq \rho^{(V)}(x,t)\frac{\delta\tilde{x}}{\delta x}. \tag{1.61}$$

In the limit as $\delta x \to 0$, and using equation (1.59), we obtain

$$\rho_0^{(V)} = \rho^{(V)}(x,t)\frac{\partial\tilde{x}}{\partial x} = \rho^{(V)}(x,t)\left[u_x(x,t) + 1\right]. \tag{1.62}$$

Solving for $\rho^{(V)}(x,t)$, we find

$$\rho^{(V)}(x,t) = \frac{\rho_0^{(V)}}{1 + u_x(x,t)} \simeq \rho_0^{(V)}[1 - u_x(x,t)], \tag{1.63}$$

where the last relation follows by assuming that the partial derivative u_x is small, so that it is sufficient to use the Taylor expansion $1/(1+z) = 1 - z + \cdots$ to first order, with $z = u_x(x,t)$. The change in density is thus given by the partial derivative of the displacement function $u(x,t)$ with respect to the position x:

$$\frac{\rho^{(V)}(x,t) - \rho_0^{(V)}}{\rho_0^{(V)}} \simeq -u_x(x,t). \tag{1.64}$$

Pressure and force

The force that acts on the gas within the slice of width $\delta\tilde{x}$ depends on the difference in pressure at the two ends of the slice. This is similar to the situation of longitudinal displacements in a spring, where the analogous quantity is the difference in tension; here, pressure plays the role of tension. The change in pressure is related to the change in density according to equation (1.58), so equation (1.64) yields

$$\frac{P(x,t) - P_0}{P_0} \simeq \gamma\frac{\rho^{(V)}(x,t) - \rho_0^{(V)}}{\rho_0^{(V)}} \simeq -\gamma u_x(x,t). \tag{1.65}$$

Applying Newton's second law

According to equation (1.56), the force $\delta\boldsymbol{F}$ on the gas within our slice of width $\delta\tilde{x}$ is given by the difference in pressure:

$$\delta\boldsymbol{F} \simeq A[P(x,t) - P(x+\delta x, t)]\,\mathbf{i}. \tag{1.66}$$

Using equation (1.65), this gives

$$\delta\boldsymbol{F} \simeq A\gamma P_0[u_x(x+\delta x, t) - u_x(x,t)]\,\mathbf{i}. \tag{1.67}$$

According to Newton's second law, the force $\delta\boldsymbol{F}$ is equal to the mass δm in the slice times the acceleration $\boldsymbol{a} = u_{tt}(x,t)\mathbf{i}$, so

$$A\gamma P_0[u_x(x+\delta x, t) - u_x(x,t)]\,\mathbf{i} \simeq A\rho_0^{(V)} u_{tt}(x,t)\,\delta x\,\mathbf{i}, \tag{1.68}$$

using equations (1.60) and (1.67). Dividing by δx and taking the limit as $\delta x \to 0$, we once more recover the wave equation

$$u_{tt}(x,t) = \frac{\gamma P_0}{\rho_0^{(V)}}\, u_{xx}(x,t), \tag{1.69}$$

which gives a speed of sound

Speed of sound in a gas

$$c = \sqrt{\frac{\gamma P_0}{\rho_0^{(V)}}}. \tag{1.70}$$

Exercise 1.24

Air is a mixture of several gases, notably oxygen and nitrogen. For such a mixture, the value $\gamma = \frac{7}{5}$ is appropriate. Consider normal conditions with pressure $P_0 = 1.013 \times 10^5 \, \text{Pa} = 1.013 \times 10^5 \, \text{N m}^{-2}$, which is normal atmospheric air pressure, and volume density $\rho_0^{(V)} = 1.2 \, \text{g l}^{-1} = 1.2 \, \text{kg m}^{-3}$.

(a) What is the speed of sound c derived from these data?

(b) At a height of $10\,000 \, \text{m}$ above sea level, the pressure falls to about $P_0 = 2.6 \times 10^4 \, \text{N m}^{-2}$, and the density is just $\rho_0^{(V)} = 0.41 \, \text{kg m}^{-3}$. What is the speed of sound here?

(c) Helium is a monatomic noble gas, for which $\gamma = \frac{5}{3}$. At normal pressure $P_0 = 1.013 \times 10^5 \, \text{N m}^{-2}$, its density is $\rho_0^{(V)} = 0.1785 \, \text{kg m}^{-3}$, so it is much lighter than air. (This is, of course, the reason why it is used in airships and balloons.) What is the speed of sound c in a helium atmosphere?

(d) As will be shown in the next chapter, the frequency ν, wavelength λ and wave speed c are related by $\lambda \nu = c$. If a sound pipe produces a wave of a certain wavelength λ, how do the frequencies in air and in helium differ from each other? What happens to your voice if you inhale helium and speak?

Solutions to Exercises in Chapter 1

Solution 1.1

Inserting $u(x,t) = a_1 u_1(x,t) + a_2 u_2(x,t)$ into the wave equation (1.1) gives

$$u_{tt}(x,t) - c^2 u_{xx}(x,t)$$
$$= a_1 \left(\frac{\partial^2 u_1}{\partial t^2}(x,t) - c^2 \frac{\partial^2 u_1}{\partial x^2}(x,t) \right) + a_2 \left(\frac{\partial^2 u_2}{\partial t^2}(x,t) - c^2 \frac{\partial^2 u_2}{\partial x^2}(x,t) \right) = 0,$$

because the derivatives are themselves linear, and u_1 and u_2 both satisfy the wave equation. So $u(x,t)$ satisfies the wave equation.

Solution 1.2

(a) We use the value $Y = 200\,\text{GPa} = 2 \times 10^{11}\,\text{N}\,\text{m}^{-2}$ from Table 1.1. The elastic modulus is given by $K = AY$, so for $A = 1\,\text{mm}^2 = 10^{-6}\,\text{m}^2$, we have $K = 2 \times 10^5\,\text{N}$, and for $A = 2\,\text{mm}^2$, we have $K = 4 \times 10^5\,\text{N}$.

(b) From part (a), the elastic modulus of the steel wire is $K = 2 \times 10^5\,\text{N}$, so the magnitude F of the force needed to extend it by 1% is $F = K/100 = 2000\,\text{N}$. For the aluminium wire we have $Y = 70\,\text{GPa}$, so the elastic modulus is $K = 7 \times 10^4\,\text{N}$, and the force is thus $F = K/100 = 700\,\text{N}$.

The masses are $m = F/g \simeq 204\,\text{kg}$ and $m = F/g \simeq 71\,\text{kg}$, respectively.

(c) The volume of the wire in its natural state is $V = Al_0 = 10^{-5}\,\text{m}^3$, thus the mass of the wire is $m = V\rho^{(V)} = 27.1\,\text{g}$, and the total mass stays the same for the extended wire. The elastic modulus of the wire is $K = AY = 7 \times 10^5\,\text{N}$. The magnitude F of the force corresponding to a strain $\varepsilon = (l - l_0)/l_0 = 1/50$ is $F = K/50 = 1.4 \times 10^4\,\text{N}$, provided that a strain of 2% is still within the linear regime.

Solution 1.3

(a) According to equation (1.13), the linear density is given by $\rho = \rho_0 l_0/l < \rho_0$, so the density is smaller in the stretched spring. The tension is given by equation (1.9) as $T = Kl/l_0 - K$. Using $l/l_0 = \rho_0/\rho$ gives $T = K\rho_0/\rho - K > 0$, thus

$$\rho = \frac{K}{K+T} \rho_0.$$

(b) Equations (1.13) and (1.9) also hold for $l < l_0$, so the resulting equation is the same. However, now $\rho > \rho_0$, as the linear density in the compressed spring is larger than in the natural state, and $T < 0$.

Solution 1.4

(a) Solving equation (1.9) for the natural length l_0 yields

$$l_0 = \frac{K}{K+T} l = \frac{100}{125} l = \frac{4}{5} l = 24\,\text{cm}.$$

(b) Using the *negative* tension in the equation above gives

$$l_0 = \frac{K}{K+T} l = \frac{100}{75} l = \frac{4}{3} l = 40\,\text{cm}.$$

(c) According to equation (1.9), we have $(l - l_0)/l_0 = T/K$. Thus, in the first case $(l - l_0)/l_0 = 25/100 = \frac{1}{4}$, and in the second case $(l - l_0)/l_0 = -25/100 = -\frac{1}{4}$.

Solution 1.5

(a) The mass follows from the relation $\rho_0 = m/l_0$:

$$m = \rho_0 l_0 = 0.1\,\text{kg} = 100\,\text{g}.$$

(b) According to Hooke's law, equation (1.8), the magnitude of the force is

$$F = K\left|\frac{l-l_0}{l_0}\right| = 0.5K = 50\,\text{N}.$$

As the spring is extended, the tension is positive, so $T = F = 50\,\text{N}$.

The linear density is given by equation (1.13) as

$$\rho = \frac{l_0}{l}\rho_0 = \frac{2}{3}\rho_0 \simeq 0.067\,\text{kg m}^{-1},$$

or by inserting the values for m and l in $\rho = m/l$.

(c) The magnitude of the force is now

$$F = K\left|\frac{l-l_0}{l_0}\right| = 0.2K = 20\,\text{N}.$$

As the spring is compressed, the tension is negative, so $T = -F = -20\,\text{N}$.

The linear density is given by

$$\rho = \frac{l_0}{l}\rho_0 = \frac{5}{4}\rho_0 = 0.125\,\text{kg m}^{-1},$$

or by using $\rho = m/l$.

Solution 1.6

(a) The slope is $u'(x) = (1-2x)/10$.

(b) The displacement $u(x)$ has an extremum where $u'(x) = 0$, which occurs when $x = \frac{1}{2}$. At the end points, $u(0) = u(1) = 0$, so $x = \frac{1}{2}$ gives the maximal displacement, namely $u(\frac{1}{2}) = \frac{1}{40}$.

(c) The second derivative is $u''(x) = -\frac{1}{5}$, which is constant. Thus the extremal values of the slope occur at $x = 0$ and $x = 1$, and are $u'(0) = \frac{1}{10}$ and $u'(1) = -\frac{1}{10}$.

(d) According to equation (1.18), the force is

$$\delta \boldsymbol{F}(x) = T\left(\frac{1-2(x+\delta x)}{10} - \frac{1-2x}{10}\right)\mathbf{j} = -\tfrac{1}{5}T\,\delta x\,\mathbf{j} = T\,u''(x)\,\delta x\,\mathbf{j}.$$

Solution 1.7

(a) The given expression follows by noting that $\rho = m/l$ and $\rho_0 = m/l_0$, so $\rho = \rho_0 l_0/l$.

(b) In Exercise 1.3, we obtained the relation $\rho = K\rho_0/(K+T)$ for the uniformly extended or compressed state. Substituting this expression for ρ into equation (1.25) yields the desired result.

Solution 1.8

(a) For $u(x,t) = A\sin(kx - \omega t)$ we find

$$u_{tt}(x,t) = -\omega^2 u(x,t), \quad u_{xx}(x,t) = -k^2 u(x,t).$$

So for $u(x,t)$ to be a solution of the wave equation (1.24), we require $\omega^2 = c^2 k^2$.

(b) For $x = 0$, we have $u(0,t) = A\sin(-\omega t) = -A\sin(\omega t)$. If this vanishes for all values of t, then either $A = 0$, which implies $u(x,t) = 0$, or $\omega = 0$, giving $u(x,t) = A\sin(kx)$, which is independent of t.

(c) If $A = 0$, we are already left with the solution $u(x,t) = 0$. If $\omega = 0$ and $c \neq 0$, the relation $\omega^2 = c^2 k^2$ implies that $k = 0$. Hence $u(x,t) = 0$ is the only solution of the form $u(x,t) = A\sin(kx - \omega t)$ that satisfies the boundary condition $u(0,t) = 0$.

Solution 1.9

(a) For $u(x,t) = A\sin(kx)\sin(\omega t)$, we again find
$$u_{tt}(x,t) = -\omega^2 u(x,t), \quad u_{xx}(x,t) = -k^2 u(x,t),$$
so we have the same condition $\omega^2 = c^2 k^2$ as in Exercise 1.8.

(b) The boundary condition at $x = 0$ is automatically satisfied, because $u(0,t) = 0$ for the given function $u(x,t)$. For $x = l$, we have $u(l,t) = A\sin(kl)\sin(\omega t) = 0$, which has to hold at any time t. Besides the choices $A = 0$ or $\omega = 0$, which both lead to the trivial solution $u(x,t) = 0$, this can be achieved by choosing k such that $\sin(kl) = 0$.

(c) As the zeros of the sine function $\sin(x)$ are at integer multiples of π, the relation $\sin(kl) = 0$ has the solutions $k_n = n\pi/l$ with integer n. As $\sin(-x) = -\sin(x)$, the solutions with k_n and k_{-n} are related by a change of sign in A. Therefore we do not need to consider negative values of n. The case $k_0 = 0$ again yields the trivial solution, leaving us with positive integer values $n = 1, 2, 3, \ldots$ for interesting solutions that correspond to a moving string. The same argument applies to the sign of ω, and we may choose $\omega_n = ck_n$ to satisfy the condition $\omega^2 = c^2 k^2$. This yields the solutions
$$u_n(x,t) = A_n \sin\left(\frac{n\pi}{l}x\right) \sin\left(\frac{n\pi c}{l}t\right).$$

(d) The nodes x_s of the solution $u_n(x,t)$ are given by the solutions of the equation $\sin(k_n x_s) = 0$. For $0 < x_s < l$, the argument $k_n x_s$ runs through the interval $0 < k_n x_s < k_n l = n\pi$, so there are $n-1$ nodes apart from the fixed end points of the string. These nodes are at $k_n x_s = m\pi$ with $m = 1, 2, \ldots, n-1$, which yields $x_s = ml/n$ with $m = 1, 2, \ldots, n-1$. Thus the labelling of the solutions $u_n(x,t)$ introduced above is such that $u_n(x,t)$ has $n-1$ nodes.

(e) We have
$$u_1(x,t) = A_1 \sin\left(\frac{\pi x}{l}\right) \sin\left(\frac{\pi c t}{l}\right), \quad u_2(x,t) = A_2 \sin\left(\frac{2\pi x}{l}\right) \sin\left(\frac{2\pi c t}{l}\right),$$
with arbitrary constants A_1 and A_2. Sketches of $u_1(x, t_j)$ and $u_2(x, t_j)$ at various instants of time t_j are shown in Figures 1.21 and 1.22, respectively.

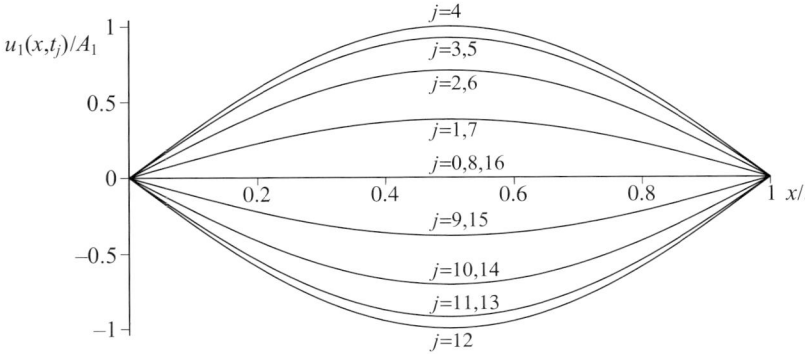

Figure 1.21 'Stroboscopic' snapshots of $u_1(x, t_j)$ at times $ct_j/l = j/8$ for $j = 0, 1, \ldots, 16$, covering one period of the motion

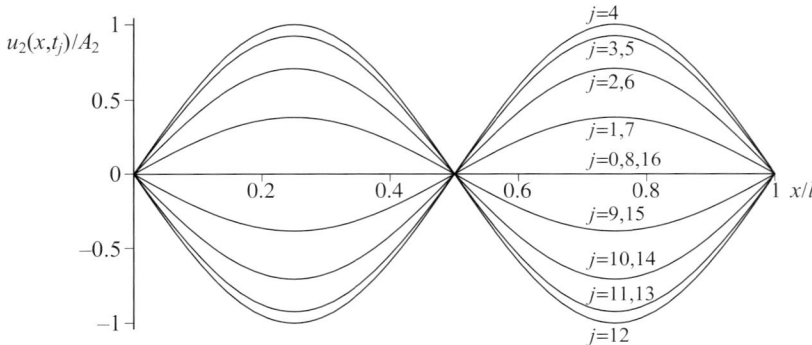

Figure 1.22 'Stroboscopic' snapshots of $u_2(x,t_j)$ at times $ct_j/l = j/16$ for $j = 0, 1, \ldots, 16$, covering one period of the motion

Solution 1.10

(a) The second partial derivatives of the complex function $u(x,t) = p(x,t) + iq(x,t)$ are

$$\frac{\partial^2 u}{\partial t^2}(x,t) = \frac{\partial^2}{\partial t^2}[p(x,t) + iq(x,t)] = \frac{\partial^2 p}{\partial t^2}(x,t) + i\frac{\partial^2 q}{\partial t^2}(x,t),$$

$$\frac{\partial^2 u}{\partial x^2}(x,t) = \frac{\partial^2}{\partial x^2}[p(x,t) + iq(x,t)] = \frac{\partial^2 p}{\partial x^2}(x,t) + i\frac{\partial^2 q}{\partial x^2}(x,t).$$

As t and x are real, and both $p(x,t)$ and $q(x,t)$ are real functions, this is also the decomposition of the second partial derivatives of $u(x,t)$ into real and imaginary parts.

So taking the real and imaginary parts of the wave equation (1.24) for $u(x,t)$ gives rise to the two wave equations

$$p_{tt}(x,t) = c^2 p_{xx}(x,t), \quad q_{tt}(x,t) = c^2 q_{xx}(x,t),$$

where we have assumed that the constant c is real.

(b) The complex conjugate of $u(x,t)$ is $u^*(x,t) = p(x,t) - iq(x,t)$. As $p(x,t)$ and $q(x,t)$ both satisfy the wave equation, so does any linear combination of them, because the wave equation is linear and homogeneous.

(c) The real and imaginary parts are

$$p(x,t) = \frac{u(x,t) + u^*(x,t)}{2}, \quad q(x,t) = \frac{u(x,t) - u^*(x,t)}{2i}.$$

Solution 1.11

(a) For $f(x,t) = A\exp[i(kx - \omega t)]$ and $g(x,t) = B\exp[i(kx + \omega t)]$, we find

$$f_{tt}(x,t) = -\omega^2 f(x,t), \quad g_{tt}(x,t) = -\omega^2 g(x,t),$$
$$f_{xx}(x,t) = -k^2 f(x,t), \quad g_{xx}(x,t) = -k^2 g(x,t).$$

So both functions satisfy the wave equation (1.24) provided that $\omega^2 = c^2 k^2$, which is the relation found in Exercises 1.8 and 1.9. Hence, if this relation is satisfied, any linear combination of the two functions is also a solution of the wave equation. So $u(x,t) = f(x,t) + g(x,t)$ satisfies the wave equation for arbitrary complex values A and B.

(b) The boundary condition gives

$$u(0,t) = A\exp(-i\omega t) + B\exp(i\omega t) = 0.$$

If $\omega = 0$, this can be fulfilled by choosing $A = -B$. For $\omega \neq 0$, the only possibility is $A = B = 0$. This can easily be seen by inserting, for instance, $t = 0$ and $t = \pi/(2\omega)$. Noting that $\exp(\pm i\pi/2) = \pm i$, this yields $A + B = 0$ and $-iA + iB = 0$, respectively, which implies $A = B = 0$.

(c) For $A = -B$, the function $u(x,t)$ becomes
$$\begin{aligned} u(x,t) &= A\left(\exp[i(kx-\omega t)] - \exp[i(kx+\omega t)]\right) \\ &= -A\exp(ikx)[\exp(i\omega t) - \exp(-i\omega t)] \\ &= -2iA[\cos(kx) + i\sin(kx)]\sin(\omega t) \\ &= 2A\sin(kx)\sin(\omega t) - 2iA\cos(kx)\sin(\omega t). \end{aligned}$$

As A is real, the real part of $u(x,t)$ is thus
$$p(x,t) = \mathrm{Re}[u(x,t)] = 2A\sin(kx)\sin(\omega t),$$
which differs from the function of Exercise 1.9 only by a factor 2, which is irrelevant as we are free to choose the coefficient A anyway.

(d) We can satisfy the boundary conditions $p(0,t) = p(l,t) = 0$ for the real part by choosing appropriate values of k, as shown in Exercise 1.9.

The imaginary part satisfies the boundary conditions $q(0,t) = q(l,t) = 0$ for $A = 0$ or $\omega = 0$. Apart from these trivial cases, it is not possible to satisfy the boundary conditions, because it requires $\cos(kx) = 0$ for $x = 0$ and $x = l$, which is impossible as $\cos 0 = 1$.

So only the trivial solution $u(x,t) = 0$ satisfies the boundary conditions $u(0,t) = u(l,t) = 0$, in accordance with the result of part (b).

Solution 1.12

(a) Since $l_0 = 1\,\mathrm{m}$, the end points are at $\tilde{x}(0) = 0$ and $\tilde{x}(1) = 2$ (in metres).

(b) The length is $l = 2\,\mathrm{m}$.

(c) The function $\tilde{x}(x) = 2x^2$ describing the deformation of the spring, and the corresponding stretching $\partial\tilde{x}/\partial x = 4x$, are shown in Figure 1.23 as functions of the position x along the spring in its natural state, with all lengths measured in metres.

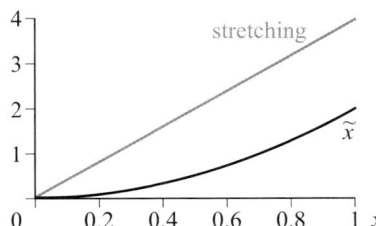

Figure 1.23 Deformation $\tilde{x}(x)$ (black) and stretching $\partial\tilde{x}/\partial x$ (grey) for the non-uniformly stretched spring of Exercise 1.12

Where the stretching is smaller than 1, the spring is compressed; where it is larger than 1, it is extended. So the spring is compressed for $x < \frac{1}{4}$ and extended for $x > \frac{1}{4}$, or equivalently for $\tilde{x} < \frac{1}{8}$ and $\tilde{x} > \frac{1}{8}$, respectively.

(d) Solving for x gives $x = \sqrt{\tilde{x}/2}$. The negative root can be ignored because $0 < x < 1$.

(e) If the stretching were uniform, the tension would be
$$T = K\frac{2l_0 - l_0}{l_0} = K = 100\,\mathrm{N},$$
and the linear density would be
$$\rho = \rho_0 l_0/(2l_0) = \rho_0/2 = 0.05\,\mathrm{kg\,m^{-1}}.$$

(f) According to equation (1.29), the tension is given by $T(x) = K(\partial\tilde{x}/\partial x) - K = K(4x - 1)$. Eliminating x, using the result of part (d), gives
$$T(x) = K(2\sqrt{2\tilde{x}} - 1) = 100(2\sqrt{2\tilde{x}} - 1)\,\mathrm{N}.$$

The tensions at the end points are $T(0) = -K = -100\,\mathrm{N}$ and $T(1) = 3K = 300\,\mathrm{N}$. At $\tilde{x} = 1$, the tension is $T(\sqrt{1/2}) = (2\sqrt{2} - 1)K \simeq 182.84\,\mathrm{N}$.

(g) The tension T, as shown in Figure 1.24, increases monotonically with $\tilde{x}$, so the tension increases along the spring. It is negative for small $\tilde{x}$, where the spring is compressed, and positive where the spring is extended.

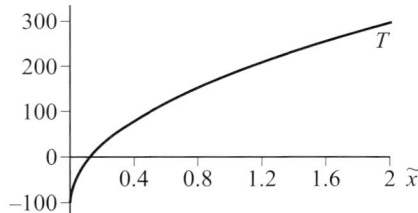

Figure 1.24 The tension in the non-uniformly stretched spring of Exercise 1.12

The tension vanishes at $\tilde{x} = \frac{1}{8}$, in accordance with the result obtained in part (c) above.

(h) The function is unrealistic at the left end, because $\partial \tilde{x}/\partial x = 4x$ vanishes for $x = 0$. So a small spring element at this end is compressed almost to length zero, which is certainly outside the range of applicability of Hooke's law for any real spring. Also, according to equation (1.33), the density $\rho(x)$ becomes infinite at $x = 0$. The maximal stretching at the other end is given by the partial derivative at $x = 1$, which is 4, so the spring would need to obey Hooke's law even when stretched to four times its natural length.

Solution 1.13

(a) According to equation (1.5) on page 14, we have $K = AY$.

The linear density ρ_0 is the mass per unit length, while $\rho^{(\mathrm{V})}$ is the mass per unit volume, so $\rho_0 = A\rho^{(\mathrm{V})}$.

(b) Inserting these relations in equation (1.46), the cross-sectional area A cancels and we obtain $c = \sqrt{Y/\rho^{(\mathrm{V})}}$, independent of the area A.

(c) Inserting the values from Table 1.1 yields $K = 2 \times 10^8 \,\mathrm{N}$, $\rho_0 = 7.86 \,\mathrm{kg\,m^{-1}}$ and $c \simeq 5 \,\mathrm{km\,s^{-1}}$. This is much faster than the speed of sound in air.

Solution 1.14

(a) Using equation (1.47) for the small element, which we assume to be uniform, we have

$$\widetilde{T}(x,t) \simeq T + K_{\mathrm{eff}} \frac{\delta \tilde{x} - \delta x}{\delta x} = T + K_{\mathrm{eff}} \left(\frac{\delta \tilde{x}}{\delta x} - 1 \right).$$

Taking the limit as $\delta x \to 0$, and using equation (1.39), we obtain

$$\widetilde{T}(x,t) = T + K_{\mathrm{eff}} \left(\frac{\partial \tilde{x}}{\partial x} - 1 \right) = T + K_{\mathrm{eff}} \, u_x(x,t).$$

(b) The force on the spring element is now

$$\delta \boldsymbol{F} = \widetilde{T}(x+\delta x, t)\,\mathbf{i} - \widetilde{T}(x,t)\,\mathbf{i} = K_{\mathrm{eff}} \left(u_x(x+\delta x, t) - u_x(x,t) \right) \mathbf{i}.$$

The mass is given by $\delta m = \rho\, \delta x$. The acceleration of the spring element is as in equation (1.43). Newton's second law now gives

$$K_{\mathrm{eff}} \left(u_x(x+\delta x, t) - u_x(x,t) \right) \mathbf{i} \simeq \rho\, \delta x\, u_{tt}(x,t)\,\mathbf{i},$$

which, upon dividing by δx and taking the limit as $\delta x \to 0$, yields

$$K_{\mathrm{eff}}\, u_{xx}(x,t) = \rho\, u_{tt}(x,t),$$

which is the wave equation (1.50) with the wave speed c as given in equation (1.51).

Solution 1.15

(a) The constant tension is $T = K(l/l_0) - K$, from equation (1.9).

(b) The constant tension $\widetilde{T}$ for length L is given by $\widetilde{T} = K(L/l_0) - K$. From the previous result, we have $l_0 = lK/(K+T)$, so

$$\widetilde{T} = K\frac{L}{l}\frac{K+T}{K} - K = (K+T)\frac{L}{l} - K = (K+T)\frac{L-l}{l} + T,$$

which gives equation (1.47) with $\delta l = L - l$ and $K_{\text{eff}} = K + T$.

Solution 1.16

(a) The solutions in Exercise 1.9 are

$$u_n(x,t) = A_n \sin\left(\frac{n\pi}{l}x\right) \sin\left(\frac{n\pi c}{l}t\right).$$

By equation (1.38), the position $\tilde{x}(x,t)$ along the deformed spring is given by $\tilde{x}(x,t) = x + u(x,t)$. A sketch for the solution $u_2(x,t)$ is shown in Figure 1.25.

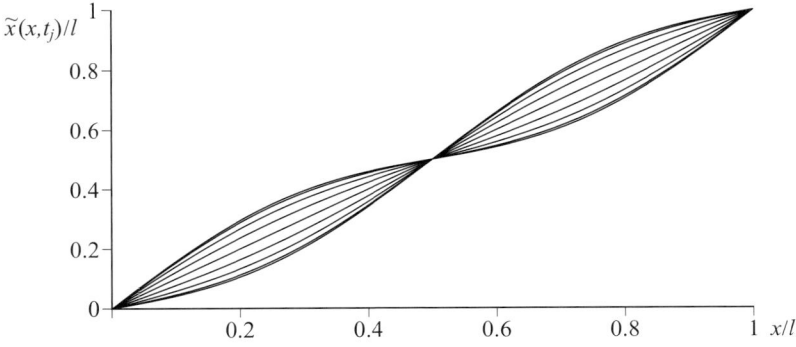

Figure 1.25 'Stroboscopic' snapshots of $\tilde{x}(x,t_j) = x + u_2(x,t_j)$ with amplitude $A_2 = l/10$, at the same time instants as shown in Figure 1.22

(b) Inserting the partial derivative

$$\frac{\partial u_n}{\partial x}(x,t) = A_n\frac{n\pi}{l}\cos\left(\frac{n\pi x}{l}\right)\sin\left(\frac{n\pi ct}{l}\right)$$

into equation (1.40) gives

$$T(x,t) = K\left[A_n\frac{n\pi}{l}\cos\left(\frac{n\pi x}{l}\right)\sin\left(\frac{n\pi ct}{l}\right)\right].$$

Note that this is the tension at the displaced position $\tilde{x}(x,t)$. It has the same sinusoidal time-dependence as the solution $u_n(x,t)$, but involves the cosine of the space coordinate $n\pi x/l$ rather than its sine. So the magnitude of the tension is maximal at those points $\tilde{x}(x_s,t)$ that correspond to the nodes of the solution $u_n(x,t)$, because $\cos(n\pi x_s/l) = \pm 1$ and the tension is maximal or minimal depending on the sign. Note that these are also the points at which $\tilde{x}(x,t) = x$.

(c) Using equations (1.33) and (1.39) gives

$$\rho(x,t) = \rho_0 \left(\frac{\partial \tilde{x}}{\partial x}(x,t)\right)^{-1} = \rho_0\left(1 + u_x(x,t)\right)^{-1}.$$

(d) Inserting the partial derivative of $u_n(x,t)$ calculated above, we obtain

$$\rho(x,t) = \frac{\rho_0}{1 - A_n\frac{n\pi}{l}\cos(\frac{n\pi x}{l})\sin(\frac{n\pi ct}{l})}.$$

Solution 1.17

(a) Using equation (1.9), the strain is $\varepsilon = T/K = 1/200$, so the string is extended by 0.5%.

(b) As we are within the linear regime, the wave speed is given by
$$c_{\text{long}} = (K+T)/\sqrt{\rho_0 K} = 2010\,\text{m s}^{-1}$$
– see equation (1.52) and the following paragraph. Note that force equals mass times acceleration, so $1\,\text{N} = 1\,\text{kg m s}^{-2}$, and we have used $\rho_0 = 0.05\,\text{kg m}^{-1}$.

(c) The wave speed for transverse waves is given by the expression in Exercise 1.7 as $c_{\text{trans}} = \sqrt{(K+T)T/(\rho_0 K)} = \sqrt{20\,100} \simeq 141.8\,\text{m s}^{-1}$. Comparison with the speed of longitudinal waves gives $c_{\text{trans}} = \sqrt{T/(K+T)}\,c_{\text{long}} \simeq 0.07\,c_{\text{long}}$. Thus the transverse wave speed is only about 7% of the longitudinal wave speed for the stretched string.

(d) As $T \to 0$, the longitudinal wave speed approaches $\sqrt{K/\rho_0} = 2000\,\text{m s}^{-1}$.

Solution 1.18

(a) As the combined spring is in equilibrium, the forces that the two parts exert on each other at the place where they are joined have to be equal in magnitude and opposite in direction. As the force acting on the end of the combined spring has magnitude F, and the combined spring is in static equilibrium, the force at the joined ends is the same, so the tension $T = F > 0$ is constant throughout the combined spring.

(b) Hooke's law applied to each part gives
$$F = K_1 \frac{L_1 - l_1}{l_1} = K_2 \frac{L_2 - l_2}{l_2}.$$
Inserting the values for K_1, K_2, l_1 and l_2, we find
$$F = 5L_1 - 10 = 5L_2 - 20.$$
Substituting $L_2 = L - L_1 = 8 - L_1$ gives
$$5L_1 - 10 = 5(8 - L_1) - 20 = 20 - 5L_1,$$
which gives $L_1 = 3\,\text{m}$, and thus $L_2 = 8 - 3 = 5\,\text{m}$.

While the absolute extension is $1\,\text{m}$ in both cases, the relative extensions are $(L_1 - l_1)/l_1 = \frac{1}{2}$ and $(L_2 - l_2)/l_2 = \frac{1}{4}$, so the first part is extended by 50%, whereas the second, stiffer part is extended by only 25%.

(c) The magnitude of the force is $F = 5L_1 - 10 = 5\,\text{N}$.

(d) If the extended spring is in equilibrium, the total force on any small spring element has to be zero, so the tension is constant even if the spring is non-uniform. However, as the example shows, the amount of stretching will vary according to the varying local stiffness of the spring.

Solution 1.19

(a) The slope is $u'(x) = \pi \cos(\pi x)/40$.

(b) The displacement $u(x)$ has a maximum or minimum where $u'(x) = 0$, thus for $x = \frac{1}{2}$. At the end points, $u(0) = u(1) = 0$, so the maximal displacement occurs at $x = \frac{1}{2}$, and is $u(\frac{1}{2}) = \frac{1}{40}$.

(c) The second derivative is $u''(x) = -\pi^2 \sin(\pi x)/40$, which vanishes for $x = 0$ and $x = 1$. The corresponding slopes are $u'(0) = \pi/40$ and $u'(1) = -\pi/40$, respectively.

(d) According to equation (1.18), the force is
$$\delta \boldsymbol{F}(x) = T\left(\frac{\pi \cos[\pi(x + \delta x)]}{40} - \frac{\pi \cos(\pi x)}{40}\right)\mathbf{j}$$
$$= \frac{\pi T}{40}\left(\cos(\pi x)[\cos(\pi \delta x) - 1] - \sin(\pi x)\sin(\pi \delta x)\right)\mathbf{j}$$
$$\simeq -\frac{\pi^2 T}{40}\sin(\pi x)\,\delta x\,\mathbf{j}$$
$$= T\,u''(x)\,\delta x\,\mathbf{j},$$

Solutions to Exercises in Chapter 1

where we have used $\cos(\alpha + \beta) = \cos\alpha\cos\beta - \sin\alpha\sin\beta$ and the Taylor expansions $\cos(\pi\,\delta x) = 1 + O(\delta x^2)$ and $\sin(\pi\,\delta x) = \pi\,\delta x + O(\delta x^3)$ to first order in δx.

Solution 1.20

(a) At the top, i.e. at $x = 0$, the elastic force has to compensate for the weight of the entire suspended spring, so the tension is $T(0) = mg$.

Along the spring, the tension decreases as the weight of the part of the spring below decreases. There is no gravitational force acting on a small element at the lower end of the spring, so the tension at $x = l_0$ is $T(l_0) = 0$.

(b) The tension $T(x)$ has to compensate for the weight of the lower part of the spring. This part has natural length $l_0 - x$, so its mass is $m(l_0 - x)/l_0 = m(1 - x/l_0)$, and its weight is $mg(1 - x/l_0) = T(x)$, which gives the desired result.

(c) Equation (1.34) gives
$$\frac{\rho(x)}{\rho_0} = \frac{K}{K + T(x)} = \frac{K}{K + mg(1 - x/l_0)}.$$

(d) Equation (1.30) gives
$$\frac{d\tilde{x}}{dx} = 1 + \frac{T(x)}{K} = 1 + \frac{mg}{K}\left(1 - \frac{x}{l_0}\right).$$

Integration with respect to x gives
$$\tilde{x}(x) = C + x + \frac{mgx}{K}\left(1 - \frac{x}{2l_0}\right).$$

The integration constant C follows from the condition $\tilde{x}(0) = 0$, which yields $C = 0$. Hence the solution is
$$\tilde{x}(x) = x + \frac{mgx}{K}\left(1 - \frac{x}{2l_0}\right).$$

(e) The length l is given by
$$l = \tilde{x}(l_0) = l_0 + \frac{mgl_0}{K}\left(1 - \tfrac{1}{2}\right) = l_0\left(1 + \frac{mg}{2K}\right).$$

(f) With $mg = 2K$, the length becomes $l = 2l_0 = 2\,\text{m}$.

The deformation, given in metres, is
$$\tilde{x}(x) = x + x(2 - x) = x(3 - x);$$

it is sketched in Figure 1.26.

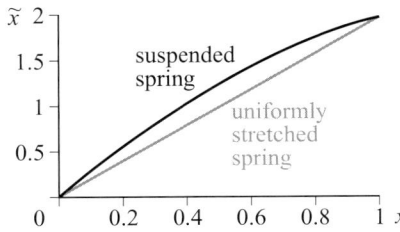

Figure 1.26 Deformation $\tilde{x}(x)$ of the suspended spring (black), compared to the corresponding function $\tilde{x}(x) = 2x$ for a uniformly stretched spring (grey)

Solving the quadratic equation $\tilde{x} = x(3 - x)$ for x, we find $x = (3 \pm \sqrt{9 - 4\tilde{x}})/2$. As $\tilde{x}(0) = 0$, the minus sign gives the appropriate solution, so
$$x = \frac{3 - \sqrt{9 - 4\tilde{x}}}{2}.$$

(g) We obtain $T(x)/K = 2 - 2x$ and $\rho(x)/\rho_0 = 1/(3-2x)$, with x measured in metres. This gives
$$\frac{T(x)}{K} = \sqrt{9-4\tilde{x}} - 1 \quad \text{and} \quad \frac{\rho(x)}{\rho_0} = \frac{1}{\sqrt{9-4\tilde{x}}}.$$
The results are shown in Figure 1.27.

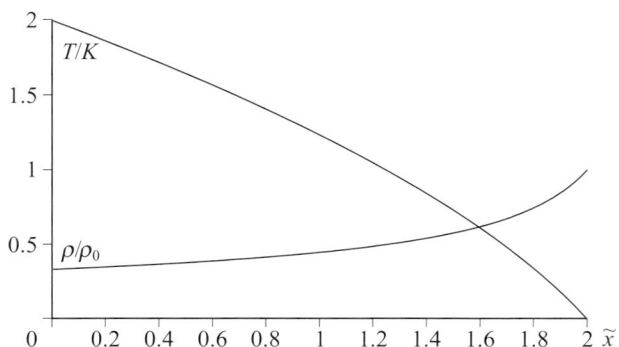

Figure 1.27 The tension ratio $T(x)/K$ and the density ratio $\rho(x)/\rho_0$ as functions of the position $\tilde{x}$ along the suspended spring

The corresponding values for a uniformly stretched spring are constant, namely $T/K = (l - l_0)/l_0 = 1$ and $\rho/\rho_0 = l_0/l = \frac{1}{2}$.

Solution 1.21

(a) There are three forces acting on an element of the string: the gravitational force $\delta \boldsymbol{W}$, and the forces $\boldsymbol{F}_1$ and $\boldsymbol{F}_2$ which are due to the tension in the string. We use $\phi_1 = \phi(x)$ and $\phi_2 = \phi(x + \delta x)$ (see Figure 1.28).

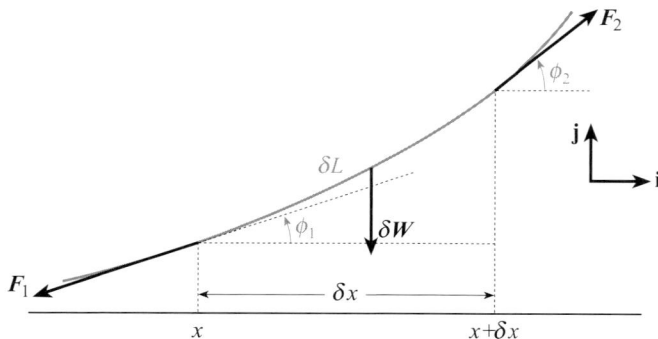

Figure 1.28 Forces on a small element of the string

The gravitational force is
$$\delta \boldsymbol{W} = -\delta m\, g\, \mathbf{j} = -\rho g\, \delta L\, \mathbf{j} \simeq -\rho g \sqrt{1 + \tan^2 \phi_1}\, \delta x\, \mathbf{j},$$
where we used Pythagoras' theorem to approximate the length δL of the string element by $\delta L^2 \simeq \delta x^2 + [\tan \phi_1\, \delta x]^2$. The gravitational force $\delta \boldsymbol{W}$ acts purely in the $-\mathbf{j}$ direction.

The forces $\boldsymbol{F}_1$ and $\boldsymbol{F}_2$ at the left- and right-hand ends of the element are given by
$$\boldsymbol{F}_1 = -T(x) \cos \phi_1\, \mathbf{i} - T(x) \sin \phi_1\, \mathbf{j},$$
$$\boldsymbol{F}_2 = T(x + \delta x) \cos \phi_2\, \mathbf{i} + T(x + \delta x) \sin \phi_2\, \mathbf{j}.$$

The net force on the string element is $\delta \boldsymbol{F} = \boldsymbol{F}_1 + \boldsymbol{F}_2 + \delta \boldsymbol{W}$. As the suspended string is in static equilibrium, the net force on the string element is zero, i.e. $\delta \boldsymbol{F} = \boldsymbol{0}$. Resolving in the $\mathbf{i}$ and $\mathbf{j}$ directions yields

Solutions to Exercises in Chapter 1

$$-T(x)\cos\phi_1 + T(x+\delta x)\cos\phi_2 = 0$$

and

$$-T(x)\sin\phi_1 + T(x+\delta x)\sin\phi_2 - \rho g \sqrt{1+\tan^2\phi_1}\,\delta x \simeq 0,$$

respectively.

(b) For the component in the **i** direction, we have

$$T(x)\cos\phi_1 = T(x+\delta x)\cos\phi_2,$$

so $T_1 = T(x)\cos[\phi(x)]$ does not change and is constant along the string. This makes sense as there are no other forces in the **i** direction. Because the string is static, the component of the tension force along the **i** direction has to be constant.

(c) We can simplify the component of $\boldsymbol{F}_1 + \boldsymbol{F}_2$ in the **j** direction as follows:

$$-T(x)\sin\phi_1 + T(x+\delta x)\sin\phi_2$$
$$= -T(x)\cos\phi_1 \tan\phi_1 + T(x+\delta x)\cos\phi_2 \tan\phi_2$$
$$= T_1(\tan\phi_2 - \tan\phi_1).$$

(d) We have $\tan\phi_1 = du/dx = u'(x)$ and $\tan\phi_2 = u'(x+\delta x)$. Inserting these in the equation for the **j** direction gives

$$T_1(u'(x+\delta x) - u'(x)) - \rho g\sqrt{1+[u'(x)]^2}\,\delta x \simeq 0.$$

Dividing by δx, taking the limit as $\delta x \to 0$ and rearranging the resulting expression gives the differential equation

$$u''(x) = k\sqrt{1+[u'(x)]^2},$$

where $k = \rho g/T_1$.

(e) The equation for $f(x)$ can be written as $f'/\sqrt{1+f^2} = k$. Integrating gives

$$\int \frac{df}{\sqrt{1+f^2}} = k\int dx,$$

and thus, using the integral provided, $\operatorname{arcsinh}(f) = k(x-a)$, where a is a constant.

(f) This yields $f(x) = \sinh[k(x-a)]$, and thus (since $f(x) = u'(x)$)

$$u(x) = \frac{1}{k}\cosh[k(x-a)] - b,$$

where b is another constant.

(g) Inserting $x = 0$ gives

$$u(0) = \frac{1}{k}\cosh(-ka) - b = \frac{1}{k}\cosh(ka) - b = 0,$$

so $b = \cosh(ka)/k$. For $x = l$, we obtain

$$u(l) = \frac{1}{k}\cosh[k(l-a)] - b = \frac{1}{k}(\cosh[k(l-a)] - \cosh(ka)) = 0,$$

which can be satisfied by choosing $a = l/2$. Thus

$$u(x) = \frac{1}{k}\left(\cosh\left[k\left(x - \frac{l}{2}\right)\right] - \cosh\left(\frac{kl}{2}\right)\right).$$

The constant $a = l/2$ gives the position of the lowest point along the string in the **i** direction, which by symmetry has to be in the middle of the interval $[0, l]$. The constant b determines the coordinate in the **j** direction of this point, because

$$u\left(\frac{l}{2}\right) = \frac{1}{k}\cosh 0 - b = \frac{1}{k} - b,$$

so $|1/k - b|$ is the maximum distance of the string from the horizontal.

(h) We have
$$1 + u'(x)^2 = 1 + f(x)^2 = 1 + \sinh^2[k(x-a)] = \cosh^2[k(x-a)],$$
so
$$L = \int_0^l dx\, \cosh[k(x - \tfrac{l}{2})] = \left[\frac{\sinh[k(x-\tfrac{l}{2})]}{k}\right]_0^l = \frac{2\sinh(\tfrac{kl}{2})}{k}.$$

From this, the constant k is the unique positive solution of the equation $kL = 2\sinh(kl/2)$, which always exists provided $L > l$. (This can be seen by realising that $kL = 2\sinh(kl/2) = 0$ for $k = 0$, and that the derivatives with respect to k are L and $l\cosh(kl/2)$, where the latter increases monotonically with k.)

(i) Within the given approximation, we have to find a positive solution of $kL = kl + k^3 l^3/24$. Dividing by k and solving for k yields
$$k = \frac{2}{l}\sqrt{6\left(\frac{L}{l} - 1\right)}.$$

The approximation is appropriate when kl is small, which is the case if $L/l - 1$ is small, which means that the string sags only slightly.

(j) We obtain $k = \tfrac{1}{3}$. (A numerical solution of the equation $3.5k = \sinh(3k)$ gives $k \simeq 0.325\,490$, so the value obtained by the approximation is only about 2.4% too large.) In the approximation, the solution reads
$$u(x) = 3\cosh\left(\frac{x-3}{3}\right) - 3\cosh 1.$$

The function $u(x)$ is minimal for $x = 3$, with $u(3) = 3 - 3\cosh 1 \simeq -1.63$, so the maximum distance to the horizontal is about $1.63\,\mathrm{m}$. The function is sketched in Figure 1.29.

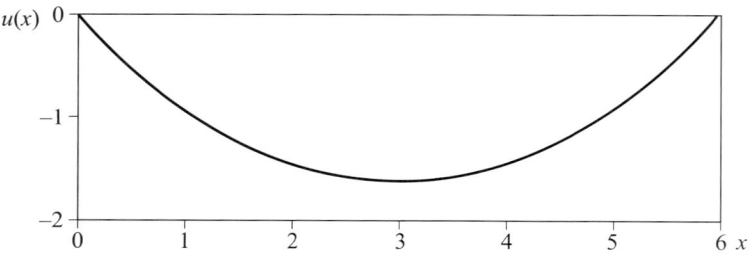

Figure 1.29 The function $u(x)$ for the suspended string

Solution 1.22

(a) The derivatives are
$$u_{tt}(x,t) = \frac{\partial}{\partial t}\left(\frac{Ac\sinh(x-ct)}{\cosh^2(x-ct)}\right) = \frac{2Ac^2\sinh^2(x-ct)}{\cosh^3(x-ct)} - \frac{Ac^2}{\cosh(x-ct)}$$
and
$$u_{xx}(x,t) = \frac{\partial}{\partial x}\left(-\frac{A\sinh(x-ct)}{\cosh^2(x-ct)}\right) = \frac{2A\sinh^2(x-ct)}{\cosh^3(x-ct)} - \frac{A}{\cosh(x-ct)},$$
so the wave equation (1.1) holds.

At time $t = 0$, this function is concentrated about $x = 0$. It is symmetric, i.e. $u(x,0) = u(-x,0)$, has a maximum at $u(0,0) = A$, and decreases rapidly with $|x|$ (see Figure 1.30).

At a time $t > 0$, the maximum has moved to $x = ct > 0$, so it moves in the positive x direction. The form of the function does not change, so we have the same behaviour apart from the shift ct.

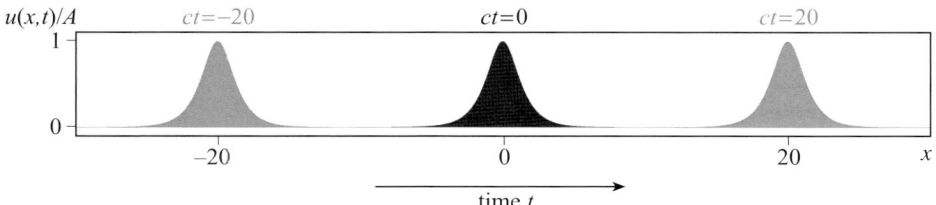

Figure 1.30 Sketch of the solution $u(x,t) = A/\cosh(x-ct)$ of the wave equation at time $t = 0$ (black) and at times $t = \pm 20/c$ (grey). The solution corresponds to a bell-shaped pulse that is moving in the positive x direction, with constant speed c.

(b) The derivatives are now
$$u_{tt}(x,t) = \frac{2Ac^2 \sinh^2(x+ct)}{\cosh^3(x+ct)} - \frac{Ac^2}{\cosh(x+ct)} = c^2 u_{xx}(x,t),$$
so this $u(x,t)$ is also a solution of the wave equation.

At time $t = 0$, this function is the same as in part (a). However, at a time $t > 0$, the maximum has moved to $x = -ct < 0$, so it now moves in the negative x direction, which makes sense because the function is obtained from the previous one by replacing t with $-t$. Again, the form of the function does not change.

Solution 1.23

(a) The change in length is
$$L - l = \int_0^l dx \left(\sqrt{1 + \left(\frac{\partial u}{\partial x}\right)^2} - 1 \right)$$
$$\leq \int_0^l dx \left(\sqrt{1+\epsilon^2} - 1 \right) = \left(\sqrt{1+\epsilon^2} - 1 \right) l.$$

Expanding the square root into a Taylor series in powers of ϵ gives
$$\frac{L-l}{l} \leq \frac{\epsilon^2}{2} + O(\epsilon^4),$$
thus the relative change in length is of second order in ϵ.

(b) If the stretching is uniform, the linear density $\tilde{\rho}$ is constant and given by
$$\tilde{\rho} = \frac{m}{L} = \rho \frac{l}{L},$$
where m denotes the total mass of the string. Hence
$$\frac{\rho - \tilde{\rho}}{\rho} = \frac{L - l}{L},$$
which is of order ϵ^2 because $L - l$ is of order ϵ^2.

For the tension, we can apply equation (1.9), for both length l and length L. This gives
$$\frac{\tilde{T}-T}{T} = \frac{K}{T}\frac{L-l_0}{l_0} - \frac{K}{T}\frac{l-l_0}{l_0} = \frac{K}{Tl_0}(L-l) = O(\epsilon^2).$$

(c) The mass δm of a small string element of length δx does not change if the element is deformed to length δL, so $\delta m = \rho\, \delta x \simeq \tilde{\rho}\, \delta L$, assuming that ρ is constant on the string element. This gives
$$\frac{\rho}{\tilde{\rho}} \simeq \frac{\delta L}{\delta x},$$
which in the limit as $\delta x \to 0$ becomes
$$\frac{\rho}{\tilde{\rho}(x)} = \frac{\partial L}{\partial x} = \sqrt{1 + \left(\frac{\partial u}{\partial x}\right)^2} \leq \sqrt{1+\epsilon^2}.$$

Hence
$$\frac{\rho - \tilde{\rho}(x)}{\rho} = 1 - \left(\frac{\partial L}{\partial x}\right)^{-1} \leq 1 - \frac{1}{\sqrt{1+\epsilon^2}} = O(\epsilon^2).$$

(d) Consider again a small string element of length δx. In the deformed string, this has a length δL which was calculated above. If we assume that the tension $\widetilde{T}$ is constant on the small string element, we obtain

$$\frac{\widetilde{T}-T}{T} \simeq \frac{K}{T}\frac{\delta L - \delta x_0}{\delta x_0} - \frac{K}{T}\frac{\delta x - \delta x_0}{\delta x_0} = \frac{K}{T}\frac{\delta L - \delta x}{\delta x_0} = \frac{K}{T}\frac{\delta x}{\delta x_0}\left(\frac{\delta L}{\delta x} - 1\right),$$

where δx_0 denotes the length of the string element in its natural state. We have $\delta x/\delta x_0 = l/l_0$, so, in the limit as $\delta x \to 0$,

$$\frac{\widetilde{T}(x)-T}{T} = \frac{Kl}{Tl_0}\left(\frac{\partial L}{\partial x} - 1\right) \leq \frac{Kl}{Tl_0}\left(\sqrt{1-\epsilon^2} - 1\right) = O(\epsilon^2).$$

(e) As $u(0,t) = u(l,t) = 0$, and as the partial derivative satisfies $|\partial u/\partial x| \leq \epsilon$, the maximum value that $|u(x)|$ can possibly attain is the value of the linear function $v(x) = \epsilon x$ at the midpoint $x = l/2$. This gives $|u(x)| \leq |v(l/2)| \leq \epsilon l/2$, as required.

Solution 1.24

(a) For the speed of sound in air, we obtain

$$c = \sqrt{\frac{\gamma P_0}{\rho_0^{(V)}}} = \sqrt{\frac{7 \times 1.013 \times 10^5}{5 \times 1.2}} \simeq 344\,\text{m}\,\text{s}^{-1}.$$

For the units, we used the definition $1\,\text{N} = 1\,\text{kg}\,\text{m}\,\text{s}^{-2}$, which corresponds to a force being the product of a mass and an acceleration.

(b) Inserting the data for a height of $10\,000\,\text{m}$ above sea level, we obtain

$$c = \sqrt{\frac{\gamma P_0}{\rho_0^{(V)}}} = \sqrt{\frac{7 \times 2.6 \times 10^4}{5 \times 0.41}} \simeq 298\,\text{m}\,\text{s}^{-1}.$$

(c) For helium, the wave speed becomes

$$c = \sqrt{\frac{\gamma P_0}{\rho_0^{(V)}}} = \sqrt{\frac{5 \times 1.013 \times 10^5}{3 \times 0.1785}} \simeq 973\,\text{m}\,\text{s}^{-1}.$$

(d) A wavelength λ corresponds to a frequency $\nu = c/\lambda$, where c is the speed of sound. For the same wavelength λ, the ratio of the frequency ν_{helium} in a helium atmosphere and ν_{air} in air is thus

$$\frac{\nu_{\text{helium}}}{\nu_{\text{air}}} = \frac{c_{\text{helium}}}{c_{\text{air}}} \simeq \frac{973}{344} \simeq 2.8.$$

So if you inhale helium and speak, your voice sounds much higher than usual, because the frequency of sound waves in helium is almost three times the frequency of sound waves of the same wavelength in air.

CHAPTER 2
Solutions of the wave equation

2.1 Introduction

In the previous chapter, we derived the one-dimensional wave equation

$$u_{tt}(x,t) = c^2 \, u_{xx}(x,t), \tag{2.1}$$

a partial differential equation of second order in both the space and time variables. It applies to various physical systems. However, in this chapter we exclusively refer to the motion of a string as a realisation of wave motion, keeping in mind that the solutions of the wave equation discussed below will also apply to other systems.

A *solution* of the wave equation is a function $u(x,t)$ that satisfies the equation for all positions x in the appropriate domain and at all times t. You have seen in Exercises 1.8 and 1.9 that there are infinitely many solutions. To describe a particular physical situation, such as the motion of a finite string, we need to find a single solution that describes the motion. This is achieved by taking into account two types of additional information: *boundary conditions* and *initial conditions*.

2.1.1 Boundary and initial conditions

Boundary conditions arise from (usually time-independent) constraints on the motion, for instance due to the geometry of the problem. For transverse waves on a stretched string, the ends of the string are usually fixed in space and not allowed to move. This imposes boundary conditions on the function $u(x,t)$ that describes the motion. For the case of a string with its ends fixed at $x = 0$ and $x = l$, the boundary conditions are $u(0,t) = u(l,t) = 0$ at all times t. This type of boundary condition, where the function $u(x,t)$ is specified at the boundary, is called a *Dirichlet condition*, named after the mathematician Johann Peter Gustav Lejeune Dirichlet (1805–1859). The wave equation together with boundary conditions constitutes a *boundary-value problem*. The solutions of the boundary-value problem are the possible motions of the *finite* string. Still, there will be infinitely many solutions that satisfy the wave equation and the boundary conditions, as we saw in Exercise 1.9 for the example of the vibrating string.

It is not always necessary to impose boundary conditions on the wave equation. Solutions of the wave equation without any boundary conditions imposed describe possible motions of an *infinite* string (meaning infinite in both positive and negative x direction). An infinite string may appear unrealistic, and you might wonder why it is mentioned at all. However, as pointed out before, wave motion occurs in many physical systems, and for some it makes sense to consider waves in a system without boundaries. For example, electromagnetic waves such as light can travel through the entire universe, and we do not need to consider boundary conditions to describe this motion. Even for finite systems, like a finite string, we can sometimes neglect part of the boundary conditions. For example, if we want to know how a pulse travelling along a string is reflected at an end, the boundary condition at the other end does not influence the reflection, so we can answer this question by treating a semi-infinite string (a string where one end extends to infinity). This approach is often used if we are interested in the dynamical (time-dependent) properties of wave motion, such as what happens when a pulse travels through an inhomogeneous (non-uniform) medium or when encountering an interface between different media. We shall discuss situations of this type later in the course.

Initial conditions specify the way in which the motion starts. The wave equation is of second order in the time variable, therefore two conditions are needed to determine the motion. For instance, we may specify the function $u(x, t_0)$ and its partial derivative $u_t(x, t_0)$ at a given initial time t_0, corresponding to the initial deformation $a(x) = u(x, t_0)$ and the initial speed $b(x) = u_t(x, t_0)$. This is analogous to the case of a particle moving according to Newton's equation of motion, where the initial position and velocity are used to specify the motion. Given the initial conditions, the wave equation determines the evolution of the initial deformation $a(x) = u(x, t_0)$ with time.

A partial differential equation with initial and boundary conditions is known as a *Cauchy problem*, in honour of the mathematician Augustin Louis Cauchy (1789–1857). If we supply sufficient information in the initial conditions, the resulting motion will be completely determined by the wave equation, so we expect that the solution of a Cauchy problem is unique.

2.1.2 Content of this chapter

In this chapter, we consider solutions of the one-dimensional wave equation. We begin with the solutions seen in some of the exercises of Chapter 1. Although there are many others, these solutions are not only particularly simple, but very important, as you will see later. After this, we consider the general solution of the wave equation. First, we introduce d'Alembert's solution, and discuss how initial- and boundary-value problems can be addressed. Then we use the method of separation of variables to solve the wave equation. This naturally leads to expressions involving infinite sums or integrals of trigonometric functions, which are known as *Fourier series* and *Fourier transforms*, respectively.

2.2 Harmonic wave motion

In Exercise 1.9 (page 23) of Chapter 1, we showed that the functions
$$u(x, t) = A \sin(kx) \sin(\omega t) \tag{2.2}$$
are solutions of the wave equation (2.1) for any value of A, provided that the positive parameters k and ω satisfy the relation
$$\omega = kc, \tag{2.3}$$
with wave speed $c > 0$.

2.2 Harmonic wave motion

Because $u(x,t)$ is given in terms of trigonometric functions, such waves are often referred to as *sinusoidal* (meaning sine-shaped) or *harmonic* (referring to sinusoidal time dependence, like for simple harmonic motion).

Without imposing additional restrictions, the solutions $u(x,t)$ describe particular motions of an infinite string. In order to obtain solutions for a finite string of length l, we need to impose the corresponding boundary conditions $u(0,t) = u(l,t) = 0$ at the fixed ends of the string. It was shown in Exercise 1.9 that the boundary conditions limit the possible values of k to integer multiples of π/l, namely $k_n = n\pi/l$ with $n = 1, 2, 3, \ldots$. The corresponding solutions

$$u_n(x,t) = A_n \sin(k_n x) \sin(\omega_n t), \quad n = 1, 2, 3, \ldots, \tag{2.4}$$

where $\omega_n = k_n c = n\pi c/l$, are called the *modes* or the *normal modes* of the finite string. The spatial part of the first six modes is depicted in Figure 2.1. Note that the function $\sin(k_n x)$ for the nth mode has $n-1$ evenly-spaced zeros for $0 < x < l$.

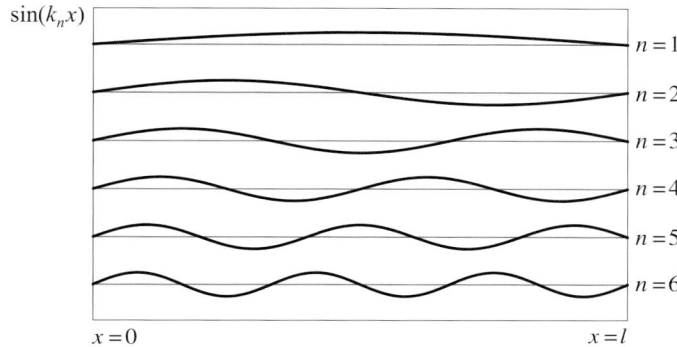

Figure 2.1 The first six modes of a transversely vibrating string. Every point along the string moves up and down sinusoidally.

For any solution of the form (2.2), the motion is *periodic* in time. This means that the wave pattern exactly repeats after a certain time τ, the *period*, has elapsed. In other words, $u(x, t+\tau) = u(x,t)$ holds at any point x and for all values of t. The period τ is related to ω by

$$\omega \tau = 2\pi, \tag{2.5}$$

because the sine function is periodic with period 2π. The inverse of the period,

$$\nu = \frac{1}{\tau} = \frac{\omega}{2\pi}, \tag{2.6}$$

is called the *frequency* of the periodic motion. It determines the number of cycles per unit time, say per second. So frequency is measured in s^{-1} (cycles per second) or Hz (hertz), where $1\,\text{Hz} = 1\,\text{s}^{-1}$, named after the physicist Heinrich Rudolf Hertz (1857–1894). You may be familiar with this unit, and with the derived units kHz, MHz or even GHz, for 10^3 Hz, 10^6 Hz and 10^9 Hz, respectively, from your microwave oven, radio and television frequencies, mobile telephone, or computer clock speed. The parameter $\omega = 2\pi\nu$ is called the *angular frequency*.

In the solution $u(x,t)$ of equation (2.2), the space variable x also enters as an argument of a sine function. Thus, at any given time t, $u(x,t)$ is periodic in x, with a period λ that is given by

$$k\lambda = 2\pi. \tag{2.7}$$

This means that $u(x + \lambda, t) = u(x, t)$ holds for all values of x at any given time t. The quantity λ is called the *wavelength*, because the wave pattern repeats after a distance λ. We shall refer to the parameter k as the *wave number*. Thus the SI unit of the wave number is m^{-1}. It follows from equations (2.3), (2.6) and (2.7) that

$$\lambda \nu = c, \tag{2.8}$$

which means that the wavelength and the frequency are inversely proportional to each other.

Sometimes the term 'wave number' is used for the inverse of the wavelength, $1/\lambda = k/2\pi$, because this counts the number of periods of the wave per unit length, say per metre.

The quantities introduced in equations (2.5)–(2.7) characterise a function periodic in space and time. You might wonder whether periodic solutions are something rather special. However, as you will see later, it is possible to analyse *any* wave motion in terms of periodic functions.

The solutions (2.2) of the wave equation contain a further parameter, the *amplitude A*, which is usually chosen to be positive. For the example of the vibrating string, the amplitude describes the maximum deviation of the string from its rest position. The value of A is, as yet, undetermined. *Any* amplitude describes a possible motion of the string; the actual amplitude depends on the initial conditions, i.e. how the string was set in motion initially.

A few remarks on the history of mathematics in music

This material is optional.

The term *harmonic* highlights the connection to music. If you release a plucked string, it will, in general, move according not to a single mode $u_n(x, t)$ of equation (2.4), but rather to a superposition of modes with different n. The mode $u_1(x, t)$ corresponds to the lowest or fundamental frequency ν_1; the corresponding wavelength is twice the length of the stretched string. Its frequency corresponds to the pitch of the tone that we hear, and is proportional to the inverse wavelength; see equation (2.8). The other modes $u_n(x, t)$ with frequencies $\nu_n = n\nu_1$ are called *higher harmonics*. The characteristic sound of a musical instrument is at least partly determined by the way in which higher harmonics are present, which means what the relative amplitudes of the various higher harmonics are. Together with some properties that depend on the way the sound is produced, and non-linear contributions that are important in some instruments, this allows one to distinguish the sound of, say, a flute from that of a violin or a trumpet, even if they play at the same pitch so that the fundamental frequency is the same.

The mathematics of music has a long history. One of the oldest detailed accounts of the role of integer and rational numbers in music is the book *Division of the canon* by Euclid (around 300 BC), but the topic reportedly goes back to the ancient Greek philosopher and mathematician Pythagoras, who lived around 570–500 BC. The problem that he faced was a practical one, namely how to tune a musical instrument. The musical scales involved rational frequency ratios, the simplest being the octave with a ratio 2:1, so the higher tone has twice the frequency of the lower tone, and the quint or fifth which corresponds to a frequency ratio 3:2. These rational numbers are the frequency ratios $\nu_2/\nu_1 = 2$ and $\nu_3/\nu_2 = 3/2$ of higher harmonics, which appear naturally in this context.

However, Pythagoras realised that a complication arises if one tries to construct consistent tonal scales with rational frequency ratios. The best known example of this is the *Pythagorean comma*. It concerns the discrepancy in the frequency ratios of seven octaves 2^7 and twelve quints $(3/2)^{12}$, which according to our tonal system correspond to the same tone (and thus the same key on a keyboard). The Pythagorean comma is the ratio $3^{12}/2^{19} = 531\,441/524\,288 \simeq 1.014$. Though small, this discrepancy rules out a consistent tonal scale based on harmonic frequency ratios which divide an octave into a number of tonal intervals, and consequently various compromise tunings were used for many centuries.

Today, we use a twelve-tone equal temperament or *well-tempered* tuning, introduced by Andreas Werckmeister in 1691. An octave is divided into twelve semitone steps of equal frequency ratios, which thus are $2^{1/12}$, the twelfth root of two. Because all semitone steps are equal, one can *transpose* music from one key to another, which means playing it at a different pitch, without changing the relations between different tones. The price one pays is that intervals between tones, apart from octaves, differ slightly from the rational values of higher harmonics. For instance, in the well-tempered tuning, a quint consists of seven semitone steps and thus a frequency ratio of $2^{7/12} \simeq 1.498$, slightly smaller than the natural quint.

(End of optional material)

Exercise 2.1

Consider the functions $u(x,t)$ of equation (2.2).

(a) Electromagnetic waves, such as radio or light waves, travel at a speed of approximately $c = 3 \times 10^8 \,\mathrm{m\,s^{-1}}$. What is the wavelength for an electromagnetic wave of frequency $\nu = 200\,\mathrm{kHz}$, which corresponds to a long-wave radio frequency? What wavelength results for $\nu = 2.45\,\mathrm{GHz}$, which corresponds to the frequency used in a microwave oven? What is the frequency for visible green light of wavelength $\lambda = 600\,\mathrm{nm} = 6 \times 10^{-7}\,\mathrm{m}$? What are the corresponding periods τ?

nm stands for nanometre.

(b) For the solutions $u_n(x,t)$ of equation (2.4), calculate the frequencies ν_n and wavelengths λ_n, and check that they satisfy relation (2.8).

Exercise 2.2

(a) Humans can hear sound in a frequency range of about 16 Hz to 20 000 Hz. The speed of sound in air at room temperature is $c \simeq 343\,\mathrm{m\,s^{-1}}$. What are the wavelengths corresponding to these frequency limits?

(b) Most pipes in a pipe organ produce sound of a fundamental wavelength λ that is twice the length l of the pipe. In a large organ, the lowest frequency is usually $\nu \simeq 16.5\,\mathrm{Hz}$. What is the length of the corresponding pipe?

(c) The largest pipe organ in the world is in the Atlantic City Convention Hall, comprising more than 32 000 pipes. Its largest pipe is approximately 64 feet in length, i.e. roughly 20 m. Calculate the basic frequency that a pipe of length $l = 20\,\mathrm{m}$ produces. Can you hear the sound produced by that pipe?

2.3 General solution for the infinite string

So far, we have discussed some example solutions (2.2) for the boundary-value problem of a finite string. We neither discussed the initial-value problem for this case, nor checked whether there exist more solutions than those given in equation (2.2).

We shall now follow a different approach, which will produce the general solution of the wave equation. In contrast to the previous solutions (2.2), which describe standing waves, this approach employs travelling wave solutions, which you can visualise as pulses travelling along a string. This works particularly well if we consider solutions on the whole real line, without boundary conditions, which correspond to motions of an infinite string. We shall therefore start from this case, derive a general solution introduced by d'Alembert, and then show how this can be used to solve the initial-value problem for the infinite string. Later, we shall see that boundary conditions can be treated as well, and the case of a semi-infinite string, fixed at one end and extending to infinity, is discussed in detail in the subsequent section. The corresponding analysis for a finite string, albeit instructive, becomes rather involved. We shall see that there are more effective methods to deal with this case.

2.3.1 D'Alembert's solution

Consider a function of the form

$$u(x,t) = F(x - ct), \tag{2.9}$$

where F is an arbitrary function of a *single* variable, assumed to be at least twice differentiable. Taking partial derivatives of $u(x,t)$ with respect to t and x gives

$$u_t(x,t) = -c\,F'(x - ct) = -c\,u_x(x,t), \tag{2.10}$$
$$u_{tt}(x,t) = c^2 F''(x - ct) = c^2\,u_{xx}(x,t), \tag{2.11}$$

so indeed $u(x,t)$ satisfies the wave equation (2.1) for *any* function F. Here, $F''(x - ct)$ denotes the second derivative of the function $F(z)$ with respect to its single argument z, evaluated at $z = x - ct$.

Both x and ct are lengths (ct being the distance a disturbance travelling at constant speed c covers in time t), so combinations such as $x \pm ct$ make sense.

What does this solution mean? The function $u(x,t)$ depends on x and t only via the combination $x - ct$. At time $t = 0$, we have $u(x,0) = F(x)$, so the function $F(x)$ describes the disturbance at time $t = 0$, for instance, the initial shape of a string. At a later time t, the solution is given by the same function F, but the argument is now $z = x - ct$. This means that the infinite string will have exactly the same shape as before, except that any given feature has moved to the right, in the positive x direction, by a distance ct. This can be seen as follows. We look at the features around a certain point x_0 at $t = 0$, where $u(x_0, 0) = F(x_0)$. At time t, we find the same feature $F(x_0)$ at a position x_t that is given by

$$x_t - ct = x_0 - c0 = x_0, \tag{2.12}$$

so $x_t = x_0 + ct$. In other words, the feature has moved to the right (in the positive x direction) by a distance ct – see Figure 2.2. Whatever the initial form of the disturbance, it moves to the right with uniform speed c.

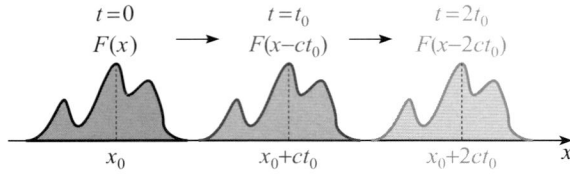

Figure 2.2 The function $F(x - ct)$ in equation (2.9) corresponds to a disturbance that moves to the right with constant speed c, retaining its form. The disturbance is shown at times $t = 0$, $t = t_0$ and $t = 2t_0$.

2.3 General solution for the infinite string

Exercise 2.3

Suppose that $u(x,t) = \exp[-(x-2t)^2]$.
(a) Verify that $u(x,t)$ solves the wave equation (2.1).
(b) Sketch the solution at $t = 0$, $t = 1$ and $t = 2$.

Exercise 2.4

Show that $u(x,t) = G(x+ct)$ is a solution of the wave equation (2.1) for any function G which is at least twice differentiable.

As shown in Exercise 2.4, the function
$$u(x,t) = G(x+ct) \qquad (2.13)$$
also satisfies the wave equation for *any* at least twice differentiable function G. What is the meaning of this solution? It is very much the same story as for the solution given by equation (2.9), but now the initial disturbance propagates to the left, in the negative x direction, with the same constant speed c. Therefore c is the characteristic speed of wave propagation in the system, justifying our notion of wave speed. In the general case, both $F(x-ct)$ and $G(x+ct)$ are present, and we simultaneously have two, possibly different, superimposed patterns of disturbance moving to the right and to the left, with the same wave speed c.

Because the wave equation is linear and homogeneous, any linear combination of the two solutions (2.9) and (2.13) is also a solution. This leads us to *d'Alembert's solution*

d'Alembert's solution
$$u(x,t) = F(x-ct) + G(x+ct), \qquad (2.14)$$
where F and G are two arbitrary, at least twice differentiable, functions of a single variable. D'Alembert proved that any solution $u(x,t)$ of the wave equation can be written in this form. This will be shown in Exercise 2.6.

Exercise 2.5

Consider the functions $u_n(x,t)$ of equation (2.4). Use the trigonometric identity $2\sin\alpha\sin\beta = \cos(\alpha-\beta) - \cos(\alpha+\beta)$ to rewrite the functions $u_n(x,t)$ in the form of equation (2.14).

Exercise 2.6

You are asked to show that any solution $u(x,t)$ of the wave equation (2.1) with $c \neq 0$ can be written in the form (2.14). To this end, introduce variables $y = x - ct$ and $z = x + ct$, and consider the function $v(y,z) = u[x(y,z), t(y,z)]$ as a function of the independent variables y and z.

(a) Express x and t in terms of y and z.
(b) Calculate $\partial x/\partial y$ and $\partial t/\partial y$.
(c) Show that
$$v_y(y,z) = \frac{1}{2}u_x - \frac{1}{2c}u_t.$$
Calculate v_{zy}, and show that the equation
$$v_{zy}(y,z) = 0$$
is equivalent to the wave equation.
(d) Hence show by integration that $v(y,z) = F(y) + G(z)$ is the general solution of this equation, where F and G are two arbitrary, at least twice differentiable, functions (compare Block 0, Exercise 2.6).
(e) Hence deduce the general solution (2.14).

According to the mixed derivative theorem cited in Block 0, Subsection 2.2.6, it does not matter in which order the two partial derivatives are taken. One may take the derivative first with respect to y and then with respect to z, or vice versa; the outcome is the same in either case.

2.3.2 Initial conditions

Having arrived at the general solution (2.14), does this immediately solve the problem of describing the motion of an infinite string? Not quite, because we still need to calculate the two functions F and G from the given initial conditions. As mentioned before, we need two sets of initial data because the wave equation is of second order in the time variable t. Given an initial disturbance $u(x,0)$ and initial velocity $u_t(x,0)$ at $t=0$, how should we split this into the sum of two functions $F(x)$ and $G(x)$?

Example 2.1

Consider a function $a(x)$ that vanishes outside the interval $[-r,r]$, as shown in Figure 2.3.

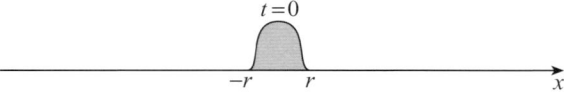

Figure 2.3 A pulse centred at the origin at time $t=0$

This depicts the initial displacement of a string at rest. What happens if the string is released from rest – will the pulse move to the left or to the right? Because the situation is symmetric, and there is no force preferring one direction over the other, we expect that the pulse splits into two 'equal' parts moving in opposite directions. Can we see this from our initial conditions and the general solution (2.14)?

We have two conditions at time $t=0$ at our disposal:

$$u(x,0) = a(x), \quad u_t(x,0) = 0. \tag{2.15}$$

The first condition states that the initial disturbance is given by the function $a(x)$. The second means that the initial velocity is zero, so the propagation is started from rest. Inserting solution (2.14), we find

$$F(x) + G(x) = a(x), \quad -cF'(x) + cG'(x) = 0. \tag{2.16}$$

From the second relation, we obtain $F'(x) = G'(x)$, hence the functions F and G differ, at most, by a constant C, i.e. $F(x) - G(x) = C$. With the first equation of (2.16), this yields

$$F(x) = \tfrac{1}{2}[a(x) + C], \quad G(x) = \tfrac{1}{2}[a(x) - C]. \tag{2.17}$$

The solution is therefore

$$u(x,t) = \tfrac{1}{2}\left[a(x-ct) + a(x+ct)\right], \tag{2.18}$$

where the constant C drops out completely. We have indeed found that the initial pulse splits into two equal parts, each having half the height of the original, as shown in Figure 2.4.

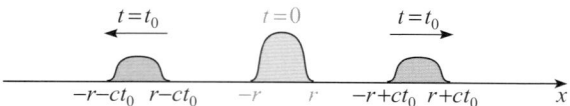

Figure 2.4 At time $t = t_0 > r/c$, the pulse has split into two parts, which move in opposite directions at the same speed c ∎

General initial conditions

Looking at the argument that led to equation (2.18), you will realise that the actual form of the function $a(x)$ does not matter. We now consider general

2.3 General solution for the infinite string

initial conditions given by two arbitrary functions $a(x)$ and $b(x)$,

$$u(x,0) = a(x), \quad u_t(x,0) = b(x), \qquad (2.19)$$

where $a(x)$ determines the initial deformation at time $t = 0$, and $b(x)$ corresponds to the initial velocity. Thus $b(x) = 0$ if we start from rest as in Example 2.1. For $b(x) \neq 0$, the deformation has an initial velocity which can vary in space. We assume that we know both functions on the whole real line.

Inserting d'Alembert's solution (2.14) yields the equations

$$F(x) + G(x) = a(x), \quad -cF'(x) + cG'(x) = b(x) \qquad (2.20)$$

for the functions F and G. The first equation determines the sum $F + G$. Integrating the second equation from an arbitrary reference point x_0, we find

$$F(x) - G(x) - C = -\frac{1}{c}\int_{x_0}^{x} dy\, b(y), \qquad (2.21)$$

where $C = F(x_0) - G(x_0)$ is a constant.

Knowing both the sum and the difference of the functions F and G, we can calculate F and G themselves:

$$F(x) = \frac{a(x)}{2} + \frac{C}{2} - \frac{1}{2c}\int_{x_0}^{x} dy\, b(y), \qquad (2.22)$$

$$G(x) = \frac{a(x)}{2} - \frac{C}{2} + \frac{1}{2c}\int_{x_0}^{x} dy\, b(y). \qquad (2.23)$$

Inserting these results into d'Alembert's solution (2.14), we obtain

$$\begin{aligned}u(x,t) &= F(x-ct) + G(x+ct) \\ &= \frac{1}{2}[a(x-ct) + a(x+ct)] - \frac{1}{2c}\int_{x_0}^{x-ct} dy\, b(y) + \frac{1}{2c}\int_{x_0}^{x+ct} dy\, b(y) \\ &= \frac{1}{2}[a(x-ct) + a(x+ct)] + \frac{1}{2c}\int_{x-ct}^{x_0} dy\, b(y) + \frac{1}{2c}\int_{x_0}^{x+ct} dy\, b(y) \\ &= \frac{1}{2}[a(x-ct) + a(x+ct)] + \frac{1}{2c}\int_{x-ct}^{x+ct} dy\, b(y). \end{aligned} \qquad (2.24)$$

The constant C has again disappeared from the solution, as has our arbitrary reference point x_0. We have thus arrived at *d'Alembert's solution of the initial-value problem*:

d'Alembert's solution of the initial-value problem

$$u(x,t) = \frac{1}{2}[a(x-ct) + a(x+ct)] + \frac{1}{2c}\int_{x-ct}^{x+ct} dy\, b(y). \qquad (2.25)$$

Equation (2.25) solves the initial-value problem posed by the wave equation (2.1) with the initial conditions of equations (2.19). If we know the initial deformation $a(x)$ and the initial velocity $b(x)$, then equation (2.25) gives us a *unique* solution of the wave equation satisfying the initial conditions.

The integration limits in equation (2.25) can easily be understood. Only the part of the initial velocity $b(y)$ for $y \in [x - ct, x + ct]$ affects the solution $u(x,t)$ at position x and time t, because c is the speed of the motion, and any disturbance outside the interval $[x - ct, x + ct]$ has not yet reached the point x at time t. So the integral in equation (2.25) covers the whole range of positions y from which one can reach the point x within time t when travelling at speed c.

We note that to obtain a twice differentiable solution, the function a needs to be at least twice differentiable, whereas it suffices for b to be differentiable once; see Exercise 2.7 below. However, one can generalise the notion of solutions, allowing for isolated singularities, for instance if we replaced the pulse shown in Figure 2.3 by a proper square pulse (with $a(x) = 1$ for $-r \le x \le r$ and $a(x) = 0$ elsewhere). Essentially, the argument is that one can approximate such cases by differentiable functions to an arbitrary accuracy. We shall therefore allow such solutions, provided that the wave equation is satisfied everywhere, except at isolated points where derivatives need not exist. In this sense, equation (2.25) also provides a solution of the wave equation for an initial square pulse given by $a(x) = 1$ for $|x| \le r$ and $a(x) = 0$ for $|x| > r$.

Exercise 2.7

(a) Calculate the first partial derivatives of the integral $\int_{x-ct}^{x+ct} dy\, b(y)$ with respect to x and t.

(b) Verify directly that the function $u(x,t)$ of equation (2.25) is a solution of the wave equation (2.1) and satisfies the initial conditions (2.19), provided that a and b are twice and once differentiable, respectively.

Exercise 2.8

Calculate d'Alembert's solution of the wave equation for initial conditions $u(x,0) = a(x) = 0$ and $u_t(x,0) = b(x) = \sin 3x$.

Exercise 2.9

Calculate d'Alembert's solution of the wave equation for initial conditions $u(x,0) = a(x) = \sin 3x$ and $u_t(x,0) = b(x) = 2x$.

Exercise 2.10

Consider the solution $u(x,t)$ of equation (2.25). For what arguments do we need to know the functions a and b in order to calculate the function $u(x_0, t_0)$ at some position x_0 and time t_0?

2.4 General solution for the semi-infinite string

What happens if we try to introduce boundary conditions? Consider the motion resulting from an initial pulse. The solution (2.25) is correct if we have an *infinite* string, so the two parts of our initial pulse can travel to the left and to the right forever. However, if the string is finite, the pulse eventually arrives at an end of the string. What happens then? Clearly, solution (2.25) does *not* obey the boundary conditions $u(0,t) = u(l,t) = 0$ for a finite string of length l, so it will describe the motion correctly only until the pulse first hits a boundary. Can we fix it? Indeed, we can.

As we shall see, when hitting a rigid boundary the pulse is *reflected*, like water waves are reflected at obstacles in the water, sound waves at walls, and light waves in a mirror. The pulse does not disappear, but it generates

2.4 General solution for the semi-infinite string

a pulse that travels in the opposite direction. This pulse then travels along the string until it again hits a boundary, is reflected again, and travels back, and so on (and so forth). So if we understand how a *single* reflection takes place, we can construct a solution for the finite string by taking into account *multiple* reflections.

This procedure is not the easiest route to find solutions for the boundary-value problem. However, the principle behind this approach is simple, and it clarifies the relation between solutions for an infinite string and for a finite string, showing why periodic solutions are important. For these reasons, we first discuss reflection at a fixed boundary and, following this, the solution constructed by multiple reflections. Reflection for other boundary conditions will be considered later in the course.

Exercise 2.11

For a finite string of length l, consider the solution of the wave equation for a pulse of width $2r < l$, which is initially at rest in the centre of the string.

(a) At what time t does an edge of the pulse first reach a boundary of the string?

(b) Considering also negative times, which correspond to a possible motion before time $t = 0$, on which time interval does the solution obey the boundary conditions $u(0, t) = u(l, t) = 0$ for the finite string?

In this section, we consider a *semi-infinite string*, ranging from $x = 0$ to infinity. The string has only one end, at $x = 0$, so the pulse can be reflected only once, which makes this case much simpler than that of a finite string. The corresponding boundary condition for a fixed end is

$$u(0, t) = 0. \tag{2.26}$$

2.4.1 A left-moving pulse

We start by considering a single pulse, moving towards the left (i.e. towards the boundary at $x = 0$). This pulse corresponds to a solution $u(x, t) = G(x + ct)$ of the wave equation. At $t = 0$, the solution is $u(x, 0) = G(x)$, so $G(x)$, for $x \geq 0$, is the initial disturbance of the string, and the boundary condition (2.26) implies $G(0) = 0$.

In this case, it will prove useful to extend the mathematical description beyond the physical range $x \geq 0$ of the actual string, and consider solutions $u(x, t)$ that solve the wave equation for all x, not just on the domain $x \geq 0$ of the semi-infinite string. We extend the function $G(x)$ to negative values of x by defining $G(x) = 0$ for $x < 0$.

> You may wonder about the interpretation of a solution $u(x, t)$ of the wave equation outside the domain, whether for a semi-infinite string or for a finite one. The motion is described by the solution in the appropriate domain; outside this domain, the functions have no *physical* meaning. Nevertheless, the solutions $u_n(x, t)$ for a finite string given in equation (2.4) are defined for all values of x, and, as you will see, this is also the case for the solutions for the semi-infinite string constructed below. This is not an accident – it is natural to choose solutions which satisfy the wave equation for *all* values of x, and this is always possible. Indeed, we obtained the solutions $u_n(x, t)$ by starting from solutions $u(x, t)$ for an infinite string, and then imposing boundary conditions, thus selecting those globally defined solutions $u(x, t)$ that satisfy the boundary conditions.

The solution $u(x,t) = G(x+ct)$ satisfies the boundary condition (2.26) at time $t = 0$. If the pulse is initially sufficiently far from the boundary so that $G(x) = 0$ for all $0 \leq x < R$ for a certain distance R, as shown in Figure 2.5, the solution $u(x,t) = G(x+ct)$ fulfils the boundary condition for a while, until the pulse reaches the boundary at $x = 0$, which happens at time $t = R/c$. Once it reaches the boundary, we need to include a reflected pulse. How do we choose the reflected pulse in order to satisfy the boundary condition (2.26)? Exercise 2.12 provides the answer.

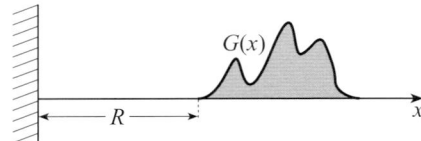

Figure 2.5 An initial pulse $G(x)$ situated at distance R from the boundary

Exercise 2.12

(a) Show that the function $u(x,t) = G(x+ct) - G(-x+ct)$, where G is an arbitrary, at least twice differentiable function, satisfies the wave equation and the boundary condition $u(0,t) = 0$ for all times t.

(b) For $G(x)$ as depicted in Figure 2.5, sketch $u(x,t)$ at $t = 0$ and at a time t with $0 < t < R/c$.

Reflection at the boundary

As shown in Exercise 2.12, the combination of two terms

$$u(x,t) = G(x+ct) - G(-x+ct) \qquad (2.27)$$

is a solution of the wave equation which satisfies the boundary condition (2.26) at any time t. The term $-G(-x+ct)$ can be interpreted as another pulse, a *mirror pulse* or *image pulse* (see Figure 2.6). It has the same shape as $G(x+ct)$, but is reflected in the horizontal and vertical axes. At $t = 0$, the mirror pulse lies outside the real string, because $-G(-x) = 0$ for $x \geq 0$.

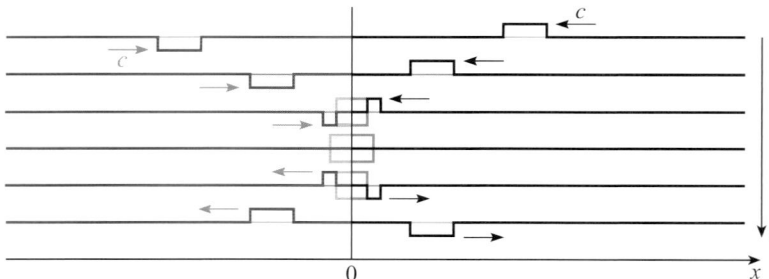

Figure 2.6 Using a mirror pulse to satisfy the boundary condition (2.26), giving rise to reflection of the pulse at the boundary

> For simplicity, Figure 2.6 shows a square-shaped pulse, which does not satisfy the condition that the function $G(x)$ is twice differentiable everywhere. However, it can be approximated arbitrarily well by continuous twice differentiable functions which satisfy the wave equation for all x.

As the real pulse moves to the left, the mirror pulse moves to the right with the same speed c, and they collide at $x = 0$. For a while, both pulses overlap close to the boundary, and the resulting displacement is given by the sum of the two pulses. Eventually, they pass through each other, then the term

2.4 General solution for the semi-infinite string

corresponding to our original pulse vanishes for $x \geq 0$, and only the mirror pulse contributes for $x > 0$. It represents the *reflected* pulse, of opposite sign, travelling to the right (see Figure 2.6).

> This is an example of the *method of images*, which uses mirror images to solve boundary-value problems. This method is frequently applied in various areas of applied mathematics, for instance in electrostatics and optics. It will also be used in the discussion of the diffusion equation later in the course.

The pulse shape shown in Figure 2.6 is used just for simplicity; the solution (2.27) works for any suitable function G. The solution involves a term that can be interpreted as a mirror pulse 'moving' outside the real string. This allows us to write down a simple expression for the solution, but in reality only the solution for $x > 0$ is part of the actual motion, which is shown in Figure 2.7.

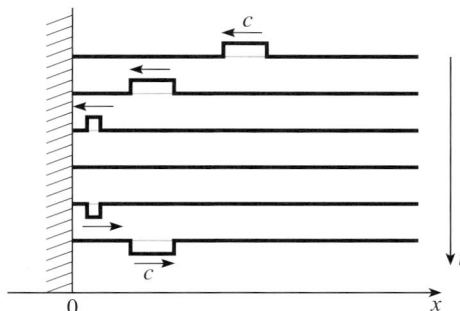

Figure 2.7 Several snapshots (top to bottom) of the reflection of a pulse at a boundary

2.4.2 A right-moving pulse

We now consider a right-moving pulse, corresponding to a solution of the wave equation of the form $F(x - ct)$, setting $F(x) = 0$ for $x \leq 0$. For $t > 0$, this pulse will never encounter the boundary at $x = 0$, so no reflection occurs. However, for $t < 0$, the function $F(x - ct)$ violates the boundary condition for some negative t. For a solution that holds at all times t, we must add a mirror pulse as well, corresponding to a reflection that occurred in the past.

Exercise 2.13

Show that the function $u(x, t) = F(x - ct) - F(-x - ct)$, where F is an arbitrary, at least twice differentiable function, satisfies the wave equation and the boundary condition $u(0, t) = 0$ for all times t.

As shown in Exercise 2.13, the combination

$$u(x, t) = F(x - ct) - F(-x - ct) \qquad (2.28)$$

satisfies the boundary condition $u(0, t) = 0$ for all times t. This situation is depicted in Figure 2.8 and represents the actual pulse $F(x - ct)$ and its mirror pulse $-F(-x - ct)$ travelling away from the boundary with velocity c.

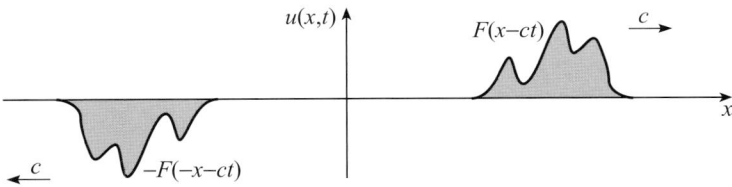

Figure 2.8 The two terms in the solution $u(x, t)$ of equation (2.28) correspond to a pulse and its mirror pulse travelling away from the origin for $t > 0$

2.4.3 General solution

Combining equations (2.27) and (2.28), we obtain the general solution for the semi-infinite string:

General solution for the semi-infinite string

$$u(x,t) = F(x-ct) - F(-x-ct) + G(x+ct) - G(-x+ct). \qquad (2.29)$$

For arbitrary, at least twice differentiable functions F and G, the function $u(x,t)$ satisfies the wave equation for all x and t, and the boundary condition $u(0,t) = 0$ for all t. As mentioned above, we extend the functions F and G to negative values of x by defining $F(x) = G(x) = 0$ for $x \leq 0$. The solution for $t=0$ becomes $u(x,0) = F(x) - F(-x) + G(x) - G(-x) = F(x) + G(x)$ for $x \geq 0$, because $F(-x) = G(-x) = 0$ for $x \geq 0$. At $t=0$, the original and mirror pulses are 'separated' in the sense that $u(x,0)$ comprises the original pulses for $x > 0$, which corresponds to the actual string, and comprises the mirror pulses for $x < 0$, because $u(x,0) = -F(-x) - G(-x)$ for $x < 0$.

Initial conditions

We can also derive the general solution of the initial-value problem for the semi-infinite string. We again specify the initial conditions by two arbitrary functions $a(x)$ and $b(x)$,

$$u(x,0) = a(x), \quad u_t(x,0) = b(x), \quad x \geq 0, \qquad (2.30)$$

setting $a(x) = b(x) = 0$ for $x \leq 0$. For $x > 0$, we then have

$$u(x,0) = F(x) + G(x) = a(x) \qquad (2.31)$$

and

$$u_t(x,0) = -cF'(x) + cG'(x) = b(x), \qquad (2.32)$$

where we have used $F(x) = G(x) = 0$ for $x \leq 0$, hence $F'(x) = G'(x) = 0$ for $x < 0$. So, for $x > 0$, we obtain precisely the same equations as in the case of the infinite string; compare equations (2.20) on page 67. Hence the solution is the same; the functions F and G are given by equations (2.22) and (2.23). Equation (2.29) then yields the solution of the initial-value problem for the semi-infinite string:

Solution of the initial-value problem for the semi-infinite string

$$u(x,t) = \frac{1}{2}[a(x-ct) + a(x+ct) - a(-x-ct) - a(-x+ct)]$$
$$+ \frac{1}{2c}\left(\int_{x-ct}^{x+ct} dy\, b(y) - \int_{-x-ct}^{-x+ct} dy\, b(y)\right). \qquad (2.33)$$

Exercise 2.14

(a) Derive equation (2.33) from equations (2.22), (2.23) and (2.29).

(b) Verify that the solution $u(x,t)$ given in equation (2.33) satisfies the initial conditions (2.30).

2.5 The finite string: solution by multiple reflection

This section contains material that is not assessed.

We can use the same approach to derive the solution for a finite string by taking account of reflections occurring at the two ends. You can think of the solution as describing pulses moving back and forth being reflected at the ends, like light between two parallel mirrors. This approach gives an instructive, intuitive understanding of the solution, but the algebra becomes rather involved and impractical, and thus is not the method of choice to solve the boundary-value problem for a finite string. While you are advised to work through this section and the exercises, you may concentrate on the main idea behind this approach rather than on the technical details, because you will not be assessed on manipulating expressions corresponding to infinitely many reflections at the boundaries.

We consider a string of length l, fixed at its ends at $x = 0$ and $x = l$. The boundary conditions are $u(0,t) = u(l,t) = 0$ for all t. We already know from the discussion of the semi-infinite string how to satisfy the boundary condition at $x = 0$.

The boundary condition at $x = l$ can be treated in the same way. For pulses travelling to the right and left described by functions $F(x - ct)$ and $G(x + ct)$, respectively, we add mirror pulses, to give

$$u(x,t) = F(x - ct) - F(2l - x - ct) \tag{2.34}$$

and

$$u(x,t) = G(x + ct) - G(2l - x + ct), \tag{2.35}$$

to satisfy the boundary condition $u(l,t) = 0$ for all t. Note that the 'mirror' is now at $x = l$, so x is related to $2l - x$, and the contributions of the original pulse and the mirror pulse cancel at $x = l$.

Exercise 2.15

Check that the functions $u(x,t)$ of equations (2.34) and (2.35) satisfy the boundary condition $u(l,t) = 0$ for all t.

2.5.1 Multiple reflections of a left-moving pulse

These elementary reflections are all we need to know to derive the solution for the finite string. For simplicity, we again start with a single pulse $G(x + ct)$, with $G(x) = 0$ for $x \leq 0$ and $x \geq l$. By reflection at both ends, this pulse gives rise to two mirror pulses, yielding

$$G(x + ct) - G(-x + ct) - G(2l - x + ct) \tag{2.36}$$

(see Figure 2.9).

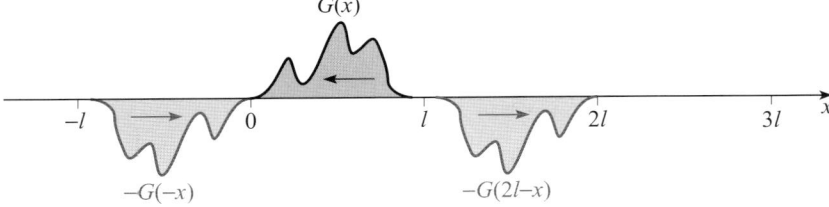

Figure 2.9 The function (2.36) at time $t = 0$

Each of the mirror pulses in turn produces one new mirror pulse. For example, by reflection at $x = 0$, $-G(2l - x + ct)$ generates a mirror pulse $G(2l + x + ct)$, while reflection at $x = l$ gives back $G[2l - (2l - x) + ct] = G(x + ct)$. Another new mirror pulse, $G(-2l + x + ct)$, is obtained by reflection of $-G(-x + ct)$ at $x = l$. Adding these two mirror pulses to the expression (2.36), we arrive at

$$G(x + ct) - G(-x + ct) - G(2l - x + ct)$$
$$+ G(2l + x + ct) + G(-2l + x + ct) \qquad (2.37)$$

(see Figure 2.10).

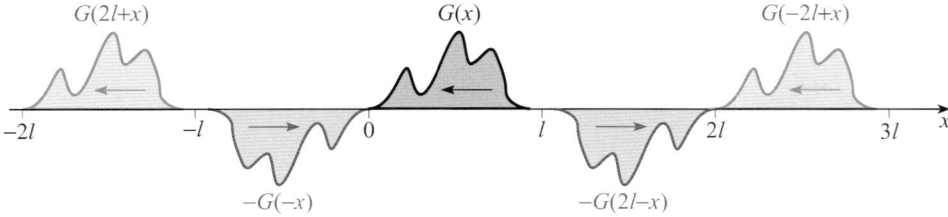

Figure 2.10 The function (2.37) at time $t = 0$

We then have to continue to add mirror pulses for the new terms. This procedure goes on indefinitely, so we obtain an infinite sequence of mirror pulses. This is like the (ideally) infinite number of reflections one sees when placing an object between two parallel mirrors. The solution $u(x,t)$ for the finite string comprises an infinite sum of terms,

$$u(x,t) = \sum_{j=-\infty}^{\infty} [G(2jl + x + ct) - G(2jl - x + ct)], \qquad (2.38)$$

where each term represents a copy of the original pulse travelling to the left and a copy of the reflected pulse travelling to the right; see Figure 2.11 for a sketch. There are no problems with convergence of the infinite sum because we assumed that the function $G(x)$ vanishes outside the string, so at most two terms of the infinite sum contribute at any position x and any time t.

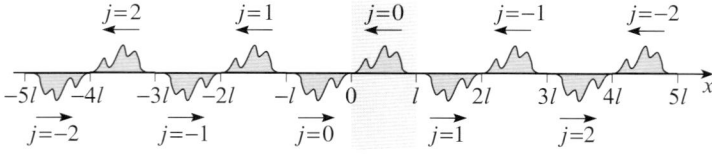

Figure 2.11 Sketch of the function $u(x,t)$ of equation (2.38) at time $t = 0$. The shaded region corresponds to the actual string.

Let us check that equation (2.38) is indeed the solution we were looking for. Clearly, $u(x,t)$ is a solution of the wave equation, because each term individually is a solution, and the wave equation is linear and homogeneous. Inserting $x = 0$ in equation (2.38) gives $u(0,t) = 0$ because each term in the infinite sum vanishes. It is slightly more difficult to verify the boundary condition for $x = l$. Inserting $x = l$ gives

$$u(l,t) = \sum_{j=-\infty}^{\infty} [G(2jl + l + ct) - G(2jl - l + ct)]$$

$$= \sum_{j=-\infty}^{\infty} G(2jl + l + ct) - \sum_{j=-\infty}^{\infty} G(2jl - l + ct), \qquad (2.39)$$

where we have split the sum into two parts.

2.5 The finite string: solution by multiple reflection

Now, noting that $2jl + l + ct = 2Jl - l + ct$ with $J = j+1$, we sum over J in the first infinite sum, obtaining

$$u(l,t) = \sum_{J=-\infty}^{\infty} G(2Jl - l + ct) - \sum_{j=-\infty}^{\infty} G(2jl - l + ct) = 0, \quad (2.40)$$

which vanishes because the sums are identical.

2.5.2 Multiple reflections of a right-moving pulse

If we start from an initially right-moving pulse $F(x - ct)$, with $F(x) = 0$ for $x \leq 0$ and $x \geq l$, the same argument leads to a solution

$$u(x,t) = \sum_{j=-\infty}^{\infty} \left[F(2jl + x - ct) - F(2jl - x - ct) \right]. \quad (2.41)$$

You are asked to verify this in the following exercise.

Exercise 2.16

Consider a function F which is at least twice differentiable and satisfies $F(x) = 0$ for $x \leq 0$ and $x \geq l$. Show that the function $u(x,t)$ of equation (2.41) is a solution of the wave equation and obeys the boundary conditions $u(0,t) = u(l,t) = 0$ for all times t.

2.5.3 General solution

In the general case, if both left- and right-moving terms are present, the solution is the sum of the expressions in equations (2.38) and (2.41):

Solution of the boundary-value problem by multiple reflections

$$u(x,t) = \sum_{j=-\infty}^{\infty} \big[F(2jl + x - ct) + G(2jl + x + ct) \\ - F(2jl - x - ct) - G(2jl - x + ct) \big]. \quad (2.42)$$

Exercise 2.17

Consider the case where $F(x) = \sin(\pi x/l)$ for $x \in [0,l]$ and $F(x) = 0$ elsewhere. Explain why the infinite sum in equation (2.41) yields $u(x,0) = \sin(\pi x/l)$ for all x.

Exercise 2.18

Calculate d'Alembert's solution (2.25) on page 67 for the initial-value problem with $u(x,0) = a(x) = \sin(2\pi x/l)$ and $u_t(x,0) = b(x) = 0$. Show that the expression you obtain satisfies the boundary conditions $u(0,t) = u(l,t) = 0$ for all times t, so it describes the motion of a finite string. Interpret this solution in terms of multiple reflections.

Initial conditions

Once more, we specify initial conditions for the finite string using two arbitrary functions: $u(x,0) = a(x)$ and $u_t(x,0) = b(x)$ for $0 \leq x \leq l$, and $a(x) = b(x) = 0$ for $x \leq 0$ and $x \geq l$. In the infinite sum (2.42) with $j = 0$, only the first two of the four terms contribute at $t = 0$, because all other terms correspond to mirror pulses that vanish on the interval $[0,l]$ for $t = 0$.

So we obtain the same relations for F and G as in equations (2.20) on page 67. This means that the functions F and G, as for infinite and semi-infinite strings, are once more given by equations (2.22) and (2.23). Hence we obtain the solution of the initial-value problem for the finite string:

Solution of the initial-value problem for the finite string

$$u(x,t) = \sum_{j=-\infty}^{\infty} \left[\frac{1}{2}\big(a(2jl + x - ct) + a(2jl + x + ct) - a(2jl - x - ct) - a(2jl - x + ct)\big) + \frac{1}{2c}\left(\int_{2jl+x-ct}^{2jl+x+ct} dy\, b(y) - \int_{2jl-x-ct}^{2jl-x+ct} dy\, b(y) \right) \right]. \quad (2.43)$$

Exercise 2.19

Derive the solution (2.43) from equations (2.22), (2.23) and (2.42).

Periodicity of solutions

We note that the solutions $u(x,t)$ for the finite string discussed above are *periodic* in t with period $\tau = 2l/c$, and periodic in x with period $2l$, so $u(x, t + 2l/c) = u(x + 2l, t) = u(x, t)$. This can be seen as follows. Consider, for simplicity, the solution given by equation (2.38). We have

$$u(x, t + 2l/c) = \sum_{j=-\infty}^{\infty} [G(2jl + x + 2l + ct) - G(2jl - x + 2l + ct)]$$

$$= \sum_{j=-\infty}^{\infty} [G(2(j+1)l + x + ct) - G(2(j+1)l - x + ct)]$$

$$= \sum_{J=-\infty}^{\infty} [G(2Jl + x + ct) - G(2Jl - x + ct)]$$

$$= u(x,t). \quad (2.44)$$

Here, we rewrote the sum with a summation index $J = j + 1$ to recover equation (2.38). The same argument applies to the solution of equation (2.41); hence the general solution that we obtain for the finite string is periodic in t. You are asked to show the periodicity in x in the following exercise.

Exercise 2.20

(a) Show that the solution $u(x,t)$ in equation (2.38) satisfies $u(x + 2l, t) = u(x, t)$. [Hint: Use a similar trick as in equations (2.39) and (2.40), splitting the sum into two parts and shifting the summation indices appropriately.]

(b) Show that $u(x + 2l, t) = u(x, t)$ also holds for the solution in equation (2.41).

2.6 Separation of variables

Using multiple reflections to account for the boundary conditions is rather tedious, because infinitely many reflections have to be taken into account.

2.6 Separation of variables

There is a more straightforward approach for some boundary-value problems. We illustrate this method by applying it to waves on a finite string with fixed ends. However, the method is very general and can be applied to many boundary-value problems associated with partial differential equations.

The wave equation (2.1) involves two independent variables: the space variable x and the time variable t. The boundary conditions for the finite string are $u(0,t) = u(l,t) = 0$, which are conditions for fixed values of x that have to hold for all times t. For solutions in the form of a product,

$$u(x,t) = f(x)\,g(t), \tag{2.45}$$

where $f(x)$ depends only on the position x along the string and $g(t)$ depends only on time t, the boundary conditions can be satisfied by choosing $f(x)$ appropriately. Solutions of a boundary-value problem of this type, and with sinusoidal time-dependence, are called *normal modes*; we shall use this as the *definition* of a normal mode. We have already encountered examples of such solutions: see equation (2.4) on page 61.

Our strategy is as follows. We try to find normal mode solutions, hoping to obtain *all* appropriate solutions by using linear combinations of these, which is allowed by the superposition principle. Using (2.45) in the wave equation (2.1) gives

$$f(x)\,g''(t) = c^2\,g(t)\,f''(x), \tag{2.46}$$

where we have used f'' and g'' to denote the second derivatives of the functions f and g. Dividing by $f(x)g(t)$ yields

$$\frac{f''(x)}{f(x)} = \frac{1}{c^2}\frac{g''(t)}{g(t)}. \tag{2.47}$$

We shall see below that no problems arise if either $f(x)$ or $g(t)$ vanishes.

All terms on the left-hand side of equation (2.47) depend only on position x, and all terms on the right-hand side are independent of x and depend only on time t. In other words, we have an equation of the form

$$\mathcal{F}(x) = \mathcal{G}(t) \tag{2.48}$$

with two functions $\mathcal{F}$ and $\mathcal{G}$. This equation has to hold for *all* values of x and t. What happens, for instance, if x changes while t is kept fixed? The right-hand side is independent of x, so it remains constant. But this means that the left-hand side has to be constant, too. Similarly, considering a change in t only, the left-hand side is constant, and so must be the right-hand side. So both sides are equal to a constant.

More formally, take the derivative of equation (2.48) with respect to x, say. This yields $\mathcal{F}'(x) = 0$, hence $\mathcal{F}(x) = C$, where C is a constant. Taking the derivative with respect to t gives $\mathcal{G}'(t) = 0$, hence $\mathcal{G}(t) = D$ with a constant D. But then equation (2.48) implies $C = D$, so $\mathcal{F}(x) = \mathcal{G}(t) = C$.

We thus obtain *two* equations,

$$\frac{f''(x)}{f(x)} = C, \quad \frac{g''(t)}{g(t)} = c^2 C, \tag{2.49}$$

which are mutually *independent*, because one involves only the variable x, and the other only the variable t.

We have *separated* the two variables of our original *partial differential equation*, obtaining *two independent ordinary differential equations* involving a new, as yet arbitrary, constant C. This approach is known as *separation of variables*, and relies on assuming a solution of the form (2.45).

> Sometimes you may find that 'normal mode' is used for the function $f(x)$, and the product $f(x)g(t)$ is referred to as a 'normal solution'.
>
> You may find various characterisations of normal modes in the literature; this definition is sufficient for our purposes.

Now that the variables have been separated, we can again multiply the resulting equations (2.49) by $f(x)$ and $g(t)$, respectively, to obtain

$$f''(x) - C f(x) = 0, \qquad (2.50)$$
$$g''(t) - c^2 C g(t) = 0. \qquad (2.51)$$

You might wonder why we first divided by $f(x)$ and $g(t)$ in equation (2.46), only to multiply by them again. The sole purpose of that transformation was to move all quantities depending on the variable t to the right-hand side of equation (2.47), and all quantities involving the variable x to the left-hand side, in order to make the separation of variables apparent. We note that equations (2.50) and (2.51) imply equation (2.46), including points where $f(x)$ or $g(t)$ vanishes, so there is no need to consider such cases separately.

> In the separation of variables, in the step from equation (2.46) to (2.47), it does not matter where we place the constant c^2. Moving this constant to the left-hand side effectively corresponds to replacing the arbitrary constant C in equations (2.49) by C/c^2.

Exercise 2.21

Consider the partial differential equation

$$w_{xx}(x,y) + 2w_{yy}(x,y) = 0.$$

Apply the separation of variables $w(x,y) = f(x)g(y)$, and derive the ordinary differential equations for f and g.

Exercise 2.22

Using $w(x,y) = f(x)g(y)$, try to separate the variables in the following partial differential equations.

(a) $w_x(x,y) + 2w_{yy}(x,y) + 3w(x,y) = 0$

(b) $w_x(x,y) + 2w_{yy}(x,y) + 3w^2(x,y) = 0$

(c) $w_x(x,y) + 2xw_{yy}(x,y) = 0$

(d) $w_x(x,y) - x^2 w_y(x,y) + 2xy w_{yy}(x,y) = 0$

(e) $w_x(x,y) + 2w_{yy}(x,y) - w_{xy}(x,y) = 0$

(f) $w_x(x,y) + 2w_{yy}(x,y) - w_{xxy}(x,y) = 0$

The previous exercises show that separation of variables will always work for homogeneous linear partial differential equations with constant coefficients that do not involve mixed derivatives. This is the case in Exercise 2.21 and in part (a) of Exercise 2.22. The partial differential equation of part (b) is non-linear, because it involves w^2. Parts (c) and (d) involve non-constant coefficients; however, as part (c) shows, there are cases where separation is still possible. Parts (e) and (f) contain mixed derivatives; again separation is possible in one case and not in the other. The method thus does not apply exclusively to linear partial differential equations with constant coefficients and without mixed derivatives. It also works if the partial differential equation essentially involves only two terms, as in part (c) and, after inserting $w(x,y) = f(x)g(y)$ and rearranging the resulting equation, in part (e).

2.6.1 Solution of the position-dependent part

We now proceed to solve the separated equations. We concentrate first on equation (2.50), which is a second-order linear differential equation for the function $f(x)$. You may already be familiar with this equation: for $C < 0$

2.6 Separation of variables

it is the equation of motion of a harmonic oscillator, so it describes simple harmonic motion. Because it is a second-order equation, the general solution is a linear combination of two independent functions.

The auxiliary equation is obtained by replacing the f by 1 and the nth derivative of f by λ^n. Here, it is $\lambda^2 = C$, which has two real solutions $\lambda = \pm\sqrt{C}$ for $C > 0$, the single solution $\lambda = 0$ for $C = 0$, and two imaginary solutions $\lambda = \pm i\sqrt{-C}$ for $C < 0$. You are asked to show in Exercise 2.23 that only solutions for $C < 0$ are of interest for the vibrating string problem, because non-trivial solutions satisfying the boundary conditions exist only in this case.

See Block 0, Subsection 1.3.2.

Exercise 2.23

Consider the second-order differential equation (2.50) for $f(x)$.

(a) What is the general real solution for $C > 0$?

(b) For $C > 0$, show that the boundary conditions $u(0,t) = u(l,t) = 0$ for the vibrating string imply that $f(x) = 0$ is the only solution.

(c) What is the general real solution for $C = 0$?

(d) For $C = 0$, show that again $f(x) = 0$ is the only solution that satisfies the boundary conditions.

Suppose now that $C < 0$. Putting $k^2 = -C > 0$, the general solution of equation (2.50) is

$$f(x) = a\cos(kx) + b\sin(kx), \tag{2.52}$$

where a and b are arbitrary real constants. An equivalent form of the general solution is discussed in the following exercise.

Exercise 2.24

(a) Verify that

$$f(x) = A\sin(kx + \psi), \tag{2.53}$$

with arbitrary real coefficients $A \geq 0$ and $-\pi/2 \leq \psi \leq \pi/2$, is a solution of the differential equation (2.50), with $k^2 = -C$.

(b) Show that this solution can be written in the form (2.52). Determine the coefficients a and b in terms of A and ψ.

(c) Use this result to express A and ψ in terms of a and b. [Hint: Use the quantity $a^2 + b^2$ to calculate A.]

Boundary conditions

For the motion of a stretched finite string with fixed ends, we need to satisfy the boundary conditions $u(0,t) = u(l,t) = 0$ at all times t. As mentioned previously, the product form $u(x,t) = f(x)g(t)$ then implies that

$$f(0) = f(l) = 0, \tag{2.54}$$

because we are not interested in the trivial solution $g(t) = 0$. Inserting $x = 0$ in solution (2.52), we obtain $a = 0$. The condition $f(l) = 0$ then yields the condition $\sin(kl) = 0$, which implies $kl = n\pi$ with integer n. Only those values of k that satisfy this condition give rise to solutions that obey the boundary conditions.

Still, we are left with an infinite set of choices for k, corresponding to solutions

$$f_n(x) = b_n \sin(k_n x), \quad k_n = \frac{n\pi}{l}, \quad n = 1, 2, 3, \ldots, \tag{2.55}$$

with arbitrary coefficients b_n. We do not need to take into account negative values of n, because $\sin(-x) = -\sin(x)$, so n and $-n$ lead to the same solution apart from a change of sign, which corresponds to a re-definition of the arbitrary constant b_n.

See also Exercise 1.9, and equation (2.4) on page 61, where we encountered the same condition.

2.6.2 Solution of the time-dependent part

The story is very much the same for equation (2.51). The differential equation for $g(t)$ is the same as that for $f(x)$, i.e. equation (2.50), apart from the constant factor c^2. Since $C = -k_n^2$, the differential equation becomes $g''(t) = \omega_n^2 g(t)$, where

$$\omega_n = k_n c = \frac{n\pi c}{l}, \quad n = 1, 2, 3, \ldots. \tag{2.56}$$

Note that the constants C in equations (2.50) and (2.51) must have the same value, so we can use the result of the position-dependent part.

The general solution is

$$g_n(t) = A_n \cos(\omega_n t) + B_n \sin(\omega_n t), \tag{2.57}$$

with arbitrary real coefficients A_n and B_n.

2.6.3 Normal mode solutions of the boundary-value problem

Now we put the two pieces together according to equation (2.45). From equations (2.55) and (2.57), we obtain the normal modes of the finite string

Normal modes of the finite string

$$u_n(x,t) = f_n(x) g_n(t) = \sin(k_n x) \left[A_n \cos(\omega_n t) + B_n \sin(\omega_n t) \right], \tag{2.58}$$

where k_n and ω_n are given in equations (2.56). Without loss of generality, we choose $b_n = 1$, because A_n and B_n are arbitrary. *Any such function $u_n(x,t)$, for any value of n and arbitrary coefficients A_n and B_n, is a solution of the wave equation that satisfies the boundary conditions $u_n(0,t) = u_n(l,t) = 0$.* Both the position- and time-dependent parts of the solutions $u_n(x,t)$ are sinusoidal.

Equation (2.58) represents an infinite set of normal mode solutions. Normal modes are sometimes referred to as *eigenmodes* or *eigenfunctions*, the corresponding values $-k_n^2$ or $-\omega_n^2$ as *eigenvalues*, and $\nu_n = \omega_n/(2\pi)$ as *eigenfrequencies* of the boundary-value problem. The infinite set of eigenvalues is called the *spectrum*.

> The notion of eigenvalue and eigenfunction is reminiscent of eigenvalues and eigenvectors of matrices, which might be a more familiar concept. Indeed, compare an eigenvalue equation $\boldsymbol{Mv} = \lambda \boldsymbol{v}$ for a matrix $\boldsymbol{M}$, where $\boldsymbol{v}$ is an eigenvector and λ is the corresponding eigenvalue, to equation (2.50) on page 78. This equation has the same formal structure: it can be written as
>
> $$\frac{d^2}{dx^2} f(x) = C f(x),$$
>
> where in place of a matrix we have a second-order differential operator d^2/dx^2, the function $f(x)$ plays the role of the eigenvector, and the constant C corresponds to the eigenvalue. In this sense, equation (2.50) is a generalisation of

2.6 Separation of variables

the matrix eigenvalue equation, where instead of vectors in a finite-dimensional space we are dealing with an infinite-dimensional space of suitable functions. Unlike in the finite-dimensional matrix case, the eigenvalues, and thus the spectrum, depend on boundary conditions. For the boundary conditions (2.54), the eigenvalues are $C = -k_n^2$ with k_n as for equation (2.55).

The number n that labels the normal mode $u_n(x,t)$ also determines the number of *nodes* of the mode. Recall that nodes are, apart from the end points, those points of the string that do not move at all during the motion. The *fundamental mode*, the lowest-frequency mode $u_1(x,t)$, has no nodes at all. The solution $u_2(x,t)$ vanishes for $x = l/2$, in the middle of the string, so it has one node. In fact, the solution $u_n(x,t)$ has $n-1$ nodes, because $f_n(x)$ has $n-1$ zeros for $0 < x < l$. In this sense, the normal modes of a string are characterised by the number of nodes.

You may have come across the notion of normal modes before, for particular motions of systems consisting of finitely many coupled harmonic oscillators, such as coupled springs. In this case, the normal modes can also be characterised by the property that the time-dependence is sinusoidal. There is indeed a close connection between these normal modes and those of the wave equation; we return to this point later in the course.

2.6.4 Superposition of normal modes and Fourier series

Each function $u_n(x,t)$ of equation (2.58) is a solution of the wave equation and satisfies the boundary conditions. The wave equation and the boundary conditions are linear and homogeneous, so any linear combination or *superposition* of the functions $u_n(x,t)$ is also a solution that satisfies the boundary conditions.

Therefore, any function

Superposition of normal modes

$$u(x,t) = \sum_{n=1}^{\infty} u_n(x,t) = \sum_{n=1}^{\infty} \sin(k_n x)[A_n \cos(\omega_n t) + B_n \sin(\omega_n t)], \quad (2.59)$$

with k_n and ω_n as given in equation (2.56), is a solution (provided that the coefficients A_n and B_n are such that the infinite sum converges in an appropriate sense), where we sum over any set of normal modes. This can be either a finite set, if all but finitely many coefficients A_n and B_n are zero, or an infinite set. In the most general case, we end up with an infinite sum of sinusoidal terms. This is an expansion of a *general* solution $u(x,t)$ in terms of the normal modes $u_n(x,t)$. *Any* solution can be expressed in this way. Such expansions of a function in a series of trigonometric functions, which arise naturally here, lead to *Fourier series*, named after Jean Baptiste Joseph Fourier, who first introduced such series when studying the heat equation, another partial differential equation which we shall consider later in the course.

When they were first introduced, at the beginning of the 19th century, Fourier series were a matter of controversy in the mathematical community. Today, they are a standard tool in mathematics, used to approximate functions. In the remainder of this chapter, we show how Fourier series and Fourier transforms (which are introduced below) can be used to solve initial-value problems for boundary conditions corresponding to finite, semi-infinite and infinite vibrating strings. The discussion of the properties of Fourier series and transforms will be the subject of the next chapter.

Initial conditions

For our solution (2.59), the initial condition $a(x) = u(x, 0)$ is given as a *Fourier sine series*

Fourier sine series

$$a(x) = \sum_{n=1}^{\infty} A_n \sin\left(\frac{\pi n x}{l}\right) \qquad (2.60)$$

with coefficients A_n. Taking the partial derivative with respect to t, we obtain

$$b(x) = u_t(x, 0) = \sum_{n=1}^{\infty} \frac{n B_n \pi c}{l} \sin\left(\frac{\pi n x}{l}\right), \qquad (2.61)$$

which is also a Fourier sine series, with coefficients $n B_n \pi c / l$. So equations (2.60) and (2.61) are representations of the functions $a(x)$ and $b(x)$ encoding the initial conditions as Fourier sine series. If we are given an initial-value problem for the finite string, which means that we know the functions $a(x)$ and $b(x)$ on the interval $[0, l]$, all we need to do is calculate the coefficients A_n and B_n from equations (2.60) and (2.61), respectively, and insert the results in equation (2.59) to obtain the solution.

Exercise 2.25

Consider the case where $B_n = 0$ for all n, so $b(x) = 0$, and $A_n = 0$ for all n except for $A_2 = 1$ and $A_5 = 2$.

(a) What is the initial form $a(x)$ of the string?

(b) Calculate the corresponding solution $u(x, t)$ of the wave equation.

(c) What is the period of the solution? In other words, for what smallest positive time τ is $u(x, t + \tau) = u(x, t)$ satisfied?

The coefficients of the Fourier sine series

To calculate the coefficients A_n and B_n, we exploit an *orthogonality* relation for sine functions:

$$\frac{2}{l} \int_0^l dx \, \sin\left(\frac{\pi m x}{l}\right) \sin\left(\frac{\pi n x}{l}\right) = \delta_{mn}, \qquad (2.62)$$

where δ_{mn} is the Kronecker delta symbol, whose values are defined by $\delta_{mm} = 1$, and $\delta_{mn} = 0$ for $m \neq n$. You are asked to derive this formula in the following exercise.

The term 'orthogonality' highlights the formal analogy to the orthogonality of vectors, with the integral in equation (2.62) playing the role of a scalar product.

Exercise 2.26

Show that the orthogonality relation (2.62) holds.

[Hint: Use the relation $2 \sin \alpha \sin \beta = \cos(\alpha - \beta) - \cos(\alpha + \beta)$, and consider separately the cases $m \neq n$ and $m = n$.]

Now, consider equation (2.60). Multiplying both sides by $\sin(\pi m x / l)$ and integrating with respect to x from 0 to l gives

$$\int_0^l dx \, a(x) \sin\left(\frac{\pi m x}{l}\right) = \int_0^l dx \sum_{n=1}^{\infty} A_n \sin\left(\frac{\pi n x}{l}\right) \sin\left(\frac{\pi m x}{l}\right)$$

$$= \sum_{n=1}^{\infty} A_n \int_0^l dx \, \sin\left(\frac{\pi m x}{l}\right) \sin\left(\frac{\pi n x}{l}\right)$$

$$= \frac{l}{2} \sum_{n=1}^{\infty} A_n \delta_{mn} = \frac{l}{2} A_m, \qquad (2.63)$$

2.6 Separation of variables

assuming that we can interchange the order of summation and integration. Replacing m by n, we obtain a formula for the Fourier coefficients

Fourier coefficients of the Fourier sine series

$$A_n = \frac{2}{l} \int_0^l dx\, a(x) \sin\left(\frac{\pi n x}{l}\right), \quad n = 1, 2, 3, \ldots, \quad (2.64)$$

which expresses the coefficients as integrals over the function $a(x)$ on the interval $[0, l]$. The coefficients B_n are given similarly by

$$B_n = \frac{2}{\pi n c} \int_0^l dx\, b(x) \sin\left(\frac{\pi n x}{l}\right), \quad n = 1, 2, 3, \ldots; \quad (2.65)$$

see Exercise 2.27 below.

Exercise 2.27

Show that the coefficients B_n in equation (2.61) are given by equation (2.65).

Exercise 2.28

Consider the motion of a stretched string of equilibrium length l which is initially released from rest with

$$a(x) = u(x, 0) = \tfrac{1}{8} \sin\left(\frac{\pi x}{l}\right) + \tfrac{1}{4} \sin\left(\frac{3\pi x}{l}\right)$$

for $0 \leq x \leq l$.

(a) Sketch (or plot, if you have access to a computer package such as Mathcad) the initial form of the string.

(b) Compare the expression for $a(x)$ with the Fourier expansion (2.60), and read off the Fourier coefficients A_n.

(c) Calculate the solution $u(x, t)$ of the initial-value problem from equation (2.59).

(d) Determine the motion $u(l/2, t)$ for the midpoint of the string. (If you have access to a suitable computer package, plot this as a function of time.)

(e) What is the time period of the solution? In other words, what is the smallest time $\tau > 0$ such that $u(x, t + \tau) = u(x, t)$?

More examples using Fourier series to solve initial-value problems will be discussed in Chapter 4.

Relation to d'Alembert's solution

The infinite Fourier series (2.59) can be expressed in the form of d'Alembert's solution (2.14) on page 65 by using trigonometric identities. With

$$2 \sin \alpha \sin \beta = \cos(\alpha - \beta) - \cos(\alpha + \beta),$$
$$2 \sin \alpha \cos \beta = \sin(\alpha - \beta) + \sin(\alpha + \beta),$$

equation (2.59) becomes

$$u(x, t) = \frac{1}{2} \sum_{n=1}^{\infty} \Big(A_n[\sin(k_n x - \omega_n t) + \sin(k_n x + \omega_n t)] + B_n[\cos(k_n x - \omega_n t) - \cos(k_n x + \omega_n t)] \Big). \quad (2.66)$$

Inserting $\omega_n = k_n c$ and splitting the sum into two parts gives

$$u(x, t) = \frac{1}{2} \sum_{n=1}^{\infty} \big(A_n \sin[k_n(x - ct)] + B_n \cos[k_n(x - ct)] \big)$$
$$+ \frac{1}{2} \sum_{n=1}^{\infty} \big(A_n \sin[k_n(x + ct)] - B_n \cos[k_n(x + ct)] \big). \quad (2.67)$$

This has the form of d'Alembert's solution (2.14), i.e. $u(x, t) = F(x - ct) + G(x + ct)$. The functions F and G are discussed in the following exercise.

Exercise 2.29

(a) Deduce the functions F and G from equations (2.14) and (2.67).

(b) Show that the functions satisfy $G(z) = -F(-z)$.

2.6.5 Semi-infinite strings and Fourier sine integrals

We have illustrated the use of separation of variables for the case of a finite vibrating string with fixed ends. The method also applies to boundary conditions that correspond to semi-infinite or infinite strings.

In the derivation of the normal mode solutions, the boundary conditions were used to show that the constant C that appears in the separated equations (2.50) and (2.51) is negative for non-trivial solutions. For semi-infinite and infinite strings, values $C \geq 0$ also lead to non-trivial solutions. However, these solutions do not correspond to vibrations, because their time-dependence is exponential in t for $C > 0$, and linear in t for $C = 0$. For physical reasons, we are not interested in these solutions, which involve motion of the entire string. We aim to describe localised disturbances of the semi-infinite or infinite string from its equilibrium position, so it suffices to concentrate on the solutions for $C < 0$. These are precisely the solutions that can be obtained from those for a finite string in the limit of infinite string length. For an infinite string, this local disturbance corresponds to the boundary condition that $u(x, t) \to 0$ as $x \to \pm\infty$.

If we consider a semi-infinite string, fixed at $x = 0$ and extending to ∞, say, then we have one boundary condition, $u(0, t) = 0$, to satisfy. This still implies that $a = 0$ in equation (2.52), so the spatial part again involves sine functions only. However, there are no restrictions on the allowed values of k, so the normal mode solution becomes

$$u_{(k)}(x, t) = \sin(kx)[A(k)\cos(kct) + B(k)\sin(kct)] \tag{2.68}$$

for any value of k; compare equation (2.58) on page 80. The solutions are labelled by k, and we think of the coefficients as functions of k. By the superposition principle, sums of solutions are also solutions. Here, because k is any real number, we can integrate over all $k \geq 0$, to give

$$u(x, t) = \int_0^\infty dk \, \sin(kx)\left[A(k)\cos(kct) + B(k)\sin(kct)\right], \tag{2.69}$$

provided that the k-dependence of the coefficients $A(k)$ and $B(k)$ is such that this integral makes sense. This corresponds to the infinite sum in equation (2.59) on page 81 for the finite string, and can be derived from equation (2.59) in the limit as the string length $l \to \infty$.

By inserting $u(x, t)$ of equation (2.69) in the wave equation, one can explicitly check that this function satisfies the wave equation, provided that differentiation and integration can be interchanged, which essentially requires that $A(k) \to 0$ and $B(k) \to 0$ sufficiently fast as $k \to \infty$.

Initial conditions

Consider now the initial condition $u(x, 0) = a(x)$. Equation (2.69) allows us to express $a(x)$ as a Fourier sine integral

Fourier sine integral

$$a(x) = \sqrt{\frac{2}{\pi}} \int_0^\infty dk \, \tilde{a}(k) \sin(kx), \tag{2.70}$$

2.6 Separation of variables

where the function $\tilde{a}(k) = \sqrt{\pi/2}A(k)$ is called the *Fourier sine transform* of the function $a(x)$, and the factor $\sqrt{\pi/2}$ is introduced for later convenience. By exchanging the order of differentiation and integration, the initial condition $u_t(x,0) = b(x)$ can also be expressed in this form:

$$b(x) = \sqrt{\frac{2}{\pi}} \int_0^\infty dk \, \tilde{b}(k) \, kc\sin(kx), \qquad (2.71)$$

with $\tilde{b}(k) = \sqrt{\pi/2}B(k)$; compare equations (2.60) and (2.61).

Given initial-value data $a(x)$ and $b(x)$ for the semi-infinite string, we need to calculate the coefficient functions $A(k)$ and $B(k)$, so we need the corresponding inverse Fourier transform. We only state the results here,

$$\tilde{a}(k) = \sqrt{\frac{2}{\pi}} \int_0^\infty dx \, a(x) \sin(kx), \qquad (2.72)$$

$$\tilde{b}(k) = \frac{1}{kc}\sqrt{\frac{2}{\pi}} \int_0^\infty dx \, b(x) \sin(kx), \qquad (2.73)$$

deferring a detailed discussion of Fourier transforms and their properties to the next chapter. Note the similarities of these expressions to equations (2.64) and (2.65) for the coefficients A_n and B_n of the Fourier sine series. This is not accidental, because one can describe an infinite string as a limiting case of a finite string, taking the limit as $l \to \infty$ in an appropriate setting.

2.6.6 Infinite strings and Fourier integrals

Infinite strings can be treated similarly, the only difference being that, in general, both $\sin(kx)$ and $\cos(kx)$ terms contribute to the solution. It turns out that it is calculationally advantageous to consider complex solutions.

Complex solutions of the wave equation

We once more consider the separated equations (2.50) and (2.51), with $C < 0$. For the spatial part, equation (2.50), the general complex solution is

$$f(x) = \alpha \exp(ikx) + \beta \exp(-ikx), \qquad (2.74)$$

with arbitrary, in general complex, constants α and β, and $k = \sqrt{-C} > 0$. An equivalent form of this solution is discussed in the following exercise.

Exercise 2.30

(a) Verify that

$$f(x) = A\exp[i(kx + \phi)] + B\exp[-i(kx - \psi)], \qquad (2.75)$$

with arbitrary real coefficients $A \geq 0$ and $B \geq 0$, and $0 \leq \phi \leq \pi$, $0 \leq \psi \leq \pi$, is a solution of the differential equation (2.50) for $k = \sqrt{-C} > 0$.

(b) Show that this solution can be written in the form (2.74). Determine the coefficients α and β in terms of A, B, ϕ and ψ.

(c) Use the results of part (b) to express A, B, ϕ and ψ in terms of α and β.

(d) Express the real and imaginary parts of $f(x)$ in terms of sine and cosine functions, and show that both can be written in the form of equation (2.52) on page 79.

Analogously, the general solution of equation (2.51) for $C < 0$ is

$$g(t) = \gamma \exp(i\omega t) + \delta \exp(-i\omega t), \tag{2.76}$$

with arbitrary complex coefficients γ and δ, and $\omega = kc > 0$. Multiplying equations (2.75) and (2.76), we arrive at normal mode solutions of the form

$$\begin{aligned}A(k)\exp[ik(x+ct)] + B(k)\exp[ik(x-ct)] \\ + D(k)\exp[-ik(x+ct)] + E(k)\exp[-ik(x-ct)],\end{aligned} \tag{2.77}$$

where the coefficients $A(k)$, $B(k)$, $D(k)$ and $E(k)$ are arbitrary. Any linear combination of these functions is a solution of the wave equation. Extending the range of k to negative values by defining $A(-k) = D(k)$ and $B(-k) = E(k)$ for $k > 0$, we can instead consider linear combinations of

$$u_{(k)}(x,t) = A(k)\exp[ik(x+ct)] + B(k)\exp[ik(x-ct)], \tag{2.78}$$

with any real value of k. As for the semi-infinite string problem, a general solution of this type can be expressed as an integral

The case $k = 0$ corresponds to a constant solution.

$$u(x,t) = \int_{-\infty}^{\infty} dk \left(A(k)\exp[ik(x+ct)] + B(k)\exp[ik(x-ct)] \right), \tag{2.79}$$

provided that the complex coefficient functions $A(k)$ and $B(k)$ are such that this integral exists.

For the initial condition $a(x) = u(x,0)$, this simplifies to the Fourier integral

Fourier integral

$$a(x) = \frac{1}{\sqrt{2\pi}} \int_{-\infty}^{\infty} dk\, \tilde{a}(k)\exp(ikx), \tag{2.80}$$

where the function $\tilde{a}(k) = \sqrt{2\pi}\,[A(k) + B(k)]$ is the (complex) *Fourier transform* of the function $a(x)$. Similarly, the initial condition $b(x) = u_t(x,0)$ becomes

$$b(x) = \frac{1}{\sqrt{2\pi}} \int_{-\infty}^{\infty} dk\, ikc\,\tilde{b}(k)\exp(ikx), \tag{2.81}$$

with $\tilde{b}(k) = \sqrt{2\pi}[A(k) - B(k)]$, provided that we can swap the order of differentiation and integration. It turns out to be convenient to introduce the factor $\sqrt{2\pi}$ in the Fourier integral. Again, we need to express $\tilde{a}(k)$ and $\tilde{b}(k)$ in terms of the initial conditions $a(x)$ and $b(x)$ in order to solve the initial-value problem. This is performed by using the theory of complex Fourier transforms.

Such Fourier transforms are a standard mathematical tool for the investigation of partial differential equations. Their properties will be discussed in detail in the next chapter.

Exercise 2.31

For $u(x,t)$ given in equation (2.79), express the real and imaginary parts in terms of sine and cosine functions.

2.7 Summary and Outcomes

In this chapter we have considered solutions of the one-dimensional wave equation. We started by characterising periodic solutions. Then we discussed d'Alembert's general solution and showed how the initial-value problem and the boundary conditions of a semi-infinite and a finite string can be satisfied. Finally, we introduced the method of separation of variables and applied it to the wave equation. This method reduces a partial differential equation to a set of ordinary differential equations, and can be applied to many partial differential equations that arise in physics. The solution of the ordinary differential equations for the finite string boundary conditions again resulted in periodic functions, the normal modes of the string. Solutions can be expanded in terms of normal modes, and these expansions naturally lead to the concept of Fourier series and Fourier transforms, which are considered in detail in the next chapter.

After working through this chapter and the exercises, you should:
- know how to characterise harmonic wave motion;
- understand what initial-value problems and boundary-value problems are;
- know d'Alembert's solution and be able to solve initial-value problems for the infinite and semi-infinite string;
- understand how boundary-value problems can be solved by taking account of reflections;
- be able to apply the method of separation of variables;
- know about normal modes and their significance;
- understand the differences between the solutions for infinite, semi-infinite and finite strings;
- understand how Fourier series and Fourier transforms arise.

2.8 Further Exercises

Exercise 2.32

Calculate the general solution of the partial differential equation
$$u_{tx}(x,t) = 5.$$

Exercise 2.33

Solve the partial differential equation
$$u_t(x,t) - 4u_x(x,t) = 2u(x,t)$$
with initial condition
$$u(x,0) = 3\exp(3x),$$
by separation of variables.

Exercise 2.34

Consider the partial differential equation
$$u_t(x,t) = u_{xx}(x,t)$$
with boundary conditions $u(0,t) = u(\pi,t) = 0$ and initial condition $u(x,0) = 3\sin x - \sin 3x$.

(a) Use separation of variables with $u(x,t) = f(x)g(t)$ to derive ordinary differential equations for f and g, and calculate the general solutions for $g(t)$ and $f(x)$.

(b) Taking into account the boundary conditions, derive the normal mode solutions for this equation.

(c) Use the initial condition to obtain the unique solution of the problem.

Exercise 2.35

Calculate d'Alembert's solution for an infinite string with initial data

$$u(x,0) = a(x) = 0, \quad u_t(x,0) = b(x) = \begin{cases} 2nc\sin(nx), & x > 0, \\ 0, & x \leq 0, \end{cases}$$

where n is a constant and c is the wave speed. [Hint: Consider the three cases $x \geq ct$, $-ct \leq x < ct$ and $x < -ct$ separately.]

Exercise 2.36

Consider a semi-infinite string on $x \geq 0$ with initial data $u(x,0) = a(x)$ and $u_t(x,0) = b(x)$ for $x \geq 0$, where

$$a(x) = 0, \quad b(x) = \begin{cases} 2nc\sin(nx), & x > 0, \\ 0, & x \leq 0, \end{cases}$$

where n is a constant and c is the wave speed.

(a) Consider a solution $u_{(k)}(x,t)$ of equation (2.68), and show that k can be chosen such that this is the solution of the wave equation that satisfies the initial conditions.

(b) For this solution, what is $u_t(x,0)$ for $x \geq 0$ and $x < 0$? Express these in terms of the function $b(x)$ that specifies the initial velocity, noting that $b(x) = 0$ for $x \leq 0$. Can you interpret the result in terms of a mirror pulse?

Exercise 2.37 *This exercise is optional.*

Consider d'Alembert's solution (2.25) for the initial-value problem of an infinite string. You are asked to show that solutions depend 'nicely' on the initial conditions, in the sense that similar initial conditions yield similar solutions.

Let $u_1(x,t)$ and $u_2(x,t)$ denote the solutions for initial conditions

$$u_1(x,0) = a_1(x), \quad \frac{\partial u_1}{\partial t}(x,0) = b_1(x),$$

and

$$u_2(x,0) = a_2(x), \quad \frac{\partial u_2}{\partial t}(x,0) = b_2(x),$$

respectively, where $a_1(x)$, $a_2(x)$ are at least twice differentiable, and $b_1(x)$, $b_2(x)$ are at least once differentiable. Assume that the initial data differ only slightly between the two cases, such that

$$|a_1(x) - a_2(x)| \leq \delta, \quad |b_1(x) - b_2(x)| \leq \epsilon$$

hold for all x. Show that this implies that the solutions differ at most by

$$|u_1(x,t) - u_2(x,t)| \leq \delta + \epsilon t$$

for any x and $t \geq 0$.

$$\left[\text{Hint: You will need to use the } \textit{triangle inequality } |x+y| \leq |x| + |y| \text{ and the result}\right.$$

$$\left.\left|\int dx\, f(x)\right| \leq \int dx\, |f(x)|.\right]$$

Solutions to Exercises in Chapter 2

Solution 2.1

(a) The wavelength of an electromagnetic wave of frequency $\nu = 200\,\mathrm{kHz} = 2 \times 10^5\,\mathrm{s}^{-1}$ is
$$\lambda = \frac{c}{\nu} \simeq \frac{3 \times 10^8}{2 \times 10^5} \simeq 1500\,\mathrm{m}.$$
The period τ is the inverse of ν, from equation (2.6). So $\tau = 5 \times 10^{-6}\,\mathrm{s}$ or $5\,\mu\mathrm{s}$. μs stands for microseconds.

The wavelength of an electromagnetic wave of frequency $\nu = 2.45\,\mathrm{GHz}$ is
$$\lambda = \frac{c}{\nu} \simeq \frac{3 \times 10^8}{2.45 \times 10^9} \simeq 0.12\,\mathrm{m},$$
i.e. about $12\,\mathrm{cm}$. The period is $\tau \simeq 4 \times 10^{-10}\,\mathrm{s}$ or about $0.4\,\mathrm{ns}$.   ns stands for nanoseconds.

Visible light of wavelength $\lambda = 600\,\mathrm{nm}$ has frequency
$$\nu = \frac{c}{\lambda} \simeq \frac{3 \times 10^8}{6 \times 10^{-7}} = 5 \times 10^{14}\,\mathrm{Hz} = 500\,\mathrm{THz},$$

$1\,\mathrm{THz} = 10^{12}\,\mathrm{Hz}$ denotes one terahertz, corresponding to 10^{12} cycles per second.

The period is now even shorter: $\tau = 2 \times 10^{-15}\,\mathrm{s} = 2\,\mathrm{fs}$.

fs stands for femtoseconds.

(b) The solutions $u_n(x,t)$ are characterised by $k_n = n\pi/l$ and $\omega_n = k_n c = n\pi c/l$. By equations (2.6) and (2.7), this yields
$$\lambda_n = \frac{2\pi}{k_n} = \frac{2l}{n}, \quad \nu_n = \frac{\omega_n}{2\pi} = \frac{nc}{2l},$$
so $\lambda_n \nu_n = c$, as required.

Solution 2.2

(a) The wavelength λ is related to the frequency ν by $\lambda = c/\nu$. With $c \simeq 343\,\mathrm{m\,s}^{-1}$, this gives $\lambda \simeq 21.4\,\mathrm{m}$ for $\nu = 16\,\mathrm{Hz}$ and $\lambda \simeq 0.017\,\mathrm{m} = 1.7\,\mathrm{cm}$ for $\nu = 20\,000\,\mathrm{Hz}$.

(b) For $\nu \simeq 16.5\,\mathrm{Hz}$, we find $l = \lambda/2 \simeq 10.4\,\mathrm{m}$.

(c) The frequency of a pipe of length $l = 20\,\mathrm{m}$ is about $8.6\,\mathrm{Hz}$, which is well below the threshold of audible frequencies. (Nevertheless, you can feel the vibration at that frequency directly, and you can also hear a mixture of other frequencies, higher harmonics, produced by the pipe.)

Solution 2.3

(a) The function $u(x,t)$ has the form $F(x - ct)$ with $F(z) = \exp(-z^2)$ and $c = 2$, so it satisfies the wave equation. Explicitly, the partial derivatives are
$$u_x(x,t) = -2(x - 2t)u(x,t),$$
$$u_{xx}(x,t) = [-2 + 4(x - 2t)^2]u(x,t),$$
$$u_t(x,t) = 4(x - 2t)u(x,t),$$
$$u_{tt}(x,t) = [-8 + 16(x - 2t)^2]u(x,t) = 4u_{xx}(x,t) = c^2 u_{xx}(x,t).$$

(b) The functions $u(x,0)$, $u(x,1)$ and $u(x,2)$ are shown in Figure 2.12.

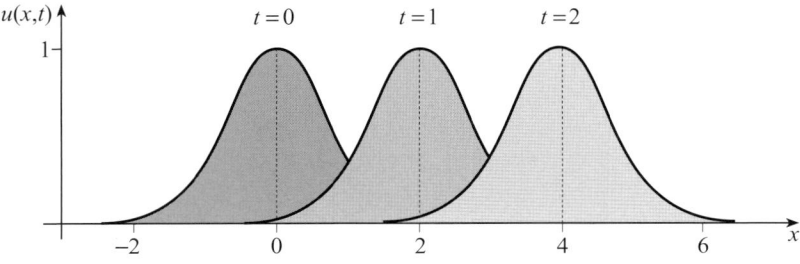

Figure 2.12 The function $u(x,t) = \exp[-(x - 2t)^2]$ at $t = 0$, $t = 1$ and $t = 2$

Solution 2.4

Differentiating $u(x,t) = G(x+ct)$ with respect to x and t gives

$$u_{xx}(x,t) = G''(x+ct) \quad \text{and} \quad u_{tt}(x,t) = c^2 G''(x+ct),$$

so $u(x,t)$ satisfies the wave equation for any function G which is at least twice differentiable.

Solution 2.5

Applying the addition formula with $\alpha = n\pi x/l$ and $\beta = n\pi ct/l$ gives

$$u_n(x,t) = \tfrac{1}{2} A_n \left[\cos\left(\tfrac{n\pi}{l}(x-ct)\right) - \cos\left(\tfrac{n\pi}{l}(x+ct)\right) \right].$$

This has the form of equation (2.14) with $F(z) = -G(z) = \tfrac{1}{2} A_n \cos(n\pi z/l)$.

Solution 2.6

(a) We have $x(y,z) = (z+y)/2$ and $t(y,z) = (z-y)/(2c)$.

(b) Taking partial derivatives yields

$$\frac{\partial x(y,z)}{\partial y} = \frac{\partial}{\partial y}\left(\frac{z+y}{2}\right) = \frac{1}{2}, \quad \frac{\partial t(y,z)}{\partial y} = \frac{\partial}{\partial y}\left(\frac{z-y}{2c}\right) = -\frac{1}{2c}.$$

(c) We have

$$v_y(y,z) = \frac{\partial}{\partial y} u[x(y,z), t(y,z)] = u_x \frac{\partial x}{\partial y} + u_t \frac{\partial t}{\partial y} = \frac{1}{2} u_x - \frac{1}{2c} u_t,$$

where we have used the chain rule and the results of part (b).

Taking the partial derivative with respect to z, we obtain

$$v_{zy}(y,z) = \frac{\partial}{\partial z}\left(\frac{1}{2} u_x - \frac{1}{2c} u_t\right)$$

$$= \frac{1}{2} u_{xx} \frac{\partial x}{\partial z} + \frac{1}{2} u_{tx} \frac{\partial t}{\partial z} - \frac{1}{2c} u_{xt} \frac{\partial x}{\partial z} - \frac{1}{2c} u_{tt} \frac{\partial t}{\partial z}$$

$$= \frac{1}{4} u_{xx} + \frac{1}{4c} u_{tx} - \frac{1}{4c} u_{xt} - \frac{1}{4c^2} u_{tt}$$

$$= \frac{1}{4c^2} \left(c^2 u_{xx} - u_{tt}\right),$$

where again we have used the chain rule, and $u_{tx} = u_{xt}$ according to the mixed derivative theorem. Therefore, assuming $c \neq 0$, the equations

$$v_{zy}(y,z) = 0, \quad \text{and} \quad c^2 u_{xx}(x,t) - u_{tt}(x,t) = 0$$

are equivalent.

(d) We can rewrite the equation as

$$\frac{\partial}{\partial z} v_y(y,z) = 0.$$

The general solution of the first-order differential equation for the function $v_y(y,z)$ is an arbitrary function of y,

$$v_y(y,z) = f(y),$$

because the solution must be constant with respect to z. This, in turn, is a first-order differential equation, which can be solved by integration to give

$$v(y,z) = F(y) + G(z),$$

where $G(z)$ is a constant with respect to y, but may be an arbitrary function of z. The function $F(y)$ is a primitive (indefinite integral) of our arbitrary function $f(y)$, thus $F' = f$, so F is also an arbitrary function of y. Note that F and G need to be twice differentiable, whereas f needs to be differentiable only once.

So $v(y, z) = F(y) + G(z)$, with arbitrary twice differentiable functions F and G, is the general solution of the wave equation.

(e) Substituting the original variables into the general solution, we obtain
$$u(x, t) = v[y(x, t), z(x, t)] = v(x - ct, x + ct) = F(x - ct) + G(x + ct),$$
which is d'Alembert's solution (2.14).

Note that we could perform the same argument with the roles of y and z interchanged, performing the two integrations in the opposite order. This would produce the same result.

Solution 2.7

(a) Denote by B a primitive of the function b, so $B' = b$. Then the integral is given by
$$\int_{x-ct}^{x+ct} dy\, b(y) = B(x + ct) - B(x - ct).$$

The partial derivatives are
$$\frac{\partial}{\partial x} \int_{x-ct}^{x+ct} dy\, b(y) = \frac{\partial}{\partial x}[B(x + ct) - B(x - ct)]$$
$$= B'(x + ct) - B'(x - ct)$$
$$= b(x + ct) - b(x - ct)$$

and
$$\frac{\partial}{\partial t} \int_{x-ct}^{x+ct} dy\, b(y) = \frac{\partial}{\partial t}[B(x + ct) - B(x - ct)]$$
$$= c[B'(x + ct) + B'(x - ct)]$$
$$= c[b(x + ct) + b(x - ct)].$$

(b) We need to calculate the second partial derivatives of the solution given in equation (2.25):
$$u_{xx}(x, t) = \frac{1}{2}[a''(x - ct) + a''(x + ct)] + \frac{1}{2c}[b'(x + ct) - b'(x - ct)]$$

and
$$u_{tt}(x, t) = \frac{c^2}{2}[a''(x - ct) + a''(x + ct)] + \frac{c}{2}[b'(x + ct) - b'(x - ct)]$$
$$= c^2\, u_{xx}(x, t).$$

So $u(x, t)$ is a solution of the wave equation. Also, $u(x, 0) = a(x)$ and
$$u_t(x, 0) = \frac{c}{2}[-a'(x - c0) + a'(x + c0)] + \frac{1}{2}[b(x + c0) + b(x - c0)] = b(x),$$
hence the initial conditions (2.19) are also satisfied.

Solution 2.8

We have $a(x) = 0$ and $b(x) = \sin(3x)$. D'Alembert's solution (2.25) gives
$$u(x, t) = \frac{1}{2c} \int_{x-ct}^{x+ct} dy\, \sin(3y)$$
$$= -\frac{1}{6c}[\cos(3x + 3ct) - \cos(3x - 3ct)]$$
$$= \frac{1}{6c}[\cos(3x - 3ct) - \cos(3x + 3ct)]$$
$$= \frac{1}{3c} \sin(3x) \sin(3ct),$$

where we have used the trigonometric identity $2 \sin \alpha \sin \beta = \cos(\alpha - \beta) - \cos(\alpha + \beta)$.

Solution 2.9

We have $a(x) = \sin(3x)$ and $b(x) = 2x$, thus d'Alembert's solution (2.25) gives

$$u(x,t) = \frac{1}{2}[\sin(3x - 3ct) + \sin(3x + 3ct)] + \frac{1}{c}\int_{x-ct}^{x+ct} dy\, y.$$

The second term gives

$$\left[\frac{y^2}{2c}\right]_{x-ct}^{x+ct} = \frac{(x+ct)^2 - (x-ct)^2}{2c} = \frac{4ctx}{2c} = 2tx,$$

so the solution becomes

$$u(x,t) = 2tx + \sin(3x)\cos(3ct),$$

where we have used the trigonometric identity $2\sin\alpha\cos\beta = \sin(\alpha+\beta) + \sin(\alpha-\beta)$.

Solution 2.10

At position x_0 and time t_0, we have

$$u(x_0, t_0) = \frac{1}{2}[a(x_0 - ct_0) + a(x_0 + ct_0)] + \frac{1}{2c}\int_{x_0 - ct_0}^{x_0 + ct_0} dy\, b(y).$$

So all we need to know are the two values $a(x_0 - ct_0)$ and $a(x_0 + ct_0)$ of the function a, and the function b on the interval $[x_0 - ct_0, x_0 + ct_0]$.

Solution 2.11

(a) At time $t = 0$, the pulse is at $l/2 - r \leq x \leq l/2 + r$. It splits into two parts that move to the left and to the right with speed c. So the left end of the left-moving part reaches $x = 0$ after travelling a distance $l/2 - r$, which takes a time $t = (l - 2r)/(2c)$. At the same time, the right end of the right-moving part of the pulse reaches $x = l$.

(b) The motion in negative time is exactly the same, because changing the sign of t just exchanges $a(x + ct)$ and $a(x - ct)$ in d'Alembert's solution (2.25). So the pulse 'left' the boundaries at time $t = -(l - 2r)/(2c)$, and hence the boundary conditions are satisfied for the time interval $-(l - 2r)/(2c) < t < (l - 2r)/(2c)$.

Solution 2.12

(a) You can differentiate $u(x,t)$ with respect to x and t to show that it satisfies the wave equation. Alternatively, notice that setting $F(z) = -G(-z)$ gives $u(x,t) = F(x - ct) + G(x + ct)$, which is the form of d'Alembert's solution of the wave equation. Inserting $x = 0$, we find $u(0,t) = G(ct) - G(ct) = 0$ at all times t.

(b) The solution $u(x,t)$ is sketched in Figure 2.13, at time $t = 0$ and at a time t, with $0 < t < R/c$, before the pulse reaches the boundary.

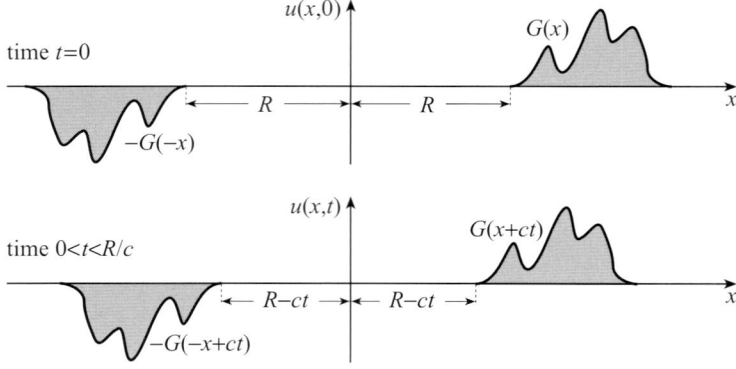

Figure 2.13 Sketch of the solution $u(x,t) = G(x + ct) - G(-x + ct)$ with $G(x)$ as in Figure 2.5, at time $t = 0$ (top) and at a time t with $0 < t < R/c$ (bottom)

Solutions to Exercises in Chapter 2

Solution 2.13

Setting $G(z) = -F(-z)$, we have $u(x,t) = F(x-ct) + G(x+ct)$, which is the form of d'Alembert's solution of the wave equation. Inserting $x = 0$, we find $u(0,t) = F(-ct) - F(-ct) = 0$ at all times t.

Solution 2.14

(a) Inserting the expressions for F and G given in equations (2.22) and (2.23), we have

$$F(x-ct) + G(x+ct) = \frac{a(x-ct)}{2} + \frac{C}{2} - \frac{1}{2c}\int_{x_0}^{x-ct} dy\, b(y)$$

$$+ \frac{a(x+ct)}{2} - \frac{C}{2} + \frac{1}{2c}\int_{x_0}^{x+ct} dy\, b(y)$$

$$= \frac{1}{2}[a(x-ct) + a(x+ct)] + \frac{1}{2c}\int_{x-ct}^{x+ct} dy\, b(y);$$

compare equation (2.24). Analogously,

$$-F(-x-ct) - G(-x+ct) = -\frac{a(-x-ct)}{2} - \frac{C}{2} + \frac{1}{2c}\int_{x_0}^{-x-ct} dy\, b(y)$$

$$- \frac{a(-x+ct)}{2} + \frac{C}{2} - \frac{1}{2c}\int_{x_0}^{-x+ct} dy\, b(y)$$

$$= -\frac{1}{2}[a(-x-ct) + a(-x+ct)]$$

$$- \frac{1}{2c}\int_{-x-ct}^{-x+ct} dy\, b(y).$$

Adding the results above yields equation (2.33).

(b) Inserting $t = 0$ gives

$$u(x,0) = a(x) - a(-x), \quad u_t(x,0) = b(x) - b(-x),$$

where the second relation follows by using the result in Exercise 2.7. These expressions agree with the initial conditions (2.30) for $x > 0$, because $a(x) = b(x) = 0$ for $x \leq 0$.

Solution 2.15

At $x = l$, equations (2.34) and (2.35) become $u(l,t) = F(l-ct) - F(2l-l-ct) = 0$ and $u(l,t) = G(l+ct) - G(2l-l+ct) = 0$, so the boundary condition is satisfied.

Solution 2.16

First, the infinite sum in equation (2.41) is well defined, because for any value of x, there are at most two values of j that give a non-vanishing contribution. This is a consequence of $F(x) = 0$ for $x \leq 0$ and $x \geq l$. Each term in the sum individually is a solution of the wave equation, so the infinite sum is also a solution.

Inserting $x = 0$ gives $u(0,t) = 0$. For $x = l$, we obtain

$$u(l,t) = \sum_{j=-\infty}^{\infty} [F(2jl + l - ct) - F(2jl - l - ct)]$$

$$= \sum_{j=-\infty}^{\infty} F(2jl + l - ct) - \sum_{j=-\infty}^{\infty} F(2jl - l - ct),$$

where we have split the sum into two parts. Performing the first sum over $J = j+1$, noting that $2jl + l - ct = 2Jl - l - ct$, gives

$$u(l,t) = \sum_{J=-\infty}^{\infty} F(2Jl - l - ct) - \sum_{j=-\infty}^{\infty} F(2jl - l - ct) = 0,$$

which vanishes because the sums are identical.

Solution 2.17

For $t = 0$, the infinite sum in equation (2.41) becomes

$$u(x, 0) = \sum_{j=-\infty}^{\infty} [F(2jl + x) - F(2jl - x)].$$

The terms $F(2jl + x)$ are $\sin(2j\pi + \pi x/l) = \sin(\pi x/l)$ for $x/l \in [-2j, -2j+1]$ and vanish elsewhere. The remaining terms $-F(2jl - x)$ are $-\sin(2j\pi - \pi x/l) = -\sin(-\pi x/l) = \sin(\pi x/l)$ for $x/l \in [2j-1, 2j]$, and vanish elsewhere. (Here we have used the identity $\sin(-x) = -\sin(x)$.) As j runs through all integers, these intervals never overlap (except at the end points of each interval), but cover the entire real line. So whatever value x has, it falls into exactly one such interval (or on the boundary of two, in which case the function is zero). Thus we find that $u(x, 0) = \sin(\pi x/l)$ for all x.

Solution 2.18

For the initial conditions $a(x) = \sin(2\pi x/l)$ and $b(x) = 0$, d'Alembert's solution in equation (2.25) gives

$$u(x, t) = \frac{1}{2}\left(\sin\left(\frac{2\pi(x-ct)}{l}\right) + \sin\left(\frac{2\pi(x+ct)}{l}\right)\right)$$

$$= \sin\left(\frac{2\pi x}{l}\right)\cos\left(\frac{2\pi ct}{l}\right).$$

Inserting $x = 0$ and $x = l$, we find that the boundary conditions $u(0, t) = u(l, t) = 0$ are automatically satisfied, so $u(x, t)$ is a solution of the boundary-value problem for all values of x and t.

As in Exercise 2.17, we may interpret this solution as the superposition of an initial disturbance described by $u(x, 0)$ for $x \in [0, l]$ with its reflections at the two ends of the string, which correspond to infinitely many mirror pulses given by $\sin(2\pi x/l)$ on any interval $[jl, (j+1)l]$, with $j = \pm 1, \pm 2, \ldots$.

Solution 2.19

From equations (2.22) and (2.23), we have

$$F(2jl + x - ct) + G(2jl + x + ct)$$
$$= \frac{a(2jl + x - ct)}{2} + \frac{C}{2} - \frac{1}{2c}\int_{x_0}^{2jl+x-ct} dy\, b(y)$$
$$+ \frac{a(2jl + x + ct)}{2} - \frac{C}{2} + \frac{1}{2c}\int_{x_0}^{2jl+x+ct} dy\, b(y)$$
$$= \frac{1}{2}[a(2jl + x - ct) + a(2jl + x + ct)] + \frac{1}{2c}\int_{2jl+x-ct}^{2jl+x+ct} dy\, b(y)$$

and

$$-F(2jl - x - ct) - G(2jl - x + ct)$$
$$= -\frac{a(2jl - x - ct)}{2} - \frac{C}{2} + \frac{1}{2c}\int_{x_0}^{2jl-x-ct} dy\, b(y)$$
$$- \frac{a(2jl - x + ct)}{2} + \frac{C}{2} - \frac{1}{2c}\int_{x_0}^{2jl-x+ct} dy\, b(y)$$
$$= -\frac{1}{2}[a(2jl - x - ct) + a(2jl - x + ct)] - \frac{1}{2c}\int_{2jl-x-ct}^{2jl-x+ct} dy\, b(y);$$

compare this with Exercise 2.14, which corresponds to the case $j = 0$. Adding the two equations and summing over all values of j yields equation (2.43).

Solution 2.20

(a) For the solution in equation (2.38), we have

$$u(x+2l,t) = \sum_{j=-\infty}^{\infty} [G(2jl+2l+x+ct) - G(2jl-2l-x+ct)]$$

$$= \sum_{j=-\infty}^{\infty} G[2(j+1)l+x+ct] - \sum_{j=-\infty}^{\infty} G[2(j-1)l-x+ct]$$

$$= \sum_{J=-\infty}^{\infty} G(2Jl+x+ct) - \sum_{J=-\infty}^{\infty} G(2Jl-x+ct)$$

$$= \sum_{J=-\infty}^{\infty} [G(2Jl+x+ct) - G(2Jl-x+ct)]$$

$$= u(x,t).$$

Because we split the sum into two parts, we can use the same summation variable J for $J = j+1$ in the first sum and $J = j-1$ in the second. Alternatively, we could have shifted one of the summation variables by 2.

(b) The solution in equation (2.41) has the same form as that in equation (2.38), apart from the sign in the argument ct. This does not affect the method used above, so $u(x+2l,t) = u(x,t)$ also holds for this solution.

Solution 2.21

Inserting the product $w(x,y) = f(x)g(y)$ into the partial differential equation gives

$$f''(x)\,g(y) + 2\,f(x)\,g''(y) = 0.$$

Dividing by $w(x,y) = f(x)g(y)$ and rearranging the result yields

$$\frac{f''(x)}{f(x)} = -\frac{2\,g''(y)}{g(y)},$$

where the left-hand side depends only on the variable x, and the right-hand side depends only on y. Thus both sides must equal a constant C, and upon multiplying with f and g again, we obtain the set of ordinary differential equations

$$f''(x) - Cf(x) = 0,$$
$$g''(y) + \frac{C}{2}g(y) = 0.$$

Solution 2.22

(a) Inserting $w(x,y) = f(x)g(y)$ gives

$$f'(x)\,g(y) + 2\,f(x)\,g''(y) + 3\,f(x)\,g(y) = 0.$$

Dividing by $w(x,y) = f(x)g(y)$ and rearranging the result yields

$$\frac{f'(x)}{f(x)} = -\frac{2\,g''(y)}{g(y)} - 3.$$

Again, the left-hand side depends just on x, and the right-hand side depends just on y, so both sides must equal a constant C. We now obtain the separated equations

$$f'(x) - Cf(x) = 0,$$
$$g''(y) + \frac{3+C}{2}g(y) = 0.$$

(b) Inserting $w(x,y) = f(x)g(y)$ gives

$$f'(x)\,g(y) + 2\,f(x)\,g''(y) + 3\,f^2(x)\,g^2(y) = 0.$$

In this case, the method fails, because the three terms do not become independent of either x or y if we divide by $f(x)g(y)$ (or indeed by any other function).

(c) Inserting $w(x,y) = f(x)g(y)$ gives
$$f'(x)\,g(y) + 2x\,f(x)\,g''(y) = 0.$$
Dividing by $xf(x)g(y)$ and rearranging the result yields
$$\frac{f'(x)}{xf(x)} = -\frac{2\,g''(y)}{g(y)}.$$
Once more, we have arranged the equation such that the left-hand side depends just on x and the right-hand side depends just on y. Thus both sides must equal a constant C, and we obtain
$$f'(x) - Cxf(x) = 0, \quad g''(y) + \frac{C}{2}g(y) = 0.$$

(d) Inserting $w(x,y) = f(x)g(y)$ gives
$$f'(x)\,g(y) - x^2\,f(x)\,g'(y) + 2xy\,f(x)\,g''(y) = 0.$$
Because of the non-constant coefficients, it is not possible to divide by a function such that all three terms depend on x or y only, so the method fails again.

(e) Inserting $w(x,y) = f(x)g(y)$ gives
$$f'(x)\,g(y) + 2\,f(x)\,g''(y) - f'(x)\,g'(y) = f'(x)\,[g(y) - g'(y)] + 2\,f(x)\,g''(y)$$
$$= 0.$$
Dividing by $f(x)[g(y) - g'(y)]$ gives
$$\frac{f'(x)}{f(x)} = -\frac{2g''(y)}{g(y) - g'(y)},$$
so again it is possible to separate the variables. The separated equations are
$$f'(x) - Cf(x) = 0, \quad g''(y) - \frac{C}{2}[g'(y) - g(y)] = 0.$$

(f) Inserting $w(x,y) = f(x)g(y)$ gives
$$f'(x)\,g(y) + 2\,f(x)\,g''(y) - f''(x)\,g'(y) = 0.$$
This involves the functions f and g and their first and second derivatives, so in contrast to the previous case the terms cannot be combined. It is not possible to cancel the dependence on either x or y simultaneously for all three terms, so the method fails.

Solution 2.23

(a) For $C > 0$, we have two real roots $\lambda = \pm\sqrt{C}$ of the auxiliary equation, so the general solution is
$$f(x) = a\exp(kx) + b\exp(-kx),$$
where $k = \sqrt{C}$, and a and b are arbitrary real constants.

(b) The boundary conditions $u(0,t) = u(l,t) = 0$ imply $f(0) = f(l) = 0$, because $u(x,t) = f(x)g(t)$ and we are not interested in solutions where $g(t) = 0$ for all times t. Inserting $f(0) = 0$ in the general solution for $C > 0$ found in part (a) gives $a + b = 0$, so $b = -a$. For $x = l$, we obtain
$$f(l) = a\exp(kl) - a\exp(-kl) = 2a\sinh(kl),$$
which vanishes only for $kl = 0$. Since $k = \sqrt{C} > 0$ (for $C > 0$), we must have $a = 0$, so the only solution is $f(x) = 0$.

(c) For $C = 0$, we have $\lambda = 0$ as the single solution of the auxiliary equation, so the general solution is a linear function
$$f(x) = a + bx,$$
with arbitrary real constants a and b.

(d) For the linear function $f(x) = a + bx$, the condition $f(0) = 0$ implies $a = 0$. Then $f(l) = bl = 0$ implies $b = 0$, so again $f(x) = 0$ is the only solution.

Solution 2.24

(a) Differentiating gives $f''(x) = -k^2 A \sin(kx+\psi) = -k^2 f(x)$, so (2.50) is satisfied with $k^2 = -C$.

(b) Using the addition formula $\sin(\alpha+\beta) = \sin\alpha\cos\beta + \cos\alpha\sin\beta$, equation (2.53) becomes
$$f(x) = A\sin(kx+\psi)$$
$$= A\cos(\psi)\sin(kx) + A\sin(\psi)\cos(kx),$$
which is of the required form (2.52) with $a = A\sin(\psi)$ and $b = A\cos(\psi)$.

(c) We find
$$a^2 + b^2 = A^2[\sin^2\psi + \cos^2\psi]$$
$$= A^2,$$
so we choose $A = \sqrt{a^2+b^2} \geq 0$. Now we can calculate ψ from
$$\tan\psi = \frac{\sin\psi}{\cos\psi} = \frac{a}{b},$$
so $\psi = \arctan(a/b)$ with $-\pi/2 < \psi < \pi/2$ if $b \neq 0$. For $b = 0$, we have $\psi = \pi/2$ for $a > 0$ and $\psi = -\pi/2$ for $a < 0$.

Solution 2.25

(a) We obtain
$$a(x) = \sum_{n=1}^{\infty} A_n \sin\left(\frac{n\pi x}{l}\right) = \sin\left(\frac{2\pi x}{l}\right) + 2\sin\left(\frac{5\pi x}{l}\right).$$

(b) From equation (2.59), the solution is given by
$$u(x,t) = \sin\left(\frac{2\pi x}{l}\right)\cos\left(\frac{2\pi ct}{l}\right) + 2\sin\left(\frac{5\pi x}{l}\right)\cos\left(\frac{5\pi ct}{l}\right).$$

(c) The periods of the two terms are given by $2\pi c\tau_2/l = 2\pi$ and $5\pi c\tau_5/l = 2\pi$, respectively. Thus $\tau_2 = l/c$ and $\tau_5 = 2l/(5c)$. The period of $u(x,t)$ is the least common multiple of the two periods, so $\tau = 2l/c = 2\tau_2 = 5\tau_5$.

Solution 2.26

For $m \neq n$, we obtain
$$\frac{2}{l}\int_0^l dx \sin\left(\frac{\pi m x}{l}\right)\sin\left(\frac{\pi n x}{l}\right)$$
$$= \frac{1}{l}\int_0^l dx \left(\cos\left(\frac{\pi(m-n)x}{l}\right) - \cos\left(\frac{\pi(m+n)x}{l}\right)\right)$$
$$= \frac{1}{l}\left[\frac{l}{\pi(m-n)}\sin\left(\frac{\pi(m-n)x}{l}\right) - \frac{l}{\pi(m+n)}\sin\left(\frac{\pi(m+n)x}{l}\right)\right]_0^l$$
$$= \frac{1}{\pi}\left(\frac{\sin[\pi(m-n)]}{m-n} - \frac{\sin[\pi(m+n)]}{m+n}\right) = 0,$$
which vanishes because $\sin(k\pi) = 0$ for any integer k.

For $m = n$, we use the same trigonometric relation for $\alpha = \beta$, which gives $2\sin^2\alpha = 1 - \cos 2\alpha$. This yields
$$\frac{2}{l}\int_0^l dx \sin^2\left(\frac{\pi n x}{l}\right) = 1 - \frac{1}{l}\int_0^l dx \cos\left(\frac{2\pi n x}{l}\right)$$
$$= 1 - \frac{1}{l}\left[\frac{l}{2\pi n}\sin\left(\frac{2\pi n x}{l}\right)\right]_0^l = 1,$$
because again the sine function vanishes at $x = 0$ and $x = l$.

Solution 2.27

Equation (2.61) has the same form as equation (2.60); the only difference is the additional factor $n\pi c/l$ in the Fourier coefficients for $b(x)$. Thus we can use equation (2.64) for $b(x)$, yielding

$$B_n \omega_n = \frac{2}{l} \int_0^l dx\, b(x) \sin\left(\frac{\pi n x}{l}\right).$$

Dividing this by $\omega_n = k_n c = \pi n c/l$ gives equation (2.65).

Solution 2.28

(a) The initial form of the string is shown in Figure 2.14.

Figure 2.14 Plot of the function $u(x, 0)$

(b) From equation (2.60), the Fourier coefficients are $A_1 = \frac{1}{8}$, $A_3 = \frac{1}{4}$ and $A_n = 0$ otherwise.

(c) With $B_n = 0$ for all n (because the string starts from rest), the solution of the initial-value problem is

$$u(x,t) = \tfrac{1}{8} \sin\left(\frac{\pi x}{l}\right) \cos\left(\frac{\pi c t}{l}\right) + \tfrac{1}{4} \sin\left(\frac{3\pi x}{l}\right) \cos\left(\frac{3\pi c t}{l}\right).$$

(d) For $x = l/2$, this becomes

$$u\left(\frac{l}{2}, t\right) = \tfrac{1}{8} \cos\left(\frac{\pi c t}{l}\right) - \tfrac{1}{4} \cos\left(\frac{3\pi c t}{l}\right),$$

because $\sin(\pi/2) = -\sin(3\pi/2) = 1$. This function is shown in Figure 2.15.

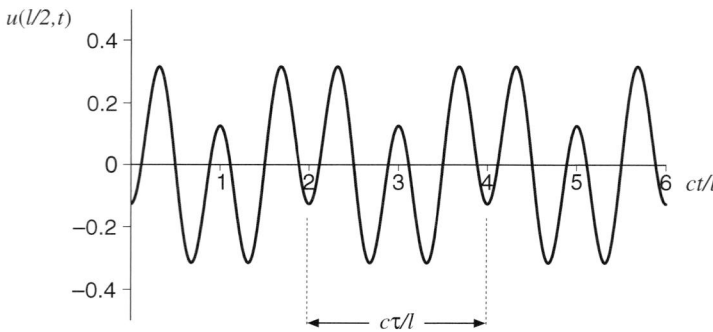

Figure 2.15 Plot of the motion of the midpoint of the string

(e) The solution $u(x,t)$ is a sum of two periodic functions of time. The periods of the two terms are given by $\pi c \tau_1/l = 2\pi$ and $3\pi c \tau_3/l = 2\pi$. Thus $\tau_3 = 2l/(3c)$ and $\tau_1 = 2l/c = 3\tau_3$. The period τ of $u(x,t)$ is the least common multiple of the two periods, so $\tau = \tau_1 = 2l/c$.

Solution 2.29

(a) Comparing equations (2.14) and (2.67) yields

$$F(z) = \frac{1}{2} \sum_{n=1}^{\infty} [A_n \sin(k_n z) + B_n \cos(k_n z)],$$

$$G(z) = \frac{1}{2} \sum_{n=1}^{\infty} [A_n \sin(k_n z) - B_n \cos(k_n z)].$$

(b) Sine and cosine are odd and even functions, respectively. Hence

$$F(-z) = \frac{1}{2}\sum_{n=1}^{\infty}[-A_n\sin(k_n z) + B_n\cos(k_n z)]$$
$$= -G(z).$$

Solution 2.30

(a) This follows from

$$f''(x) = A(ik)^2\exp[i(kx+\phi)] + B(-ik)^2\exp[-i(kx-\psi)]$$
$$= -k^2 f(x)$$

and $k^2 = -C$.

(b) Rewriting equation (2.75), we obtain

$$f(x) = A\exp(i\phi)\exp(ikx) + B\exp(i\psi)\exp(-ikx),$$

which is of the required form (2.74) with

$$\alpha = A\exp(i\phi) = A[\cos\phi + i\sin\phi],$$
$$\beta = B\exp(i\psi) = B[\cos\psi + i\sin\psi].$$

(c) We have $A = |\alpha| \geq 0$, $B = |\beta| \geq 0$, $\phi = \text{Arg}(\alpha)$ and $\psi = \text{Arg}(\beta)$, as long as $\text{Re}(\alpha) \neq 0$ and $\text{Re}(\beta) \neq 0$.

The argument, with $-\pi < \text{Arg}(z) \leq \pi$, of a complex number z was defined in Block 0, Subsection 1.2.7.

(d) We start from the form of $f(x)$ given above. Using $\exp(i\theta) = \cos\theta + i\sin\theta$, and expanding the products, the real part of $f(x)$ becomes

$$A[\cos\phi\cos(kx) - \sin\phi\sin(kx)] + B[\cos\psi\cos(kx) + \sin\psi\sin(kx)]$$
$$= [A\cos\phi + B\cos\psi]\cos(kx) - [A\sin\phi - B\sin\psi]\sin(kx)$$
$$= a\cos(kx) + b\sin(kx),$$

which has the form of equation (2.52), with real constants

$$a = A\cos\phi + B\cos\psi,$$
$$b = -A\sin\phi + B\sin\psi.$$

Analogously, the imaginary part is given by

$$A[\cos\phi\sin(kx) + \sin\phi\cos(kx)] + B[\sin\psi\cos(kx) - \cos\psi\sin(kx)]$$
$$= [A\sin\phi + B\sin\psi]\cos(kx) + [A\cos\phi - B\cos\psi]\sin(kx)$$
$$= a\cos(kx) + b\sin(kx),$$

where the constants a and b are now given by

$$a = A\sin\phi + B\sin\psi,$$
$$b = A\cos\phi - B\cos\psi.$$

Solution 2.31

Using $\exp[ik(x \pm ct)] = \cos[k(x \pm ct)] + i\sin[k(x \pm ct)]$, and noting that the coefficient functions $A(k) = \text{Re}[A(k)] + i\,\text{Im}[A(k)]$ and $B(k) = \text{Re}[B(k)] + i\,\text{Im}[B(k)]$ are complex, the real and imaginary parts of $u(x,t)$ are obtained as

$$\text{Re}[u(x,t)] = \int_{-\infty}^{\infty} dk \Big(\text{Re}[A(k)]\cos[k(x+ct)] - \text{Im}[A(k)]\sin[k(x+ct)]$$
$$+ \text{Re}[B(k)]\cos[k(x-ct)] - \text{Im}[B(k)]\sin[k(x-ct)]\Big)$$

and

$$\text{Im}[u(x,t)] = \int_{-\infty}^{\infty} dk \Big(\text{Im}[A(k)]\cos[k(x+ct)] + \text{Re}[A(k)]\sin[k(x+ct)]$$
$$+ \text{Im}[B(k)]\cos[k(x-ct)] + \text{Re}[B(k)]\sin[k(x-ct)]\Big),$$

respectively.

Solution 2.32

We can think of this equation as

$$\frac{\partial}{\partial t} u_x = 5.$$

The general solution for the function u_x is

$$u_x(x,t) = 5t + f(x),$$

where f is an arbitrary function of x. The general solution of this equation is

$$u(x,t) = 5tx + F(x) + G(t),$$

where F is a primitive (indefinite integral) of f, so $F' = f$, and G is an arbitrary function of t. As f is arbitrary, F is also arbitrary, and the general solution is $u(x,t) = 5tx + F(x) + G(t)$ with arbitrary functions F and G.

Note that you could have integrated with respect to x first. The order is arbitrary because $u_{xt} = u_{tx}$.

Note that $u(x,t) = F(x) + G(t)$ is the general solution of the homogeneous equation $u_{tx}(x,t) = 0$.

Solution 2.33

Putting $u(x,t) = f(x)g(t)$ gives

$$f(x)\,g'(t) - 4\,f'(x)\,g(t) = 2\,f(x)\,g(t).$$

Dividing by $f(x)g(t)$ and rearranging yields

$$\frac{4f'(x)}{f(x)} + 2 = \frac{g'(t)}{g(t)}.$$

The left-hand side depends only on x, and the right-hand side depends only on t, so we obtain two ordinary differential equations

$$f'(x) = \frac{C-2}{4} f(x), \quad g'(t) = C\,g(t),$$

for f and g, where C is an arbitrary constant. Their general solutions are

$$f(x) = c_1 \exp\left(\frac{(C-2)\,x}{4}\right), \quad g(t) = c_2 \exp(Ct),$$

with arbitrary constants c_1 and c_2. The initial condition gives $f(x)g(0) = 3\exp(3x)$, thus $c_1 c_2 = 3$ and $(C-2)/4 = 3$, which yields $C = 14$. Thus the solution is

$$u(x,t) = 3\exp(3x + 14t).$$

Solution 2.34

(a) Inserting $u(x,t) = f(x)g(t)$ in the partial differential equation, and dividing by $u(x,t) = f(x)g(t)$, we obtain

$$\frac{f''(x)}{f(x)} = \frac{g'(t)}{g(t)} = C.$$

The solution of the differential equation $g'(t) = Cg(t)$ is

$$g(t) = A\exp(Ct),$$

with an arbitrary constant A.

As shown in Exercise 2.23, only negative values of C yield non-trivial solutions to boundary-value problems of this type, so we set $k^2 = -C > 0$. The solution of the differential equation for f is then

$$f(x) = a\cos(kx) + b\sin(kx),$$

where a and b are arbitrary constants.

(b) The boundary conditions imply $f(0) = 0$ and $f(\pi) = 0$, so $a = 0$ and $\sin(\pi k)=0$. This means that $k = 1, 2, 3, \ldots$ label the non-trivial normal mode solutions of the boundary-value problem, which are

$$u(x,t) = B_k \sin(kx) \exp(-k^2 t)$$

with arbitrary B_k and $k = 1, 2, 3, \ldots$.

(c) The general solution is the linear combination
$$u(x,t) = \sum_{k=1}^{\infty} B_k \sin(kx) \exp(-k^2 t).$$
Inserting the initial condition gives
$$u(x,0) = \sum_{k=1}^{\infty} B_k \sin(kx) = 3\sin x - \sin 3x,$$
which implies $B_1 = 3$, $B_3 = -1$ and $B_k = 0$ otherwise, so the solution of the problem is
$$u(x,t) = 3\sin x \exp(-t) - \sin 3x \exp(-9t).$$

Solution 2.35

From equation (2.25), with $a(x) = 0$, and $b(x) = 2nc\sin(nx)$ for $x > 0$ and $b(x) = 0$ for $x \leq 0$, we obtain
$$u(x,t) = \frac{1}{2c} \int_{x-ct}^{x+ct} dy\, b(y)$$
$$= n \int_{\max\{x-ct,0\}}^{\max\{x+ct,0\}} dy\, \sin(ny).$$
For $x \geq ct$, we thus obtain
$$u(x,t) = n \int_{x-ct}^{x+ct} dy\, \sin(ny)$$
$$= \big[-\cos(ny)\big]_{x-ct}^{x+ct}$$
$$= \cos[n(x-ct)] - \cos[n(x+ct)]$$
$$= 2\sin(nx)\sin(nct).$$
For $-ct \leq x < ct$, the result becomes
$$u(x,t) = n \int_0^{x+ct} dy\, \sin(ny)$$
$$= \big[-\cos(ny)\big]_0^{x+ct}$$
$$= 1 - \cos[n(x+ct)].$$
Finally, for $x < -ct$, the solution is $u(x,t) = 0$.

Solution 2.36

(a) For the functions $u_{(k)}(x,t)$ of equation (2.68), the initial data imply $A(k) = 0$ and
$$B(k)\, kc\sin(kx) = 2nc\sin(nx)$$
for $x > 0$. This can be satisfied by choosing $k = n$ and $B(k) = 2$, so the appropriate solution is
$$u(x,t) = 2\sin(nx)\sin(nct).$$

(b) For $x \geq 0$, we recover the initial condition
$$u_t(x,0) = 2nc\sin(nx) = b(x).$$
For $x < 0$, we obtain the same form,
$$u_t(x,0) = 2nc\sin(nx) = -2nc\sin(-nx) = -b(-x),$$
but now expressed in terms of $b(-x)$, because $-x$ is positive. So the result is
$$u_t(x,0) = b(x) - b(-x),$$
which you can interpret as the initial velocity for $x \geq 0$ and its mirror image for $x < 0$; compare part (b) of Exercise 2.14.

Solution 2.37

The solutions for the initial data are

$$u_1(x,t) = \frac{1}{2}[a_1(x-ct) + a_1(x+ct)] + \frac{1}{2c}\int_{x-ct}^{x+ct} dy\, b_1(y),$$

$$u_2(x,t) = \frac{1}{2}[a_2(x-ct) + a_2(x+ct)] + \frac{1}{2c}\int_{x-ct}^{x+ct} dy\, b_2(y).$$

For their difference, we obtain

$$\begin{aligned}|u_1(x,t) - u_2(x,t)| &= \left|\frac{1}{2}[a_1(x-ct) - a_2(x-ct)] + \frac{1}{2}[a_1(x+ct) - a_2(x+ct)]\right.\\
&\quad \left. + \frac{1}{2c}\int_{x-ct}^{x+ct} dy\, [b_1(y) - b_2(y)]\right|\\
&\leq \frac{1}{2}|a_1(x-ct) - a_2(x-ct)| + \frac{1}{2}|a_1(x+ct) - a_2(x+ct)|\\
&\quad + \frac{1}{2c}\left|\int_{x-ct}^{x+ct} dy\, [b_1(y) - b_2(y)]\right|,\end{aligned}$$

where we have used the triangle inequality. The first two terms can be bounded as

$$\frac{1}{2}|a_1(x\pm ct) - a_2(x\pm ct)| \leq \frac{\delta}{2}$$

because the bound on the difference, $|a_1(x) - a_2(x)| \leq \delta$, is independent of x. For the integral, we have

$$\begin{aligned}\frac{1}{2c}\left|\int_{x-ct}^{x+ct} dy\, [b_1(y) - b_2(y)]\right| &\leq \frac{1}{2c}\int_{x-ct}^{x+ct} dy\, |b_1(y) - b_2(y)|\\
&\leq \frac{1}{2c}\int_{x-ct}^{x+ct} dy\, \epsilon = \frac{2ct\epsilon}{2c} = \epsilon t.\end{aligned}$$

Combining the estimates, we obtain $|u_1(x,t) - u_2(x,t)| \leq \delta + \epsilon t$, as required.

CHAPTER 3
Fourier series and Fourier transforms

3.1 Introduction

Fourier series and Fourier transforms are powerful tools which find applications in many areas of mathematics. In this course they will prove very useful for finding solutions to linear differential equations; they will be used to solve both the wave equation and the diffusion equation. The necessary properties of Fourier series and transforms will be developed in this chapter, so that you can see how the mathematical structure works, before seeing it applied in different contexts.

You have already seen examples of Fourier series and Fourier transforms in connection with solutions of the wave equation. A solution of the wave equation in the form of a Fourier series, and one in the form of a Fourier transform, were given in Chapter 2, Section 2.6.

The Fourier series and transforms introduced in the previous chapter were of the trigonometric (sine and cosine) type. This chapter will switch to considering complex exponential Fourier series and Fourier transforms. The theorems that make Fourier transforms so useful are expressed more simply and elegantly when we use the complex exponential form.

This chapter will also introduce the concept of the convolution of two functions. Convolutions are most often encountered when dealing with problems involving probability, and they will play an important role in studying the diffusion equation in Block II. The convolution is introduced at this point because problems involving convolution are often solved by Fourier transform methods, using the convolution theorem, which is derived in this chapter.

After studying this chapter you should be able to determine Fourier series and transforms for a wide range of functions. You should also have an appreciation of two applications of convolutions, and of how to use the convolution theorem. Section 3.6 shows how linear differential equations are simplified by taking Fourier transforms. The practical use of this method is best demonstrated by its applications which will be described later in the course.

3.2 Fourier series

3.2.1 The trigonometric Fourier series

The most easily understood application of the Fourier series is as an approximation to a periodic function. Let $f(x)$ be a real periodic function with period L (that is, $f(x+L) = f(x)$ for all x). We might hope to express it as a sum of sine and cosine functions which also have period L. A rather general way to do this is to write

$$f(x) = \tfrac{1}{2}a_0 + \sum_{n=1}^{\infty}\left(a_n \cos\left(\frac{2\pi n x}{L}\right) + b_n \sin\left(\frac{2\pi n x}{L}\right)\right) \quad (3.1)$$

and attempt to find a set of real coefficients a_n and b_n such that the infinite sum is equal to the function $f(x)$. The series on the right-hand side is called a *Fourier series*. The formulae for determining the *Fourier coefficients* a_n and b_n will be discussed later. This Fourier series is a generalisation of the sine and cosine series introduced in Chapter 2.

Writing the first coefficient as $\tfrac{1}{2}a_0$ rather than a_0 is a commonly used convention, because it simplifies writing a general expression for the cosine coefficients a_n.

As more and more terms are added to the *partial sums* of the Fourier series, i.e.

$$f_N(x) = \tfrac{1}{2}a_0 + \sum_{n=1}^{N}\left(a_n \cos\left(\frac{2\pi n x}{L}\right) + b_n \sin\left(\frac{2\pi n x}{L}\right)\right), \quad (3.2)$$

we hope that the functions $f_N(x)$ will approach the given function $f(x)$. Depending on our choice of the function $f(x)$, we may find that the partial sums *converge* to $f(x)$ (that is, $\lim_{N\to\infty} f_N(x) = f(x)$) for all real values of x, or for a subset of the real line. For practical calculations, the infinite series is approximated by a partial sum with a sufficiently large value of N. If $f(x)$ is a continuous periodic function with only finitely many maxima and minima in each period, it can be shown that the Fourier series converges to $f(x)$ for all x.

Figure 3.1 illustrates this application of Fourier series by comparing the function $f(x) = |\sin x|$ (which is periodic with period π) with partial sums of its Fourier series. In this case, the *Fourier coefficients* are

$$a_n = \frac{4}{\pi(1-4n^2)}, \quad b_n = 0. \quad (3.3)$$

The Fourier coefficients for $|\sin x|$ are obtained in Exercise 3.26, which you will be able to attempt after Section 3.2.

Figure 3.1 compares the function $f(x)$ with two partial sums (3.2) for $N=1$ and $N=4$. The partial sums converge rapidly to the function $f(x)$ as N increases.

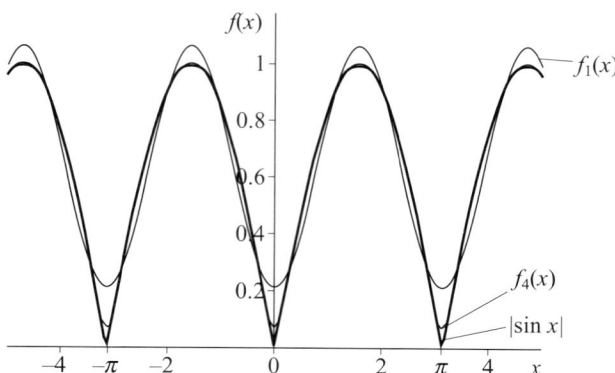

Figure 3.1 The periodic function $|\sin x|$, plotted together with two partial sums of its Fourier series

3.2 Fourier series

You have already encountered another case of approximating a specified function $f(x)$ by an infinite series, namely the Taylor series, of the form $f(x) = \sum_{n=0}^{\infty} A_n x^n$. The Taylor series is often called a *power series*, because the function is approximated by a sum of powers of the variable x. Similarly, the Fourier series is often called a *trigonometric series*, because it is a sum of trigonometric functions. The two series usually converge for different sets of values of x.

The Fourier series can be used even when the function $f(x)$ is not periodic. In this case the Fourier series cannot converge to the function everywhere, but it may converge to the function over an interval of length L. The interval may extend from some arbitrarily chosen point x_0 to $x_0 + L$, and the values of the Fourier coefficients will depend upon the choice of x_0. An example is shown in Figure 3.2, which compares the function $f(x) = x^2$ with partial sums of its Fourier series (for $N = 2$ and $N = 15$), in the case where we have chosen $L = 2$ and $x_0 = -1$. In this case, the Fourier coefficients are

$$a_0 = \tfrac{2}{3}, \quad a_n = \frac{4(-1)^n}{\pi^2 n^2} \; (n \neq 0), \quad b_n = 0. \tag{3.4}$$

Fourier coefficients for this example are obtained in Exercise 3.27.

The partial sums are periodic functions with period 2, but they converge to the non-periodic function $f(x)$ for $-1 \leq x \leq 1$.

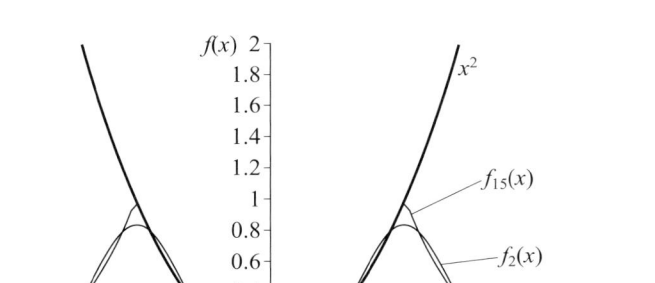

Figure 3.2 The function x^2, plotted together with two partial sums of its Fourier series on the interval $[-1, 1]$

In cases such as this, where $f(x)$ is not a periodic function, the Fourier series is a *periodic extension* of $f(x)$. To be more precise, let us denote the periodic extension of $f(x)$ by $f(x, L)$, which is defined by $f(x, L) = f(x)$ when $x_0 < x < x_0 + L$ and by the relation $f(x + L, L) = f(x, L)$ for all other points except $x = x_0 + nL$ for $n = 0, \pm 1, \pm 2, \ldots$. Then the value of $f(x, L)$ for all points where it is defined is given by the Fourier series.

Many linear differential and partial differential equations which occur as models for physical problems have solutions which may be expressed in terms of trigonometric functions. Solutions of these differential and partial differential equations are often in the form of Fourier series. An obvious example is the wave equation, in which case the Fourier series has a physical interpretation as a sum (or *superposition*) of waves. A Fourier series solution of the heat equation (diffusion equation) will be discussed later in the course.

> Fourier series were first introduced by Jean Baptiste Joseph Fourier (1758–1830) in order to solve the heat equation. This work first appeared in 1807, and in its final form in 1822 as a book entitled *La théorie analytique de la chaleur*.

3.2.2 The complex exponential Fourier series

We shall now introduce an alternative form of Fourier series, the *complex exponential Fourier series*. The advantage of the complex exponential form is that the expressions for the Fourier coefficients are often simpler, and are usually derived in a more transparent way. When we take the step from Fourier series to Fourier transforms, these advantages make the complex exponential form of the Fourier transform much easier to handle. (In fact, the 'trigonometric' Fourier transforms, which were introduced briefly in Chapter 2, will not be considered further in this course.)

The complex exponential Fourier series is obtained from (3.1) using Euler's formula

$$\exp(ix) = \cos x + i \sin x. \tag{3.5}$$

Using this formula, the cosine and sine functions may be written as

$$\cos x = \frac{1}{2}[\exp(ix) + \exp(-ix)]$$

and

$$\sin x = \frac{1}{2i}[\exp(ix) - \exp(-ix)].$$

The Fourier series may then be written in the form

$$f(x) = \sum_{n=-\infty}^{\infty} c_n \exp(2\pi i n x / L), \tag{3.6}$$

These coefficients will be related to the a_n and b_n of (3.1) in Exercise 3.1.

where the c_n are, in general, a set of complex coefficients, even if $f(x)$ is a real-valued function. Note that because the cosine and sine terms are composed of complex exponentials with arguments having opposite signs, the sum over positive n is replaced by a sum over all n. Because this form of Fourier series is more compact and has simpler properties, it will be used extensively in the remainder of this course.

> An equation equivalent to (3.5) was first written down in 1710 by Roger Cotes (1682–1716), a colleague of Newton. It was later given a more precise interpretation by de Moivre and Euler. It is often referred to as Euler's formula.

Exercise 3.1

Show that the coefficients a_n, b_n of (3.1) and the c_n of (3.6) are related as follows:

$$\begin{aligned} c_n &= \tfrac{1}{2}(a_n - ib_n) \quad (n > 0), \\ c_{-n} &= \tfrac{1}{2}(a_n + ib_n) \quad (n > 0), \\ c_0 &= \tfrac{1}{2}a_0. \end{aligned} \tag{3.7}$$

The result of Exercise 3.1 shows that the complex exponential Fourier coefficients are purely real when the Fourier series contains only cosines, and purely imaginary when it contains only sines.

Exercise 3.2

(a) Assuming that the chain rule is valid for complex-valued functions, show that

$$\frac{d}{dx}\exp(ikx) = ik\exp(ikx). \tag{3.8}$$

(b) What is the indefinite integral of $\exp(ikx)$ with respect to x?

The simplicity of this expression is one of the reasons why the complex exponential form for Fourier series (and transforms) is preferred when dealing with differential equations.

3.2 Fourier series

Exercise 3.3
Show that for any integer n,

$$\exp(i\pi n) = (-1)^n \quad \text{and} \quad \exp(i\pi n/2) = i^n. \quad (3.9)$$

These results will be useful later for simplifying expressions for the coefficients of many Fourier series.

Exercise 3.4
(a) Show that the complex conjugate of $\exp(ix)$ is $[\exp(ix)]^* = \exp(-ix)$.

(b) Show that the function $f(x)$ given by (3.6) is real if $c_{-n} = c_n^*$. (For $n = 0$, this relation states that $c_0 = c_0^*$, thus c_0 is real.)

3.2.3 Calculating Fourier coefficients

In order to use Fourier series, it is necessary to have a method for calculating Fourier coefficients. Here we shall calculate the coefficients c_n which appear in the complex exponential Fourier series (3.6). The following result is crucial (here n and m are both integers):

$$\frac{1}{L}\int_{x_0}^{x_0+L} dx \exp(-2\pi i m x/L)\exp(2\pi i n x/L)$$

$$= \frac{1}{L}\int_{x_0}^{x_0+L} dx \exp[2\pi i(n-m)x/L] = \delta_{nm}, \quad (3.10)$$

$L = \text{period}$.

where δ_{nm} is the *Kronecker delta symbol*, defined as follows:

$$\delta_{nm} = \begin{cases} 1, & n = m, \\ 0, & n \neq m. \end{cases} \quad (3.11)$$

Exercise 3.5
Check that the integral in (3.10) has the value given by (3.11). [Hint: Treat the case $n = m$ separately from $n \neq m$.]

In order to understand how the Fourier coefficients are obtained, consider the following, where on the right-hand side we have replaced the function $f(x)$ by its Fourier series (3.6):

$$\frac{1}{L}\int_{x_0}^{x_0+L} dx\, f(x)\exp(-2\pi i n x/L)$$

$$= \frac{1}{L}\int_{x_0}^{x_0+L} dx \exp(-2\pi i n x/L) \sum_{m=-\infty}^{\infty} c_m \exp(2\pi i m x/L). \quad (3.12)$$

The left-hand side contains the integer n, so the sum in the Fourier series has to run over an index which is given another name, in this case m. Now we rearrange this expression and exchange the order of summation and integration:

$$\frac{1}{L}\int_{x_0}^{x_0+L} dx\, f(x)\exp(-2\pi i n x/L)$$

$$= \frac{1}{L}\int_{x_0}^{x_0+L} dx \sum_{m=-\infty}^{\infty} c_m \exp[2\pi i(m-n)x/L]$$

$$= \sum_{m=-\infty}^{\infty} c_m \left(\frac{1}{L}\int_{x_0}^{x_0+L} dx \exp[2\pi i(m-n)x/L]\right)$$

$$= \sum_{m=-\infty}^{\infty} c_m \delta_{nm} = c_n. \quad (3.13)$$

Here equation (3.10) was used to substitute for the integrals in the penultimate line. In the final step we use the definition of the Kronecker delta symbol: the only term in the final line which is not equal to zero is the $m = n$ term, so the sum is simply a single Fourier coefficient, c_n.

We now have an expression for the Fourier coefficients c_n appearing in the series

$$f(x) = \sum_{n=-\infty}^{\infty} c_n \exp(2\pi i n x/L). \tag{3.6}$$

The Fourier coefficients c_n are given by

$$c_n = \frac{1}{L} \int_{x_0}^{x_0+L} dx\, f(x) \exp(-2\pi i n x/L). \tag{3.14}$$

The cosine and sine coefficients a_n, b_n in (3.1) can be obtained from these using the expressions quoted in Exercise 3.1.

This will be considered in detail in Example 3.1.

It can be shown that if $f(x)$ is a continuous periodic function, then the Fourier series given by (3.6) and (3.14) does converge to $f(x)$. In cases where $f(x)$ has discontinuities, there are complicated issues concerning the nature of the convergence of the series, but it can be shown that it does converge to the function $f(x)$ at all points except those where $f(x)$ is discontinuous (provided that there are only a finite number of discontinuities or maxima and minima in any finite interval).

There are also related issues concerning the interpretation of Fourier series. For example, the function $|\sin x|$ plotted in Figure 3.1 has a discontinuous first derivative, whereas all of the terms in the series have continuous first derivatives. Any finite sum of continuous functions is continuous, but taking the number of terms to infinity can, somewhat surprisingly, produce a discontinuous function. Questions about convergence of Fourier series stimulated controversies which led to a refinement of real analysis in the 19th century.

Example 3.1

We wish to determine the coefficients of the trigonometric Fourier series (3.1) for a function defined on the interval from x_0 to $x_0 + L$. Use the results of Exercise 3.1 to show that the coefficients of the trigonometric Fourier series for the function $f(x)$ with period L are given by

$$a_n = \frac{2}{L} \int_{x_0}^{x_0+L} dx\, f(x) \cos(2\pi n x/L), \quad n \geq 0, \tag{3.15}$$

$$b_n = \frac{2}{L} \int_{x_0}^{x_0+L} dx\, f(x) \sin(2\pi n x/L), \quad n \geq 1. \tag{3.16}$$

Solution

By considering $c_{-n} + c_n$ and $c_{-n} - c_n$, the results of Exercise 3.1 can be rewritten as

$$a_n = c_{-n} + c_n, \quad n \geq 0, \tag{3.17}$$
$$ib_n = c_{-n} - c_n, \quad n \geq 1. \tag{3.18}$$

We now substitute the expression for c_n given by equation (3.14) into equations (3.17) and (3.18). From equation (3.17), we have for $n \geq 0$

$$a_n = \frac{1}{L} \int_{x_0}^{x_0+L} dx\, f(x) \left(\exp(2\pi i n x/L) + \exp(-2\pi i n x/L) \right)$$

$$= \frac{2}{L} \int_{x_0}^{x_0+L} dx\, f(x) \cos(2\pi n x/L),$$

where we have used the identity $\exp(ix) + \exp(-ix) = 2\cos x$; we have reached equation (3.15). From equation (3.18), we have for $n \geq 1$

$$ib_n = \frac{1}{L} \int_{x_0}^{x_0+L} dx\, f(x) \left(\exp(2\pi inx/L) - \exp(-2\pi inx/L)\right)$$
$$= \frac{2i}{L} \int_{x_0}^{x_0+L} dx\, f(x) \sin(2\pi nx/L),$$

where we have used the identity $\exp(ix) - \exp(-ix) = 2i\sin x$. Dividing through by i leads directly to equation (3.16). ∎

Example 3.2

Consider the periodic function shown in Figure 3.3. For obvious reasons, this is called the *square wave function*; it can be expressed in symbolic form as

$$f(x) = \text{sign}(\sin x),$$

where

$$\text{sign}(x) = \begin{cases} -1 & \text{for } x > 0, \\ 0 & \text{for } x = 0, \\ -1 & \text{for } x < 0. \end{cases}$$

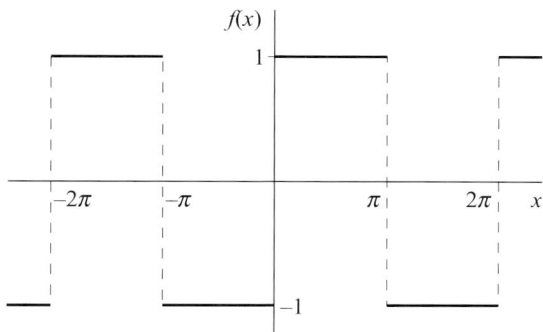

Figure 3.3 The square wave function. (For clarity, dashed vertical lines are shown at the points of discontinuity, although the function is defined to be equal to zero when x is a multiple of π.)

What is the period, L, of $f(x)$? Show that the Fourier coefficients are

$$c_n = \begin{cases} \dfrac{[(-1)^n - 1]i}{\pi n}, & n \neq 0, \\ 0, & n = 0. \end{cases}$$

Solution

The period is $L = 2\pi$. In equation (3.14), the choice of x_0 is arbitrary: here we take $x_0 = -\pi$. The integral is evaluated by dividing the region of integration into two parts: one from $-\pi$ to 0, where $f(x) = -1$, and one from 0 to π, where $f(x) = +1$. For $n \neq 0$,

$$c_n = \frac{1}{2\pi} \int_{-\pi}^{0} dx\, (-1) \exp(-inx) + \frac{1}{2\pi} \int_{0}^{\pi} dx\, (+1) \exp(-inx)$$
$$= \frac{1}{2\pi in} \left[\exp(-inx)\right]_{-\pi}^{0} - \frac{1}{2\pi in} \left[\exp(-inx)\right]_{0}^{\pi}$$
$$= \frac{1}{2\pi in} [1 - \exp(\pi in)] - \frac{1}{2\pi in}[\exp(-i\pi n) - 1]$$
$$= \frac{i}{\pi n} \left[(-1)^n - 1\right].$$

Here we have used a result from Exercise 3.3, $\exp(i\pi n) = (-1)^n$.

For $n = 0$, c_0 is the integral of $f(x)$ from $-\pi$ to π. This is clearly zero, since f is odd.

The coefficients are all purely imaginary, implying that when this Fourier series is expressed in terms of trigonometric functions, only the sine terms are present. (That this is true is explained in the discussion immediately following Exercise 3.1.) ∎

There are other functions which are called 'square waves'. Later, we shall use the Fourier coefficients of a square wave with an *uneven mark-space ratio*, considered in the following example.

Example 3.3

Let $f(x)$ be the periodic function with period L, so $f(x + L) = f(x)$, whose values on the interval $-L/2 \leq x \leq L/2$ are given by

$$f(x) = \begin{cases} 1 & \text{for } 0 \leq |x| < a, \\ \frac{1}{2} & \text{for } |x| = a, \\ 0 & \text{for } a < |x| \leq L/2. \end{cases} \quad (3.19)$$

This function is said to have an uneven mark-space ratio when $a \neq L/4$; an example is shown in Figure 3.4.

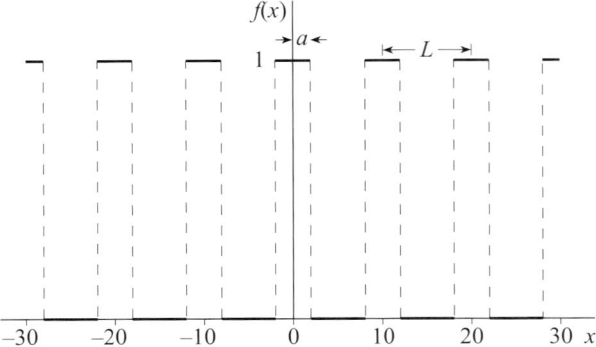

Figure 3.4 A square wave with an uneven mark-space ratio: here $L = 10$ and $a = 2$

Show that the Fourier coefficients for $f(x)$ are given by

$$c_n = \begin{cases} \dfrac{1}{\pi n} \sin(2\pi n a/L), & n \neq 0, \\ 2a/L, & n = 0. \end{cases} \quad (3.20)$$

Solution

The Fourier coefficients are again given by equation (3.14), but here $x_0 = -L/2$. Substituting for $f(x)$ from equation (3.19) gives

$$c_n = \frac{1}{L} \int_{-L/2}^{L/2} dx\, f(x) \exp(-2\pi i n x/L)$$

$$= \frac{1}{L} \int_{-a}^{a} dx \exp(-2\pi i n x/L)$$

$$= \frac{1}{L} \left[\frac{-L}{2\pi i n} \exp(-2\pi i n x/L) \right]_{-a}^{a}$$

$$= \frac{1}{2\pi i n} \left[\exp(2\pi i n a/L) - \exp(-2\pi i n a/L) \right]$$

$$= \frac{1}{\pi n} \sin(2\pi n a/L)$$

for $n \neq 0$, and

$$c_0 = \frac{1}{L} \int_{-L/2}^{L/2} dx\, f(x) = \frac{1}{L} \int_{-a}^{a} dx = 2a/L.$$

3.2 Fourier series

The function $(\sin x)/x$ is often written $\operatorname{sinc} x$, and is referred to as the *sinc function*. At $x = 0$, it is defined to be equal to 1 (because $\lim_{x \to 0} \sin x / x = 1$):

'sinc' is pronounced 'sink'.

$$\operatorname{sinc} x = \begin{cases} \dfrac{\sin x}{x} & \text{for } x \neq 0, \\ 1 & \text{for } x = 0. \end{cases} \qquad (3.21)$$

With this definition, the expressions for the Fourier coefficients c_n ($n \neq 0$) and c_0 can be combined into a single formula, valid for all integers n:

$$c_n = \frac{2a}{L} \operatorname{sinc}\left(\frac{2\pi n a}{L}\right). \qquad \blacksquare \qquad (3.22)$$

Exercise 3.6

Determine the Fourier coefficients for the *triangular wave function*, shown in Figure 3.5. This function may be specified by noting that it is even, so $f(x) = f(-x)$, periodic, so $f(x + 2\pi) = f(x)$, and takes the value $f(x) = x$ in the interval from $x = 0$ to $x = \pi$.

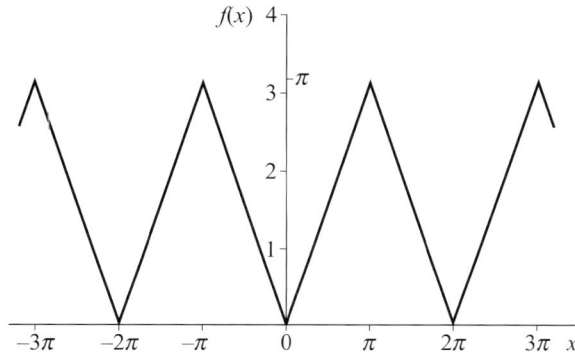

Figure 3.5 The triangular wave function

Show that the Fourier coefficients are

$$c_n = \begin{cases} \dfrac{(-1)^n - 1}{\pi n^2}, & n \neq 0, \\ \pi/2, & n = 0. \end{cases}$$

[Hint: Again, you should split the region of integration into two parts. Unlike previous examples, both integrals now require integration by parts.]

If you would like additional practice in calculating Fourier coefficients, the Further Exercises section contains two more examples (for the functions shown in Figures 3.1 and 3.2).

Exercise 3.7

The derivative of a Fourier series can normally be obtained by differentiating the series term by term. If the Fourier series for $f(x)$ is given by (3.6), show that the corresponding Fourier series for the derivative $f'(x)$ is

$$f'(x) = \sum_{n=-\infty}^{\infty} \frac{2\pi i n}{L} c_n \exp(2\pi i n x / L), \qquad (3.23)$$

provided that this series converges. (This is an easy exercise: just differentiate the general term in the series.)

Note that the coefficients of the derivative of the Fourier series decrease more slowly than those of the original series as $|n| \to \infty$, which means that it is possible for the Fourier series for $f(x)$ to converge while the Fourier series for the derivative $f'(x)$ does not converge. This happens at points where the function $f(x)$ is continuous but non-differentiable, so that $f'(x)$ does not exist. For example, for $f(x)$ in Exercise 3.6, these are the points $x = n\pi$ with $n = 0, \pm 1, \pm 2, \ldots$.

Exercise 3.8

Note that the square wave function in Figure 3.3 is the derivative of the triangular wave function in Figure 3.5 (except at multiples of π, where the derivative of the triangular wave is undefined). Check that the Fourier coefficients of the square wave are obtained from those of the triangular wave by multiplication by in.

Exercise 3.9

A Fourier series with $L = 2\pi$ can also be written in the form

$$f(x) = A_0 + \sum_{n=1}^{\infty} A_n \sin(nx - \phi_n),$$

where the coefficients A_n (called *amplitudes*), and parameters ϕ_n (called *phases*) are real numbers. Determine formulae giving the complex coefficients c_n in terms of the A_n and ϕ_n.

[Hint: Use Euler's formula (3.5) to express the sine function in this expression in terms of complex exponentials.]

3.3 Fourier transforms

3.3.1 Introduction and definition

The Fourier transform can be regarded as an extension of the concept of the Fourier series (3.6). In many cases, a function $f(x)$ may be represented in the form

$$f(x) = \frac{1}{\sqrt{2\pi}} \int_{-\infty}^{\infty} dk \exp(ikx) \tilde{f}(k), \qquad (3.24)$$

where $\tilde{f}(k)$ is called the *Fourier transform* of $f(x)$. In this representation, the sum over n in the Fourier series (3.6) is replaced by an integral over k, and the discrete set of Fourier coefficients c_n is replaced by a function $\tilde{f}(k)$. Equation (3.24) is an implicit definition of the $\tilde{f}(k)$ corresponding to the function $f(x)$. A formula for calculating $\tilde{f}(k)$ from $f(x)$ will be given shortly.

The Fourier transform does not exist for all functions. In this chapter we shall assume that the function $f(x)$ approaches zero as $x \to \pm\infty$. This course will emphasise the applications of Fourier transforms, and will not consider the conditions for the Fourier transform to exist.

This section starts by introducing a general formula for the Fourier transform, then calculates some specific examples, and concludes by explaining how the formula is obtained. Later sections will describe the properties that make the Fourier transform such a powerful tool.

3.3 Fourier transforms

We shall show that the Fourier transform of the function $f(x)$ is given by

$$\tilde{f}(k) = \frac{1}{\sqrt{2\pi}} \int_{-\infty}^{\infty} dx \exp(-ikx) f(x). \tag{3.25}$$

Compare this with formula (3.24) giving $f(x)$ in terms of $\tilde{f}(k)$. Note that the formulae for the Fourier transform (mapping from f to $\tilde{f}$, equation (3.25)) and its inverse (mapping from $\tilde{f}$ to f, equation (3.24)) differ only by a sign. This close relation between the Fourier transform and its inverse transform is a powerful feature, because results which apply to the Fourier transform can be adapted to its inverse transform.

We make some remarks about the Fourier transform.

- The Fourier transform is very closely related to the Fourier series. The Fourier series (3.6) is a sum of terms of the form $A(k)\exp(ikx)$, where k and n are related by $k = 2\pi n/L$, and $A(k) = c_n$. If L is very large, then the separation of successive values of k, $\Delta k = 2\pi/L$, is very small. In this case the sum can be approximated by an integral. This observation forms the basis of the derivation of the Fourier transform in Subsection 3.3.3.

- The function $A\exp(ikx)$ occurs in descriptions of wave motion: we have seen that the real part of this function can represent the vertical displacement of a string at some instant of time. The number k is called the *wave number*, and A is the *complex amplitude* of the wave. A more general wave might be written in the form of a superposition:

$$f(x) = \sum_{n=1}^{N} A(k_n) \exp(ik_n x),$$

where $\{k_n : n = 1, \ldots, N\}$ is a set of wave numbers. In situations where N is large, the sum of the k_n may be often be approximated by an integral over k, that is, by a Fourier transform.

- It should be emphasised that the Fourier transform is a mapping from one function f to another function $\tilde{f}$. From a mathematical point of view, the name of the variable which is integrated over in equations (3.25) and (3.24) is irrelevant. The variables used in the Fourier transform $\tilde{f}$ and the function f are sometimes referred to as *conjugate variables* (terminology that has no relation to complex conjugates). Thus in (3.25) and (3.24), k is the conjugate variable to x.

  In applications of Fourier transforms, the names of variables reflect the quantities that they represent: for example, x is often used for a distance, and t for a time. The function $\exp[i(kx - \omega t)]$ is often used to represent a wave with wave number k and angular frequency ω. For this reason, the symbol k is often used as the conjugate variable to x, so that if the function f is a function of a position variable x, then its Fourier transform $\tilde{f}$ will often be written as a function of a variable named k, i.e. as $\tilde{f}(k)$. Similarly, if f is a function of time, its Fourier transform will typically be written $\tilde{f}(\omega)$.

- Many different definitions of the Fourier transform can be found in different textbooks. For example, some books use the relation

$$f(x) = \int_{-\infty}^{\infty} dy \exp(2\pi i x y) \tilde{f}(y)$$

to define the Fourier transform $\tilde{f}$ of a function f. The useful properties of Fourier transforms continue to hold true when alternative definitions are used, but some of the factors appearing in formulae are different (e.g. $1/\sqrt{2\pi}$ may have to be replaced by $1/2\pi$ or 1). When reading other texts, you must be careful to note how the Fourier transform is defined.

3.3.2 Some examples of Fourier transforms

Now we consider some examples of Fourier transforms. The three examples discussed here are the most commonly used Fourier transforms. Two of them are introduced through an example and an exercise, which you are recommended to work through.

Example 3.4

Use equation (3.25) to show that the Fourier transform of the function $f(x) = \exp(-|x|)$ is

$$\tilde{f}(k) = \frac{1}{\sqrt{2\pi}} \frac{2}{1+k^2}.$$

Figure 3.8 on page 121 includes graphs of this f and $\tilde{f}$. The function $1/(1+k^2)$ occurs frequently in applied mathematics, and is often called the *Lorentzian function*, named after Hendrik Antoon Lorentz (1853–1928), a Dutch physicist.

Solution

We split the integral into two parts, one for $x > 0$, where $|x| = x$, and the other for $x < 0$, where $|x| = -x$:

$$\tilde{f}(k) = \frac{1}{\sqrt{2\pi}} \int_{-\infty}^{\infty} dx \exp(-ikx) \exp(-|x|)$$

$$= \frac{1}{\sqrt{2\pi}} \int_{0}^{\infty} dx \exp[-(1+ik)x] + \frac{1}{\sqrt{2\pi}} \int_{-\infty}^{0} dx \exp[(1-ik)x]$$

$$= \frac{1}{\sqrt{2\pi}} \left(\left[\frac{-1}{1+ik} \exp[-(1+ik)x] \right]_{0}^{\infty} + \left[\frac{1}{1-ik} \exp[(1-ik)x] \right]_{-\infty}^{0} \right)$$

$$= \frac{1}{\sqrt{2\pi}} \left(\frac{1}{1+ik} + \frac{1}{1-ik} \right)$$

$$= \frac{1}{\sqrt{2\pi}} \frac{2}{1+k^2}. \quad \blacksquare$$

Exercise 3.10

The *characteristic function* or *top-hat function* on the interval from $-a$ to a is denoted by $\chi_a(x)$ and defined by

$$\chi_a(x) = \begin{cases} 1 & \text{for } |x| \leq a, \\ 0 & \text{for } |x| > a. \end{cases} \tag{3.26}$$

The characteristic function on the interval from -1 to $+1$ will be denoted by $\chi(x)$, so $\chi(x) = \chi_1(x)$.

Show that the Fourier transform of $\chi(x)$ is

$$\tilde{\chi}(k) = \sqrt{\frac{2}{\pi}} \frac{\sin k}{k} \tag{3.27}$$

for $k \neq 0$, and that $\tilde{\chi}(0) = \sqrt{2/\pi}$.

This Fourier transform may also be expressed in terms of the sinc function, discussed in Example 3.3: $\tilde{\chi}(k) = \sqrt{2/\pi}\,\text{sinc}(k)$. Figure 3.8 shows graphs of $\chi(x)$ and $\sqrt{2/\pi}\,\text{sinc}(k)$.

3.3 Fourier transforms

The final example is the Fourier transform of a Gaussian function,

$$f(x) = \exp(-x^2/2). \tag{3.28}$$

Its Fourier transform is

$$\tilde{f}(k) = \exp(-k^2/2) \tag{3.29}$$

– i.e. this function is its own Fourier transform. This result is one of several very useful properties of Gaussian functions. In Section 3.4 we shall discuss the rules which enable us to find the Fourier transform of the general Gaussian function from this specific case.

We cannot give a rigorous derivation of this result here, but below we give some insight into why (3.29) is true. First, recall from Block 0 the following formula for the Gaussian integral:

$$\int_{-\infty}^{\infty} dx \exp(-\alpha x^2) = \sqrt{\frac{\pi}{\alpha}}. \tag{3.30}$$

Note that $-x^2/2 - ikx = -(x+ik)^2/2 - k^2/2$, so the expression for the Fourier transform can be written as

$$\tilde{f}(k) = \frac{1}{\sqrt{2\pi}} \int_{-\infty}^{\infty} dx \exp(-ikx) \exp(-x^2/2)$$

$$= \frac{1}{\sqrt{2\pi}} \exp(-k^2/2) \int_{-\infty}^{\infty} dx \exp[-(x+ik)^2/2].$$

By changing the variable to $u = x - a$ and using (3.30), the integral

$$\int_{-\infty}^{\infty} dx \exp[-(x-a)^2/2] = \sqrt{2\pi} \tag{3.31}$$

is seen to be independent of a for any real value of a. If this result is also true for complex values of a, then equation (3.29) follows (after setting $a = -ik$). It can be shown that the integral is independent of a when the integrand is a Gaussian function, but demonstrating this fact requires some knowledge of complex analysis.

3.3.3 Derivation of the Fourier transform

In this subsection we shall sketch the derivation of the formula for the Fourier transform of a function $f(x)$ that approaches zero as $|x| \to \infty$. The derivation of the expression for $\tilde{f}(k)$ will not be assessed, but you are expected to be able to use the result, equation (3.25). The Fourier transform is closely related to the Fourier series. We have seen (in Subsection 3.2.1) that a Fourier series can be used to represent a non-periodic function on an interval of length L. Here we consider the Fourier series for the periodic extension of $f(x)$ on an interval of length L, with its centre at $x = 0$. We call this periodic extension $f(x, L)$, and for any x in the interval $[-L/2, L/2]$ we assume that $f(x) = f(x, L)$. (This assumption is valid for all continuous functions $f(x)$.) It follows that for any x, $f(x, L)$ approaches $f(x)$ as $L \to \infty$.

We shall see that as the length L of this interval is taken to infinity, the sum over Fourier coefficients is approximated by an integral, which will correspond to the integral in equation (3.24).

Before showing that this is plausible for rather general functions, let us consider how it works for a specific example: we shall obtain the Fourier transform of the characteristic function $\chi_a(x)$ introduced in Exercise 3.10.

Its periodic extension has already been introduced: the uneven mark-space ratio square wave considered in Example 3.3 is periodic with period L, and is equal to the characteristic function inside the interval $[-L/2, L/2)$ (except at the discontinuities). The Fourier coefficients c_n are given by equation (3.20), leading to the Fourier series

$$f(x, L) = \frac{1}{\sqrt{2\pi}} \sum_{n=-\infty}^{\infty} \left(\frac{2\pi}{L}\right) \sqrt{\frac{2}{\pi}} a \operatorname{sinc}(2\pi n a/L) \exp(2\pi i n x/L), \quad (3.32)$$

where $\operatorname{sinc}(x) = (\sin x)/x$. Now put $\Delta k = 2\pi/L$, and define the variable $k_n = 2\pi n/L = n\Delta k$ so that equation (3.32) becomes

$$f(x, L) = \frac{1}{\sqrt{2\pi}} \sum_{n=-\infty}^{\infty} \Delta k \sqrt{\frac{2}{\pi}} a \operatorname{sinc}(k_n a) \exp(i k_n x). \quad (3.33)$$

At this point we recall that the definite integral may be defined by means of a limit as

$$\lim_{\delta x \to 0} \sum_{\{a < x_n < b\}} \delta x\, f(x_n) = \int_a^b dx\, f(x), \quad (3.34)$$

where $x_n = n\delta x$ and the notation $\{a < x_n < b\}$ means 'the set of all x_n such that $a < x_n < b$' (see Figure 3.6). (There are classes of functions for which this definition of the definite integral is not suitable, but these are not normally encountered in applied mathematics.)

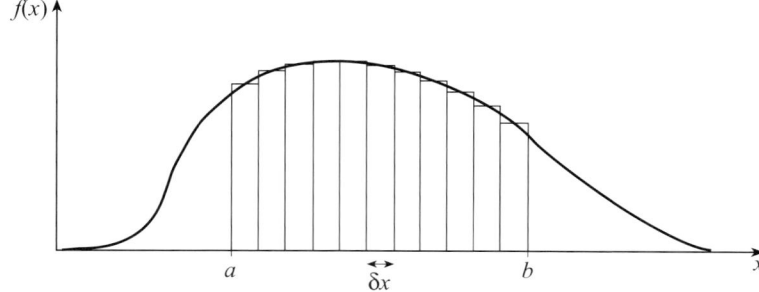

Figure 3.6 The area under this curve from a to b is the definite integral of $f(x)$ between $x = a$ and $x = b$. It can be expressed as the limit of a sum of areas of small rectangles of width δx as $\delta x \to 0$.

We apply this to equation (3.33), with Δk and k_n playing the roles of δx and x_n, respectively, and note that the limit $\Delta k \to 0$ corresponds to the infinite interval limit, $L \to \infty$. As this limit is taken, $\Delta k \to 0$ and we have $f(x, L) \to f(x)$, so

$$f(x) = \frac{1}{\sqrt{2\pi}} \int_{-\infty}^{\infty} dk \sqrt{\frac{2}{\pi}} a \operatorname{sinc}(ka) \exp(ikx). \quad (3.35)$$

Comparing this expression with (3.24), we see that the Fourier transform of $f(x) = \chi_a(x)$ is given by

$$\tilde{f}(k) = \sqrt{\frac{2}{\pi}} a \operatorname{sinc}(ka). \quad (3.36)$$

This is in agreement with the result of Exercise 3.10, obtained (for the case where $a = 1$) by using formula (3.25) for the Fourier transform. Thus we have seen that the Fourier transform of $f(x)$ may be obtained from the Fourier coefficients of the periodic extension $f(x, L)$.

3.4 Some properties of Fourier transforms

Now we consider the general case of a function $f(x)$ for which $f(x) \to 0$ as $|x| \to \infty$. Given a function $f(x)$, we consider its periodic extension $f(x, L)$ with period L, and using (3.6) we write the Fourier series of this periodic extension in the form

$$f(x, L) = \frac{1}{\sqrt{2\pi}} \sum_{n=-\infty}^{\infty} \left(\frac{2\pi}{L}\right) \left(\frac{Lc_n}{\sqrt{2\pi}}\right) \exp(2\pi i n x / L)$$

$$= \frac{1}{\sqrt{2\pi}} \sum_{n=-\infty}^{\infty} \Delta k \, C(k_n, L) \exp(i k_n x), \qquad (3.37)$$

where $\Delta k = 2\pi/L$ and $k_n = n \Delta k$. The coefficients $C(k_n, L)$ are obtained using (3.14) as

$$C(k_n, L) = \frac{Lc_n}{\sqrt{2\pi}} = \frac{1}{\sqrt{2\pi}} \int_{-L/2}^{L/2} dx \, f(x) \exp(-i k_n x). \qquad (3.38)$$

We now take the limit $L \to \infty$ in relations (3.37) and (3.38). Since $\Delta k \to 0$, the sum in equation (3.37) becomes an integral. Also, $f(x, L) \to f(x)$ as $L \to \infty$, so the integral may be written in the form

$$f(x) = \frac{1}{\sqrt{2\pi}} \int_{-\infty}^{\infty} dk \, \lim_{L \to \infty} C(k, L) \exp(ikx). \qquad (3.39)$$

Comparing this with the definition of the Fourier transform (3.24), and using equation (3.38), we see that

$$\tilde{f}(k) = \lim_{L \to \infty} C(k, L) = \frac{1}{\sqrt{2\pi}} \int_{-\infty}^{\infty} dx \, f(x) \exp(-ikx). \qquad (3.40)$$

This agrees with equation (3.25).

It should be emphasised that in this discussion we have not provided all the steps needed to justify the taking of the limit $L \to \infty$, so this should be regarded as a plausibility argument rather than a full proof, which would be beyond the scope of this course.

3.4 Some properties of Fourier transforms

This section will introduce some elementary properties of Fourier transforms, via a series of worked examples and exercises. A common use of these results is to obtain Fourier transforms of a wide range of functions from a small number of elementary examples. These exercises are followed by Table 3.2, showing elementary *Fourier transform pairs*, $f(x)$ and $\tilde{f}(k)$, based upon the examples introduced in Subsection 3.3.2. Together with the properties described earlier, these will be sufficient to evaluate most of the examples encountered in this course. Further exercises will illustrate methods for finding more general Fourier transforms from the examples in this table.

Most of the applications of the Fourier transform arise from two further properties, discussed in Sections 3.5 and 3.6.

It is recommended that you attempt all of the exercises in this section, in order to become accustomed to manipulating Fourier transforms.

Exercise 3.11

Linearity. Here we check that taking the Fourier transform is a linear operation. Consider a linear combination of functions $f_1(x)$, $f_2(x)$, with constant coefficients a_1, a_2, i.e. $F(x) = a_1 f_1(x) + a_2 f_2(x)$. Check that the Fourier transform of $F(x)$ is

$$\tilde{F}(k) = a_1 \tilde{f}_1(k) + a_2 \tilde{f}_2(k), \qquad (3.41)$$

where $\tilde{f}_1$ and $\tilde{f}_2$ are the Fourier transforms of f_1 and f_2, respectively.

Example 3.5

Scaling. Consider the effect of scaling the coordinate x by a factor α (a positive real number). Show that the Fourier transform of the function $F(x) = f(x/\alpha)$ is

$$\tilde{F}(k) = \alpha \, \tilde{f}(\alpha k), \qquad (3.42)$$

where $\tilde{f}$ is the Fourier transform of f.

Solution

To determine the Fourier transform of $F(x) = f(x/\alpha)$, we change the variable of integration to $u = x/\alpha$:

$$\begin{aligned}
\tilde{F}(k) &= \frac{1}{\sqrt{2\pi}} \int_{-\infty}^{\infty} dx \exp(-ikx) \, f(x/\alpha) \\
&= \frac{\alpha}{\sqrt{2\pi}} \int_{-\infty}^{\infty} du \exp(-ik\alpha u) \, f(u) \\
&= \alpha \, \tilde{f}(\alpha k).
\end{aligned}$$

(We assumed $\alpha > 0$ when making the change of variable; more generally, $\tilde{F}(k) = |\alpha| \, \tilde{f}(\alpha k)$.) ∎

Example 3.5 shows that stretching the x-axis by a factor α corresponds to making a contraction of the k-axis by the same factor (see Figure 3.7).

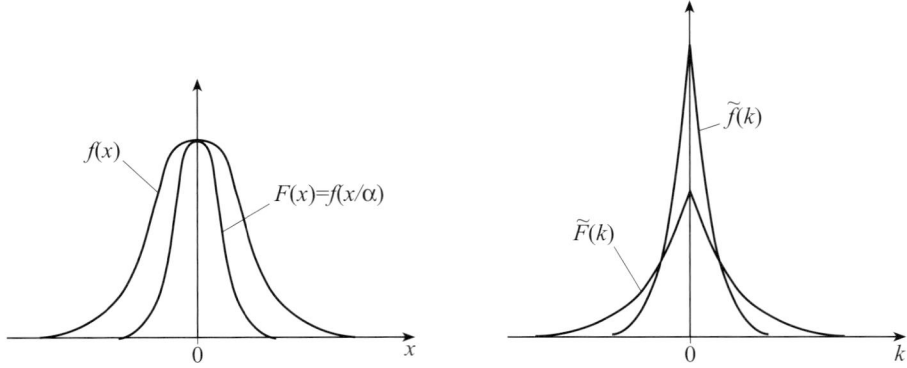

Figure 3.7 The scaling property of Fourier transforms, as considered in Example 3.5; in this example, $\alpha < 1$

Example 3.6

Translation. Consider the effect of translating or shifting a function $f(x)$ along the x-axis by a distance X. The translated function is given by $F(x) = f(x - X)$. Note that the graph of F is obtained by shifting the graph of f along the x-axis to the right (if $X > 0$) by a distance X. Show that the Fourier transform of this function $F(x)$ is

$$\tilde{F}(k) = \exp(-ikX) \, \tilde{f}(k), \qquad (3.43)$$

where $\tilde{f}$ is the Fourier transform of f.

3.4 Some properties of Fourier transforms

We can describe this result by saying that multiplication of $\tilde{f}$ by $\exp(-ikX)$ is the *image* of the operation of translation of f by a displacement X.

Solution

To determine the Fourier transform of $F(x) = f(x - X)$, we change the variable of integration to $u = x - X$:

$$\tilde{F}(k) = \frac{1}{\sqrt{2\pi}} \int_{-\infty}^{\infty} dx \exp(-ikx) f(x - X)$$
$$= \frac{1}{\sqrt{2\pi}} \int_{-\infty}^{\infty} du \exp[-ik(u + X)] f(u)$$
$$= \exp(-ikX) \tilde{f}(k). \quad \blacksquare$$

If the Fourier transform of a function f is the function $\tilde{f}$, what is the Fourier transform of the function $\tilde{f}$? To answer this question, attempt the following exercise.

Exercise 3.12

Inversion. Show that the Fourier transform of $\tilde{f}(x)$ is $f(-k)$.

This result implies that if you know that $\tilde{f}$ is the Fourier transform of f, you also have the Fourier transform of $\tilde{f}$. These are often described as forming a *Fourier transform pair*. In words, the Fourier transform of the Fourier transform is obtained from the original function by changing the sign of the argument. This is summarised by the following, in which $A \xrightarrow{\text{FT}} B$ means 'take the Fourier transform of function A to give function B':

$$\text{if } f(x) \xrightarrow{\text{FT}} \tilde{f}(k), \quad \text{then } \tilde{f}(k) \xrightarrow{\text{FT}} f(-x). \tag{3.44}$$

In the case of an even function, applying the Fourier transform twice returns you to the original function:

$$\text{if } f(x) = f(-x) \text{ and } f(x) \xrightarrow{\text{FT}} \tilde{f}(k), \quad \text{then } \tilde{f}(k) \xrightarrow{\text{FT}} f(x). \tag{3.45}$$

Exercise 3.13

Symmetry. Show that the Fourier transform of an even real-valued function is also an even real-valued function (provided that the Fourier transform exists). What is the corresponding result for the Fourier transform of an odd real-valued function?

You should be prepared to think of using the properties considered in the exercises above for manipulations involving Fourier transforms. The remaining exercises of this subsection illustrate some applications of these results.

Exercise 3.14

What is the Fourier transform of $F(x) = \exp(iKx)f(x)$, where K is a constant?

Show that the Fourier transform of $g(x) = f(x)\cos(\lambda x)$ is

$$\tilde{g}(k) = \tfrac{1}{2}[\tilde{f}(k + \lambda) + \tilde{f}(k - \lambda)],$$

where $\tilde{f}(k)$ is the Fourier transform of the function $f(x)$.

[Hint: Recall that $\cos(\theta) = \tfrac{1}{2}[\exp(i\theta) + \exp(-i\theta)]$.]

Note that there is a near symmetry here: in Example 3.6, we saw that multiplication by $\exp(-ikX)$ is the image of translation by X, and Exercise 3.14 showed that translation by K is the image of multiplication by $\exp(iKx)$.

The properties that we have considered thus far are summarised in Table 3.1.

Table 3.1 Summary of some properties of Fourier transforms

Property	Function	Fourier transform
Linearity	$a\,f(x) + b\,g(x)$	$a\,\tilde{f}(k) + b\,\tilde{g}(k)$
Scaling ($\alpha \neq 0$)	$f(x/\alpha)$	$|\alpha|\,\tilde{f}(\alpha k)$
Translation	$f(x-a)$	$\exp(-ika)\,\tilde{f}(k)$
Translation in k	$\exp(iax)\,f(x)$	$\tilde{f}(k-a)$
Inversion	$\tilde{f}(x)$	$f(-k)$

For this course, only three basic Fourier transform pairs will be used. These three pairs are those considered as examples in Subsection 3.3.2: for convenience, they are listed in Table 3.2. Other Fourier transforms encountered in this course can be obtained from these with the help of the properties listed above, together with the *convolution theorem* (to be introduced in Section 3.5) or results on Fourier transforms of derivatives and integrals (Section 3.6).

Table 3.2 Some Fourier transform pairs

$f(x)$	$\tilde{f}(k)$
$\exp(-|x|)$	$\sqrt{\dfrac{2}{\pi}}\,\dfrac{1}{1+k^2}$
$\exp(-x^2/2)$	$\exp(-k^2/2)$
$\chi(x)$	$\sqrt{\dfrac{2}{\pi}}\,\dfrac{\sin k}{k}$

Figure 3.8 shows graphs of the elementary Fourier transform pairs listed in Table 3.2. Note that because all of the functions are even (i.e. $f(x) = f(-x)$), all of these pairs are *dual* (in the sense that the Fourier transform of a function in either column is the function in the other column (recall equation (3.45)).

> Computer algebra packages (such as Mathcad or Maple) can often find Fourier transforms; if you have access to one of these, it could be helpful, but bear in mind that there can be many equivalent ways to present a given result, and these packages may sometimes present formulae in forms different to those quoted in our answers to exercises.
>
> Also, mathematical tables (such as *Tables of integrals, series and products*, by I. S. Gradshteyn and I. M. Ryzhik, 5th English edition, edited by A. Jeffrey, Academic Press, New York, 1994) list large numbers of functions for which the Fourier transforms are known. Often these are expressed in terms of the 'special functions' of mathematical physics, such as Bessel functions. If you visit a library that has a copy, you might find it interesting to look at these tables; but we would not suggest investing in this book, as it is a tool for specialists.

3.4 Some properties of Fourier transforms

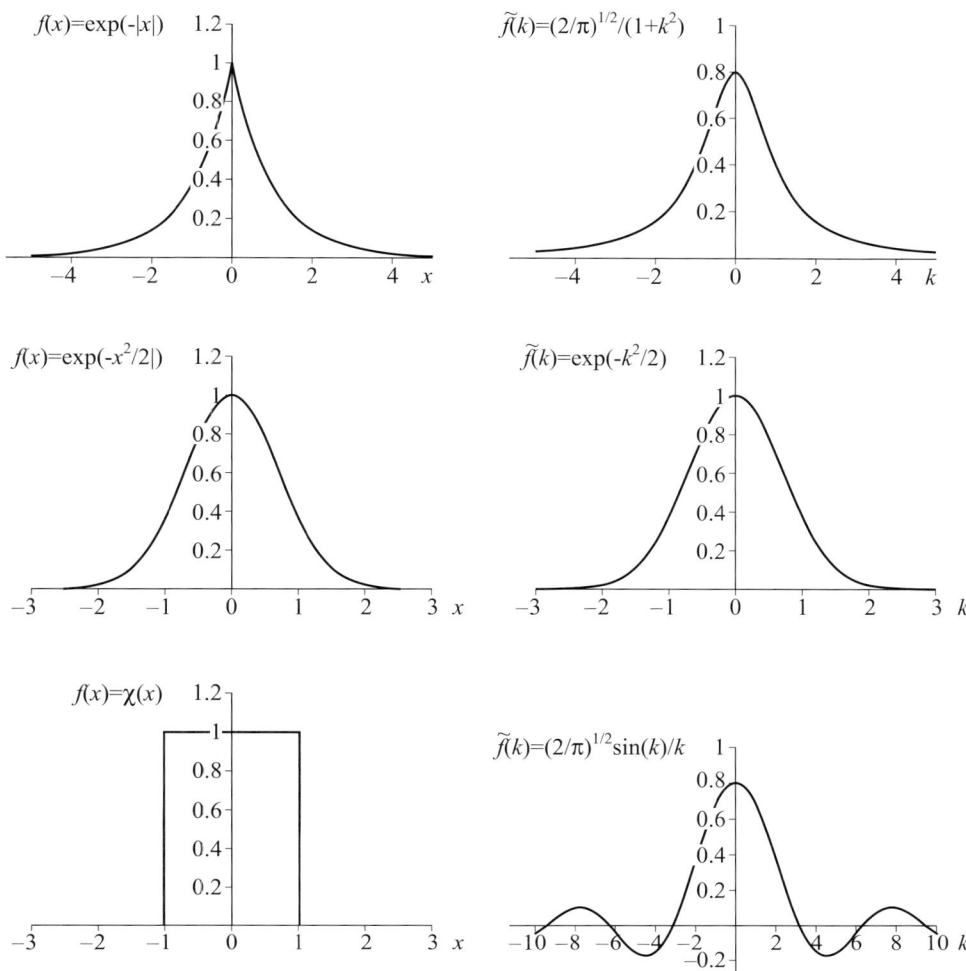

Figure 3.8 Plots of three elementary Fourier transform pairs

The following exercises will give you more practice in using the rules summarised in Table 3.1.

Exercise 3.15

Show that the Fourier transform of $g(x) = Af(\alpha x - \alpha x_0)$ (where x_0 and A are constants, and α is a positive number) is

$$\tilde{g}(k) = \frac{A}{\alpha} \exp(-ikx_0) \, \tilde{f}(k/\alpha).$$

Exercise 3.16

Use the scaling rule (see Example 3.5) to write down Fourier transforms of the following functions (where a is a positive constant):

$$f_1(x) = \exp(-a|x|), \quad f_2(x) = \exp(-ax^2), \quad f_3(x) = \chi(2x/a).$$

Exercise 3.17

Use the rule for inversion of Fourier transforms (3.45) to determine the Fourier transform of

$$f(x) = \frac{\sin x}{x},$$

then apply the scaling rule to obtain the Fourier transform of

$$F(x) = \frac{\sin(ax)}{x}$$

(where a is a positive constant).

Exercise 3.18

What are the Fourier transforms of the functions
$$f(x) = \exp[-a(x-x_0)^2], \quad g(x) = \exp(iKx)\exp(-ax^2),$$
where a is a positive constant, and x_0 and K are real numbers?

3.5 Convolutions and the convolution theorem

Given two functions $f(x)$ and $g(x)$, a third function $h(x)$ can be defined, which is called the *convolution* of f and g:

$$h(x) = \int_{-\infty}^{\infty} du\, f(x-u)\, g(u). \tag{3.46}$$

This is a rather complicated looking construction, but the convolution of two functions occurs in many contexts in applied mathematics; we shall describe two examples (blurred images, and time-delay processes) below. Several other examples of convolutions will be given later in the course.

We use the notation $h = f \otimes g$ to indicate that h is the convolution of f and g. The convolution may not be defined in all cases; we restrict attention to functions $f(x)$ and $g(x)$ that approach zero rapidly as $|x| \to \infty$. The reason for discussing convolution in a chapter on Fourier transforms is because of the *convolution theorem*, which often enables problems involving convolutions to be solved by Fourier transforming.

3.5.1 Blurred images

One example of a convolution is a blurred image produced by an out-of-focus camera, and this example can give a useful mental picture when thinking about problems involving convolution. For simplicity, we discuss this as a one-dimensional example, in which the image is formed on a line, with coordinate x, rather than in a plane with coordinates (x,y). Here the function $g(u)$ represents the light intensity at position u for a perfect image, and $f(x-u)$ is the light intensity at x which is produced by the camera if the perfect image is a point of light at u: the function f therefore describes the blurring of the image (see Figure 3.9). The intensity of light in the observed image is $h(x)$. The contribution to the intensity at x from u is proportional to the perfect-image intensity there, $g(u)$, multiplied by the amount of this intensity that spreads to the position x, namely $f(x-u)$. The observed intensity of the image at x is the sum of all the intensities at each point u, and is obtained by integrating over u. Equation (3.46) is therefore a reasonable model for the light intensity at x when the ideal image would be described by $g(x)$.

Because of a fault in its assembly, the Hubble space telescope produced blurred images until adjustments were made by an astronaut. Until the repairs were made, the images were 'enhanced' to counteract the effect of the blurring. A two-dimensional version of the convolution equation (3.46) was used to model the relation between the blurred images transmitted to Earth from the Hubble telescope, and the true images.

3.5 Convolutions and the convolution theorem

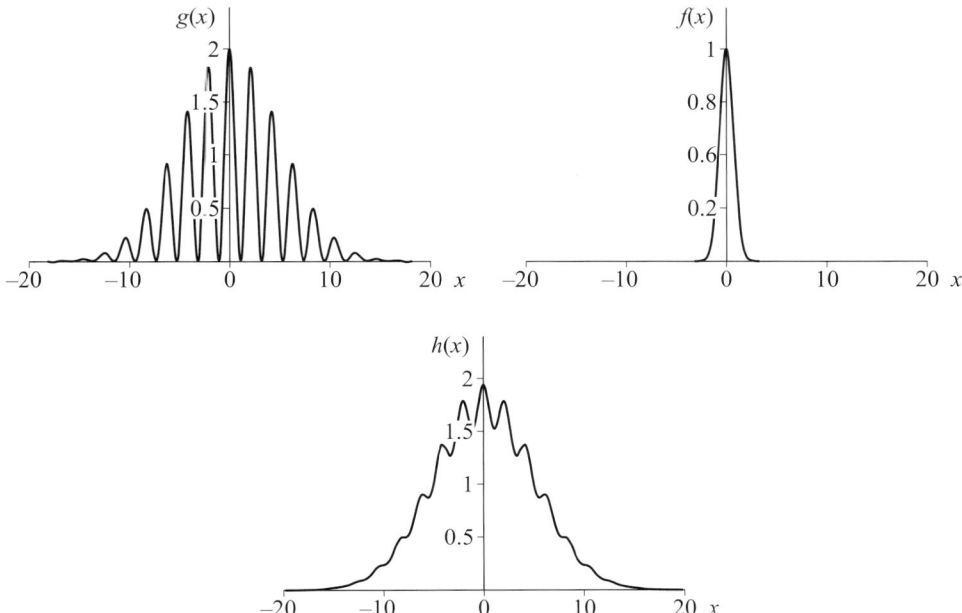

Figure 3.9 Convolution can be a model for a blurred or out-of-focus image. Here $g(x)$ is the light intensity as a function of position x for a perfect image, and $f(x)$ is the light intensity observed at position x when a perfect camera would show a point of light at the origin. The image produced by a real camera, with its imperfect or out-of-focus lens, is modelled by the convolution $h = f \otimes g$.

The functions f and g used in Figure 3.9 are
$$g(x) = \exp(-x^2/50)[1 + \cos 3x], \quad f(x) = \exp(-x^2).$$

Have a look at the form of all three curves, and think about how they are related to the definition of the convolution. In particular, note that the rapidly varying part of the function $g(x)$ is suppressed after convolution with $f(x)$, whereas the slowly varying part remains nearly unchanged. In this example, the rapidly oscillating part of the function represents fine detail in the ideal image g which is lost in the blurred image h.

3.5.2 Time-delay processes

Another example where convolutions arise is in modelling *time-delay processes*, such as in the following situation. People arrive at a tourist attraction, and spend a variable amount of time looking around before visiting the souvenir shop. The number of people going through the entrance between time k and time $k+1$ minutes is $n_1(k)$. The number of people arriving at the souvenir shop between k and $k+1$ minutes is $n_2(k)$. The probability that a person will spend l minutes (to the nearest minute) looking at the attraction is $P(l)$ (note that $P(l) = 0$ if $l < 0$). A model for the relationship between the functions $n_1(t)$ and $n_2(t)$ is then

$$n_2(k) = \sum_{l=-\infty}^{\infty} P(l)\, n_1(k-l). \tag{3.47}$$

(Note that only positive values of l contribute to the sum, because $P(l) = 0$ when l is negative.) This expression is already similar in structure to the convolution as defined in equation (3.46), and n_2 is said to be the *discrete convolution* of P and n_1.

If $n_1(k)$, $n_2(k)$ and $P(l)$ are slowly varying functions of k and l, then these functions may be approximated by functions of continuous time variables t and τ, and the sum may be approximated by an integral of the form

$$n_2(t) = \int_{-\infty}^{\infty} d\tau\, P(\tau)\, n_1(t-\tau). \tag{3.48}$$

This expression for $n_2(t)$ is a convolution integral. (The function $P(\tau)$ must be zero for negative values of τ, because the probability of spending a negative amount of time in the shop must be zero.) Figure 3.10 illustrates the relation between $n_1(t)$ and $n_2(t)$ (for some artificially constructed data).

Note that the time dependence of the time-delayed quantity, $n_2(t)$, looks 'smoother' than that of the original number, $n_1(t)$. This is analogous to the blurring effect in the previous example of a convolution.

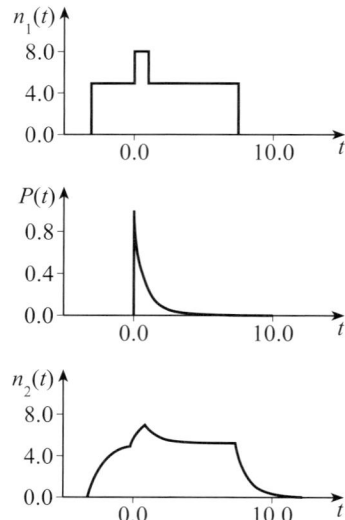

Figure 3.10 The number of people arriving at a tourist attraction per unit time is $n_1(t)$. The probability of spending a time between t and $t + \delta t$ looking round the attraction is $P(t)\delta t$: in this illustration, $P(t) = \exp(-t)$. The rate at which people arrive at the souvenir shop, $n_2(t)$, is given by the convolution $n_2 = P \otimes n_1$. In this illustration, all three time axes are in units of hours, not minutes.

Exercise 3.19

Show that

$$f \otimes g = g \otimes f \tag{3.49}$$

(i.e. the convolution operation is commutative). [Hint: Change the variable of integration.]

3.5.3 The convolution theorem

The Fourier transform of the convolution $h = f \otimes g$ is expressed very simply in terms of the Fourier transforms of the component functions f and g, as the following calculation shows:

3.5 Convolutions and the convolution theorem

$$\begin{aligned}
\tilde{h}(k) &= \frac{1}{\sqrt{2\pi}} \int_{-\infty}^{\infty} dx\, h(x) \exp(-ikx) \\
&= \frac{1}{\sqrt{2\pi}} \int_{-\infty}^{\infty} dx\, \exp(-ikx) \int_{-\infty}^{\infty} du\, f(x-u)\, g(u) \\
&= \frac{1}{\sqrt{2\pi}} \int_{-\infty}^{\infty} du\, g(u) \int_{-\infty}^{\infty} dx\, \exp(-ikx)\, f(x-u) \\
&= \frac{1}{\sqrt{2\pi}} \int_{-\infty}^{\infty} du\, g(u) \exp(-iku) \int_{-\infty}^{\infty} dx\, \exp[-ik(x-u)]\, f(x-u) \\
&= \frac{1}{\sqrt{2\pi}} \int_{-\infty}^{\infty} du\, g(u) \exp(-iku) \int_{-\infty}^{\infty} dy\, f(y) \exp(-iky) \\
&= \sqrt{2\pi}\, \tilde{g}(k)\, \tilde{f}(k).
\end{aligned} \tag{3.50}$$

(In the third step, the order of the integrations was reversed, and in the fifth step the substitution $y = x - u$ was used.) This result is known as the *convolution theorem*. In words, the Fourier transform of the convolution $f \otimes g$ is $\sqrt{2\pi}$ times the product of the Fourier transforms of the functions f and g:

$$h = f \otimes g \quad \Longleftrightarrow \quad \tilde{h} = \sqrt{2\pi}\, \tilde{f}\, \tilde{g}. \tag{3.51}$$

After Fourier transforming, convolution, a difficult operation involving evaluation of an integral, is replaced by multiplication, a very simple operation. This means that problems in which convolutions appear are often solved by application of the Fourier transform.

Figure 3.11 shows the Fourier transforms of the three functions plotted in Figure 3.9. Note that the third function is $\sqrt{2\pi}$ times the product of the first two functions, as is implied by the convolution theorem. The Fourier transform of g has three peaks: the central one is the Fourier transform of the Gaussian function, and the other two arise from the Gaussian function multiplied by the cosine function, and are shifted images of this Fourier transform as implied by Exercise 3.14. The magnitudes of the two 'satellite' peaks are reduced in the Fourier transform of h, representing the fact that the rapidly varying features (represented by contributions to the Fourier integral (3.24) with larger values of k) are suppressed when g is convoluted with f.

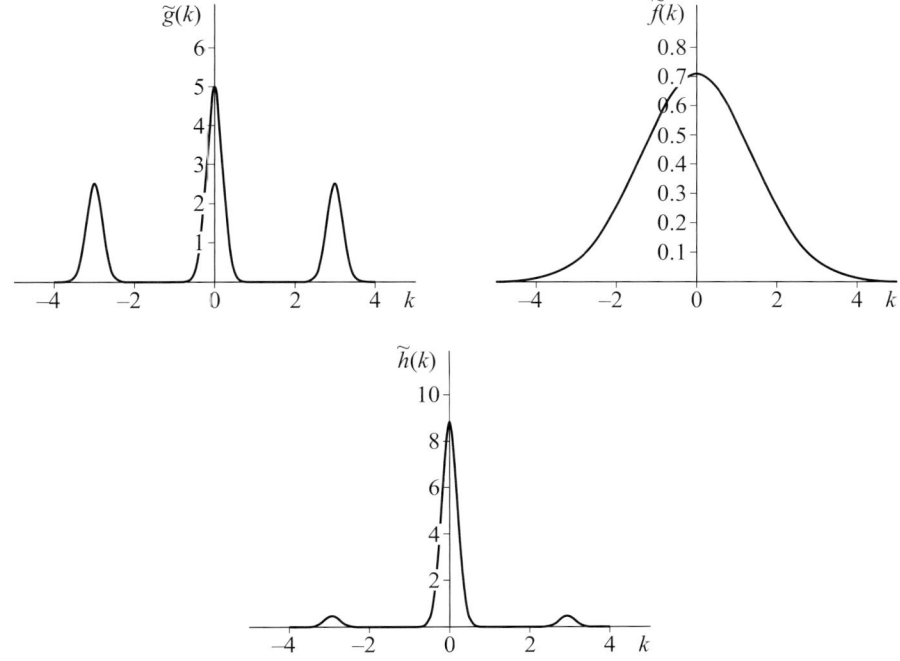

Figure 3.11 The Fourier transforms of the functions in Figure 3.9

The convolution theorem gives a hint about how a true image, g, can be recovered from a blurred image, h. If the blurred image is modelled by $h = f \otimes g$ (where f describes the blurring effect, as in Subsection 3.5.1), the convolution theorem indicates that the Fourier transforms are related by $\tilde{h} = \sqrt{2\pi}\, \tilde{f}\, \tilde{g}$. We Fourier transform both h and f, and obtain the Fourier transform of g:

$$\tilde{g}(k) = \frac{\tilde{h}(k)}{\sqrt{2\pi}\, \tilde{f}(k)}.$$

In principle we can then determine $g(x)$ by inverse Fourier transforming $\tilde{g}(k)$. In practice, this 'deconvolution' does not work smoothly, because there are values of k for which $\tilde{f}(k)$ is very small. When we divide $\tilde{h}(k)$ by $\tilde{f}(k)$, small errors in $\tilde{h}(k)$ can be greatly magnified. The Hubble space telescope pictures were enhanced by a 'deconvolution' process, but this required much more than a direct application of the convolution theorem.

Example 3.7

Calculate the convolution $F = f_1 \otimes f_2$ of two Gaussian functions

$$f_1(x) = \exp(-x^2/2), \quad f_2(x) = \exp(-\alpha x^2/2),$$

where $\alpha > 0$. Use the table of standard Fourier transforms and the scaling rule to determine the Fourier transform of F.

Solution

The convolution of $f_1(x) = \exp(-x^2/2)$ and $f_2(x) = \exp(-\alpha x^2/2)$ is

$$F(x) = \int_{-\infty}^{\infty} du \exp\bigl(-[(x-u)^2/2 + \alpha u^2/2]\bigr)$$

$$= \int_{-\infty}^{\infty} du \exp[-\Phi(u)/2],$$

where $\Phi(u) = A(u-B)^2 + C$, with $A = \alpha + 1$, $B = x/A$, $C = \alpha x^2/(\alpha + 1)$. The reason for writing the exponent in this form is that the integral with respect to u is found using a standard Gaussian integral:

$$F(x) = \int_{-\infty}^{\infty} du \exp[-\Phi(u)/2]$$

$$= \exp(-C/2) \int_{-\infty}^{\infty} du \exp\bigl[-A(u-B)^2/2\bigr]$$

$$= \sqrt{2\pi/A}\, \exp(-C/2).$$

(The last step used the substitution $v = u - B$ and the expression for the integral of the Gaussian function.) Inserting the values of A and C, we see that F is a Gaussian function:

$$F(x) = \sqrt{\frac{2\pi}{\alpha + 1}}\, \exp\left[-\frac{\alpha x^2}{2(\alpha + 1)}\right].$$

Using the second result in Exercise 3.16 with $a = \tfrac{1}{2}\alpha/(\alpha+1)$, we see that the Fourier transform of this Gaussian function is also a Gaussian function:

$$\tilde{F}(k) = \sqrt{\frac{2\pi}{\alpha}}\, \exp\left[-\frac{k^2(\alpha+1)}{2\alpha}\right]. \quad \blacksquare$$

Thus we have seen that the convolution of two Gaussian functions is another Gaussian function.

Exercise 3.20

Determine the Fourier transforms of the functions f_1 and f_2 in Example 3.7, and use the convolution theorem to write down the Fourier transform of their convolution F. Confirm that you obtain the same result for the Fourier transform of F as in Example 3.7.

Exercise 3.21

Use the convolution theorem to calculate the convolution of two Lorentzian functions

$$F(x) = \int_{-\infty}^{\infty} dy \, \frac{1}{1 + (x-y)^2} \frac{1}{a^2 + y^2}.$$

[Hints: Use the table of Fourier transforms and the scaling relation to find the Fourier transforms of the Lorentzian functions. Then use the convolution theorem to determine the Fourier transform of F. Finally, invert this Fourier transform to determine F itself.]

Exercise 3.22

Calculate the convolution of the characteristic function $\chi(x)$ (as defined in Exercise 3.10) with itself. Calculate the Fourier transform of $F = \chi \otimes \chi$ using the convolution theorem.

[Hint: Consider the region where the integrand is non-zero.]

3.6 Fourier transforms of derivatives

Perhaps the most useful property of Fourier transforms is that there is a very simple relationship between the Fourier transform $\tilde{f}(k)$ of a function $f(x)$ and the Fourier transform of its derivative, $f'(x)$. You will see that the Fourier transform of the first derivative is $ik\tilde{f}(k)$, and that higher derivatives follow a simple pattern. You will see that these properties can be used to obtain solutions to certain types of differential equations.

Consider the Fourier transform of the derivative of the function $f(x)$. This Fourier transform will be called $\tilde{f}_1(k)$:

$$\tilde{f}_1(k) = \frac{1}{\sqrt{2\pi}} \int_{-\infty}^{\infty} dx \, f'(x) \exp(-ikx). \tag{3.52}$$

Integrating by parts,

$$\tilde{f}_1(k) = \frac{1}{\sqrt{2\pi}} \left[\exp(-ikx) f(x) \right]_{-\infty}^{\infty} + ik\tilde{f}(k). \tag{3.53}$$

The first term on the right-hand side is equal to zero if the function $f(x)$ approaches zero in the limits $x \to \pm\infty$, in which case

$$\tilde{f}_1(k) = ik\tilde{f}(k). \tag{3.54}$$

Applying (3.54) repeatedly, this result can be generalised immediately to higher derivatives. The Fourier transform of $d^n f/dx^n$ is

$$\tilde{f}_n(k) = (ik)^n \tilde{f}(k), \tag{3.55}$$

provided that the function $d^{n-1}f/dx^{n-1}$ approaches zero as $|x| \to \pm\infty$. The conclusion is that differentiation, an operation of calculus, is replaced by multiplication, an algebraic operation, on taking the Fourier transform.

Note that *for this section only*, $\tilde{f}_n$ means the Fourier transform of $d^n f/dx^n$, for any n; elsewhere, it would mean the Fourier transform of a function f_n.

This is one of the features that makes the Fourier transform a powerful tool: the Fourier transform of a linear constant-coefficient differential equation is an algebraic equation, which is much easier to solve. The required solution may then be recovered by an inverse Fourier transform. Because linear constant-coefficient differential and partial differential equations arise in describing many natural phenomena, this is a very useful method. The approach is illustrated in Figure 3.12 and used in Exercise 3.25. Later in the course, the method will be used to obtain solutions of the wave equation and the heat/diffusion equation.

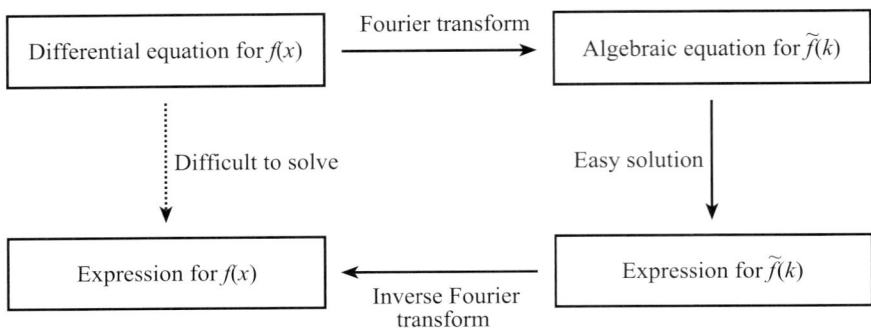

Figure 3.12 Illustrating the use of Fourier transforms to solve differential equations. This approach is directly applicable to linear constant-coefficient differential equations, but Fourier transforms can also be useful for other types of differential equations.

Because integration is the reverse of differentiation, there is a result corresponding to (3.54) for the Fourier transform of the integral of a function. If F is the indefinite integral (or primitive) of f (that is, $f = F' = dF/dx$), then their Fourier transforms are expected to be related by $\tilde{F}(k) = \tilde{f}(k)/(ik)$. This approach can sometimes be useful, but there may be difficulties in applying it. For example, because of the division by k, $|\tilde{F}(k)|$ may approach infinity as $k \to 0$, and this can cause problems when calculating the inverse Fourier transform.

Exercise 3.23

Calculate the derivative of the Lorentzian function $f(x) = 1/(1 + x^2)$. Determine the Fourier transform of $f(x)$ and its derivative using (3.54).

Exercise 3.24

Show that the Fourier transform of $xf(x)$ is $i\tilde{f}'(k)$, where $\tilde{f}'(k)$ is the derivative with respect to k of the Fourier transform of f.

$$\left[\text{Hint: Note that } -ix\exp(-ikx) = \frac{d}{dk}\exp(-ikx).\right]$$

Note the near symmetry between the results in Exercise 3.24 and equation (3.54): multiplication by one variable is the 'Fourier image' of differentiation with respect to the conjugate variable.

Exercise 3.25

Previously, the Fourier transform of the Gaussian function was obtained by an incomplete argument, which rested on an assumption about integration over complex variables (see equation (3.31)). This exercise provides stronger evidence for accepting that the Fourier transform of a Gaussian function is also a Gaussian function.

If a function f satisfies the differential equation

$$\frac{df}{dx} + xf = 0,$$

show that its Fourier transform $\tilde{f}(k)$ satisfies the same differential equation, i.e.

$$\frac{d\tilde{f}}{dk} + k\tilde{f} = 0.$$

Solve the first of these differential equations for $f(x)$, showing that it is solved by the Gaussian function $f(x) = \exp(-x^2/2)$. Hence show that this Gaussian function is its own Fourier transform.

3.7 Summary of Fourier transforms

Table 3.3 summarises the properties of Fourier transforms, and lists the elementary Fourier transform pairs, for use as a quick reference.

Fourier transform:

$$\tilde{f}(k) = \frac{1}{\sqrt{2\pi}} \int_{-\infty}^{\infty} dx \exp(-ikx) f(x).$$

Inverse Fourier transform:

$$f(x) = \frac{1}{\sqrt{2\pi}} \int_{-\infty}^{\infty} dk \exp(ikx) \tilde{f}(k).$$

Table 3.3 Summary of Fourier transforms for quick reference

Property	Function	Fourier transform
Linearity	$a\,f(x) + b\,g(x)$	$a\,\tilde{f}(k) + b\,\tilde{g}(k)$
Scaling ($\alpha \neq 0$)	$f(x/\alpha)$	$\|\alpha\|\,\tilde{f}(\alpha k)$
Translation	$f(x-a)$	$\exp(-ika)\,\tilde{f}(k)$
Translation in k	$\exp(iax)\,f(x)$	$\tilde{f}(k-a)$
Inversion	$\tilde{f}(x)$	$f(-k)$
Convolution	$f(x) \otimes g(x)$	$\sqrt{2\pi}\,\tilde{f}(k)\,\tilde{g}(k)$
Differentiation	$f'(x)$	$ik\,\tilde{f}(k)$
Multiplication by x	$x\,f(x)$	$i\,\tilde{f}'(k)$

Fourier transform pairs	$f(x)$	$\tilde{f}(k)$
Exponential/Lorentzian	$\exp(-\|x\|)$	$\sqrt{\frac{2}{\pi}}\,\frac{1}{1+k^2}$
Gaussian/Gaussian	$\exp(-x^2/2)$	$\exp(-k^2/2)$
Characteristic/sinc	$\chi(x)$	$\sqrt{\frac{2}{\pi}}\,\frac{\sin k}{k}$

3.8 Outcomes

After studying this unit you should be able to:
- understand the trigonometric Fourier series and its use for both periodic and non-periodic functions;
- understand the complex exponential Fourier series;
- calculate the complex exponential Fourier coefficients c_n for a given function;
- calculate the trigonometric Fourier coefficients a_n and b_n for a given function;
- understand the Fourier transform;
- determine the Fourier transforms for a few simple and commonly encountered functions;
- understand some basic properties of Fourier transforms, i.e. linearity, scaling, translations, and inversion and its consequences for even real-valued functions;
- use the basic properties (linearity, scaling, etc.) to determine Fourier transforms of more complicated functions given the Fourier transforms of a few simpler functions;
- understand what is meant by the convolution of two functions;
- understand the convolution theorem;
- take Fourier transforms of derivatives.

3.9 Further Exercises

The following two exercises provide further practice in determining Fourier coefficients.

Exercise 3.26

Determine the Fourier coefficients for the function $f(x) = |\sin x|$, which was plotted in Figure 3.1. Show that the general Fourier coefficient is

$$c_n = \frac{2}{\pi(1 - 4n^2)}.$$

Exercise 3.27

Determine the Fourier coefficients for the periodic function which takes the value $f(x) = x^2$ in the interval $[-1, 1]$, for which two partial sums were plotted in Figure 3.2. Show that the general Fourier coefficient is

$$c_n = \begin{cases} \frac{2(-1)^n}{\pi^2 n^2}, & n \neq 0, \\ \frac{1}{3}, & n = 0. \end{cases}$$

Exercise 3.28

The function $f(x)$ is defined as follows:
$$f(x) = \begin{cases} +1 & \text{for } 0 < x \leq 1, \\ -1 & \text{for } -1 \leq x < 0, \\ 0 & \text{otherwise.} \end{cases}$$

Sketch a graph of this function. Express $f(x)$ as the difference of two translated characteristic functions. Use the Fourier transform of the characteristic function, together with the elementary properties of Fourier transforms discussed in Section 3.4, to determine the Fourier transform of $f(x)$.

Exercise 3.29

By direct calculation, determine the function $F(x)$ which is minus the convolution of the characteristic function on the interval $[-\frac{1}{2}, \frac{1}{2}]$ with itself, i.e.

$$F = -\chi_{1/2} \otimes \chi_{1/2}.$$

Show that the function $f(x)$ considered in Exercise 3.28 is the derivative of this function $F(x)$ (except at the three points where F is non-differentiable).

Use the convolution theorem to determine the Fourier transform of F. Hence, using the results of Section 3.6, determine the Fourier transform of $f(x)$. Compare your answer with that for Exercise 3.28.

3.10 Appendix: Multi-dimensional Fourier transforms

The Fourier transform can be extended in a natural way to functions of two or more variables, and the two-dimensional case will be used later. You do not need to study this material; it is provided as a reference if you wish to see how the two-dimensional version is obtained and if you are interested in higher-dimensional generalisations. Only equations (3.56) and (3.57) are used later, not the discussion forming the remainder of this appendix.

A function of two variables $f(x_1, x_2)$ has a Fourier transform $\tilde{f}(k_1, k_2)$ which is defined in the equation

$$f(x_1, x_2) = \frac{1}{2\pi} \int_{-\infty}^{\infty} dk_1 \int_{-\infty}^{\infty} dk_2 \exp[i(k_1 x_1 + k_2 x_2)] \tilde{f}(k_1, k_2). \quad (3.56)$$

The Fourier transform itself is given by

$$\tilde{f}(k_1, k_2) = \frac{1}{2\pi} \int_{-\infty}^{\infty} dx_1 \int_{-\infty}^{\infty} dx_2 \exp[-i(k_1 x_1 + k_2 x_2)] f(x_1, x_2). \quad (3.57)$$

These relations are analogous to the single-variable case.

We show that (3.57) is the inverse of (3.56).

Consider an intermediate function $g(k_1, x_2)$ which is the Fourier transform of f with respect to only one variable (x_1, say). Then

$$f(x_1, x_2) = \frac{1}{\sqrt{2\pi}} \int_{-\infty}^{\infty} dk_1 \, g(k_1, x_2) \exp(ik_1 x_1),$$

where the standard formula for the Fourier transform gives

$$g(k_1, x_2) = \frac{1}{\sqrt{2\pi}} \int_{-\infty}^{\infty} dx_1 \, f(x_1, x_2) \exp(-ik_1 x_1).$$

Now write $g(k_1, x_2)$ in terms of its Fourier transform with respect to the second variable, x_2, which is taken to be $\tilde{f}(k_1, k_2)$,

$$g(k_1, x_2) = \frac{1}{\sqrt{2\pi}} \int_{-\infty}^{\infty} dk_2 \, \tilde{f}(k_1, k_2) \exp(ik_2 x_2),$$

and write down the corresponding inverse. The resulting equations can be combined to express $f(x_1, x_2)$ in terms of $\tilde{f}(k_1, k_2)$ and vice versa, resulting in equations (3.56) and (3.57).

In d dimensions, the variables x_i and k_i may be written as vectors: $\boldsymbol{x} = (x_1, x_2, \ldots, x_d)$ and $\boldsymbol{k} = (k_1, k_2, \ldots, k_d)$. The corresponding expressions relating the function $f(\boldsymbol{x})$ to its transform $\tilde{f}(\boldsymbol{k})$ are

$$f(\boldsymbol{x}) = \frac{1}{(2\pi)^{d/2}} \int d\boldsymbol{k} \exp(i\boldsymbol{k} \cdot \boldsymbol{x}) \, \tilde{f}(\boldsymbol{k}), \tag{3.58}$$

$$\tilde{f}(\boldsymbol{k}) = \frac{1}{(2\pi)^{d/2}} \int d\boldsymbol{x} \exp(-i\boldsymbol{k} \cdot \boldsymbol{x}) \, f(\boldsymbol{x}), \tag{3.59}$$

where

$$\boldsymbol{k} \cdot \boldsymbol{x} = \sum_{i=1}^{d} k_i x_i \tag{3.60}$$

is the usual scalar product of vectors. The integral over $d\boldsymbol{x}$ stands for the multiple integral

$$\int d\boldsymbol{x} = \int_{-\infty}^{\infty} dx_1 \int_{-\infty}^{\infty} dx_2 \cdots \int_{-\infty}^{\infty} dx_d, \tag{3.61}$$

and the integral over $d\boldsymbol{k}$ is defined in the same way.

Exercise 3.30

Show that the Fourier transform of $F(x_1, x_2) = f(x_1) g(x_2)$ is the product of Fourier transforms $\tilde{F}(k_1, k_2) = \tilde{f}(k_1) \tilde{g}(k_2)$.

Most of the properties of Fourier transforms extend in a natural way to functions of two or more variables. As an example, let us consider the convolution theorem. Convolution of functions of two variables is defined by a natural extension of (3.46): the convolution of $f(x_1, x_2)$ and $g(x_1, x_2)$ is

$$h(x_1, x_2) = \int_{-\infty}^{\infty} du_1 \int_{-\infty}^{\infty} du_2 \, f(x_1 - u_1, x_2 - u_2) \, g(u_1, u_2). \tag{3.62}$$

Exercise 3.31

Generalise the convolution theorem to functions of two variables.

Solutions to Exercises in Chapter 3

Solution 3.1

Using the relations
$$\cos x = \frac{1}{2}[\exp(ix) + \exp(-ix)]$$
and
$$\sin x = \frac{1}{2i}[\exp(ix) - \exp(-ix)],$$
we obtain
$$f(x) = \tfrac{1}{2}a_0 + \sum_{n=1}^{\infty} a_n \cos(2\pi nx/L) + \sum_{n=1}^{\infty} b_n \sin(2\pi nx/L)$$
$$= \tfrac{1}{2}a_0 + \tfrac{1}{2}\sum_{n=1}^{\infty}(a_n - ib_n)\exp(2\pi inx/L) + \tfrac{1}{2}\sum_{n=1}^{\infty}(a_n + ib_n)\exp(-2\pi inx/L)$$
$$= \tfrac{1}{2}a_0 + \tfrac{1}{2}\sum_{n=1}^{\infty}(a_n - ib_n)\exp(2\pi inx/L) + \tfrac{1}{2}\sum_{n=-\infty}^{-1}(a_{-n} + ib_{-n})\exp(2\pi inx/L)$$
$$= \sum_{n=-\infty}^{\infty} c_n \exp(2\pi inx/L).$$

We use $-i = 1/i$ in simplifying the second line.

Comparing the coefficients of the sums in the last two lines gives the result quoted in the exercise.

Solution 3.2

(a) We may use the relation $\exp(ikx) = \cos(kx) + i\sin(kx)$ to obtain (using the chain rule)
$$\frac{d}{dx}\exp(ikx) = -k\sin(kx) + ik\cos(kx)$$
$$= ik[\cos(kx) + i\sin(kx)]$$
$$= ik\exp(ikx).$$

However, it will help you to calculate more rapidly if you get used to working with the exponential function directly. The relation $d\exp(z)/dz = \exp(z)$ remains valid when z is a complex number. Using the chain rule gives
$$\frac{d}{dx}\exp(ikx) = \frac{d(ikx)}{dx}\left.\frac{d\exp(z)}{dz}\right|_{z=ikx}$$
$$= ik\exp(ikx).$$

The notation
$$df(z)/dz|_{z=y}$$
means 'differentiate f with respect to z and substitute $z = y$ in the resulting function'.

(b) The indefinite integral (primitive) of $\exp(ikx)$ is $\exp(ikx)/ik$ (which can be checked by differentiation).

Solution 3.3

For integer values of n,
$$\exp(i\pi n) = \cos(\pi n) + i\sin(\pi n)$$
$$= (-1)^n$$
and
$$\exp(i\pi n/2) = \big[\exp(i\pi/2)\big]^n$$
$$= \big[\cos(\pi/2) + i\sin(\pi/2)\big]^n$$
$$= i^n.$$

Solution 3.4

(a) The complex conjugate of $\exp(ix)$ is
$$[\exp(ix)]^* = [\cos x + i \sin x]^* = \cos x - i \sin x = \exp(-ix).$$

(b) If $c_{-n} = c_n^*$, then
$$f(x) = \sum_{n=-\infty}^{\infty} c_n \exp(2\pi i n x/L)$$
$$= \sum_{m=-\infty}^{\infty} c_{-m} \exp(-2\pi i m x/L)$$
$$= \sum_{m=-\infty}^{\infty} c_m^* [\exp(2\pi i m x/L)]^*$$
$$= \sum_{m=-\infty}^{\infty} [c_m \exp(2\pi i m x/L)]^* = f^*(x),$$

$f^*(x)$ is shorthand notation for $[f(x)]^*$.

so f is a real-valued function. Note that the substitution $m = -n$ was used, and that the name of the variable being summed over is irrelevant.

Alternatively, assuming the relation $c_{-n} = c_n^*$, you could split up the summation as follows:
$$f(x) = c_0 + \sum_{n=1}^{\infty} c_n \exp(2\pi i n x/L) + \sum_{n=1}^{\infty} [c_n \exp(2\pi i n x/L)]^*.$$

The function $f(x)$ is real because c_0 is real and the two sums are complex conjugates of each other.

Solution 3.5

Let $N = n - m$. For $N \neq 0$, we have
$$\int_{x_0}^{x_0+L} dx \exp(2\pi i N x/L) = \frac{L}{2\pi i N} \left[\exp(2\pi i N x/L)\right]_{x_0}^{x_0+L}$$
$$= \frac{L}{2\pi i N} \left[\exp(2\pi i N(x_0+L)/L) - \exp(2\pi i N x_0/L)\right]$$
$$= \frac{L \exp(2\pi i N x_0/L)[\exp(2\pi i N) - 1]}{2\pi i N} = 0.$$

The last step uses the fact that $\exp(2\pi i N) = 1$ for integer N.

When $N = 0$, the integrand is unity, so the integral is equal to the length of the interval, namely L. Hence the expression on the right-hand side of equation (3.10) is 1 when $N = 0$ (i.e. when $m = n$), and 0 when $N \neq 0$ (i.e. when $m \neq n$).

Solution 3.6

We take $L = 2\pi$ and $x_0 = -\pi$. Again, for a function defined piecewise on intervals, the approach is to split the region of integration into segments. For $n \neq 0$, we have
$$c_n = \frac{1}{2\pi} \int_{-\pi}^{0} dx\, (-x) \exp(-inx) + \frac{1}{2\pi} \int_{0}^{\pi} dx\, x \exp(-inx)$$
$$= \frac{1}{2\pi} \int_{0}^{\pi} dx\, x [\exp(inx) + \exp(-inx)],$$

where the change of variable $u = -x$ was used in the first integral. Now, integrating by parts,
$$\int_{0}^{\pi} dx\, x \exp(inx) = \left[\frac{x \exp(inx)}{in}\right]_{0}^{\pi} - \frac{1}{in} \int_{0}^{\pi} dx \exp(inx)$$
$$= \frac{\pi(-1)^n}{in} - \frac{1}{(in)^2}[(-1)^n - 1],$$

where we have used the fact that $\exp(i\pi n) = (-1)^n$.

We obtain the Fourier coefficient c_n by combining this integral with another, differing only by the sign of n. The terms proportional to $1/n$ cancel, and we obtain

$$c_n = \frac{(-1)^n - 1}{\pi n^2}, \quad n \neq 0.$$

For $n = 0$, we have

$$c_0 = \frac{-1}{2\pi}\int_{-\pi}^{0} dx\, x + \frac{1}{2\pi}\int_{0}^{\pi} dx\, x = \frac{\pi}{2}.$$

The coefficients are all purely real, implying that when this Fourier series is expressed in terms of trigonometric functions, only the cosine terms are present.

Solution 3.7

We have

$$f'(x) = \sum_{n=-\infty}^{\infty} c_n \frac{d}{dx}\exp(2\pi i n x/L)$$

$$= \sum_{n=-\infty}^{\infty} \frac{2\pi i n}{L} c_n \exp(2\pi i n x/L).$$

Solution 3.8

The derivative of the triangular wave is $+1$ for $0 < x < \pi$, and -1 for $\pi < x < 2\pi$. Also, the derivative must be a periodic function with the same period, namely $L = 2\pi$. The derivative is therefore identical to the square wave (except at points where x is an integer multiple of π, where the derivative of the triangular wave is undefined). The Fourier coefficients for the triangular wave are (from Exercise 3.6) $c_n = [(-1)^n - 1]/\pi n^2$ ($n \neq 0$), and $c_0 = \pi/2$. Those for the square wave are therefore expected to be

$$c_n = \frac{2\pi i n}{L}\frac{(-1)^n - 1}{\pi n^2} = \frac{2i}{Ln}[(-1)^n - 1]$$

for $n \neq 0$, and $c_0 = 0$. Substituting $L = 2\pi$, these c_n are in agreement with those calculated for the square wave in Example 3.2.

Solution 3.9

We have

$$f(x) = A_0 + \sum_{n=1}^{\infty} A_n \sin(nx - \phi_n)$$

$$= A_0 + \sum_{n=1}^{\infty} \frac{A_n}{2i}\left(\exp[i(nx - \phi_n)] - \exp[-i(nx - \phi_n)]\right)$$

$$= A_0 + \sum_{n=1}^{\infty} \frac{A_n}{2i}\exp(-i\phi_n)\exp(inx) - \sum_{n=-\infty}^{-1} \frac{A_{-n}}{2i}\exp(i\phi_{-n})\exp(inx)$$

$$= \sum_{n=-\infty}^{\infty} c_n \exp(inx),$$

where

$$c_0 = A_0,$$

$$c_n = \begin{cases} \dfrac{A_n}{2i}\exp(-i\phi_n) & (n > 0), \\ \dfrac{-A_{-n}}{2i}\exp(i\phi_{-n}) & (n < 0). \end{cases}$$

Solution 3.10

For $k \neq 0$,

$$\tilde{f}(k) = \frac{1}{\sqrt{2\pi}} \int_{-1}^{1} dx \exp(-ikx)$$
$$= \frac{-1}{\sqrt{2\pi}ik}[\exp(-ik) - \exp(ik)]$$
$$= \frac{1}{\sqrt{2\pi}ik}[\cos(k) + i\sin(k) - \cos(-k) - i\sin(-k)]$$
$$= \sqrt{\frac{2}{\pi}} \frac{\sin k}{k}.$$

When $k = 0$,

$$\tilde{f}(k) = \frac{1}{\sqrt{2\pi}} \int_{-1}^{1} dx = \sqrt{\frac{2}{\pi}}.$$

Solution 3.11

We assume a standard result for integrals (in which P and Q are constants):

$$\int_a^b dx \, [P\,f(x) + Q\,g(x)] = P \int_a^b dx \, f(x) + Q \int_a^b dx \, g(x).$$

Applying this to the Fourier transform, we obtain

$$\tilde{F}(k) = \frac{1}{\sqrt{2\pi}} \int_{-\infty}^{\infty} dx \exp(-ikx) [a_1 f_1(x) + a_2 f_2(x)]$$
$$= \frac{a_1}{\sqrt{2\pi}} \int_{-\infty}^{\infty} dx \exp(-ikx) f_1(x) + \frac{a_2}{\sqrt{2\pi}} \int_{-\infty}^{\infty} dx \exp(-ikx) f_2(x)$$
$$= a_1 \tilde{f}_1(k) + a_2 \tilde{f}_2(k).$$

Solution 3.12

Starting with equation (3.25),

$$\tilde{f}(k) = \frac{1}{\sqrt{2\pi}} \int_{-\infty}^{\infty} dx \exp(-ikx) f(x),$$

and using the substitution $u = -x$, we find

$$\tilde{f}(k) = \frac{1}{\sqrt{2\pi}} \int_{-\infty}^{\infty} du \exp(iku) f(-u).$$

This expression is in the same form as equation (3.24), apart from differences in the names of variables. Changing the names of the variables ($k \to x$ and $u \to k$), this becomes

$$\tilde{f}(x) = \frac{1}{\sqrt{2\pi}} \int_{-\infty}^{\infty} dk \exp(ikx) f(-k).$$

Comparing with the standard expression for the Fourier transform, equation (3.24), this shows that $f(-k)$ is the Fourier transform of $\tilde{f}(x)$.

Solution 3.13

If $f(x) = f(-x)$, then

$$\tilde{f}(-k) = \frac{1}{\sqrt{2\pi}} \int_{-\infty}^{\infty} dx \exp(ikx) f(x)$$
$$= \frac{1}{\sqrt{2\pi}} \int_{-\infty}^{\infty} du \exp(-iku) f(-u)$$
$$= \frac{1}{\sqrt{2\pi}} \int_{-\infty}^{\infty} du \exp(-iku) f(u) = \tilde{f}(k).$$

The second equality uses a change of variable $u = -x$, and the third uses $f(u) = f(-u)$.

Solutions to Exercises in Chapter 3

If f is also a real-valued function, satisfying $f(x) = [f(x)]^*$, then

$$[\tilde{f}(k)]^* = \frac{1}{\sqrt{2\pi}} \int_{-\infty}^{\infty} dx \, [\exp(-ikx)]^* f(x) = \tilde{f}(-k).$$

So if f is both even and real-valued, then $[\tilde{f}(k)]^* = \tilde{f}(-k) = \tilde{f}(k)$, i.e. $\tilde{f}(k)$ is real-valued.

If f is odd, $f(-x) = -f(x)$, then the same arguments show that $\tilde{f}$ is also odd, $\tilde{f}(k) = -\tilde{f}(-k)$. If f is also a real-valued function, then we have $[\tilde{f}(k)]^* = \tilde{f}(-k) = -\tilde{f}(k)$, so $\tilde{f}$ is purely imaginary.

Solution 3.14

The Fourier transform of $F(x) = \exp(iKx) f(x)$ is

$$\tilde{F}(k) = \frac{1}{\sqrt{2\pi}} \int_{-\infty}^{\infty} dx \exp(-ikx) \left[\exp(iKx) f(x)\right]$$
$$= \frac{1}{\sqrt{2\pi}} \int_{-\infty}^{\infty} dx \exp[-i(k-K)x] f(x) = \tilde{f}(k-K),$$

so translation in the Fourier transform variable is the image of multiplication by an oscillatory exponential (i.e. an exponential whose argument is the conjugate variable times a purely imaginary constant).

The Fourier transform of $g(x) = \cos(\lambda x) f(x) = \frac{1}{2}[\exp(i\lambda x) + \exp(-i\lambda x)] f(x)$ is given by applying this result to each term.

Solution 3.15

Define $f_1(x) = f(\alpha x)$ and $f_2(x) = f_1(x - x_0)$, so that

$$g(x) = A f(\alpha(x - x_0)) = A f_1(x - x_0) = A f_2(x).$$

Then, using the result of Example 3.6, the Fourier transform of f_2 is

$$\tilde{f}_2(k) = \exp(-ikx_0) \, \tilde{f}_1(k).$$

Using the result of Example 3.5 (with α replaced by $1/\alpha$), the Fourier transform of f_1 is

$$\tilde{f}_1(k) = \frac{1}{\alpha} \tilde{f}(k/\alpha).$$

The Fourier transform of g is therefore

$$\tilde{g}(k) = A \tilde{f}_2(k) = A \exp(-ikx_0) \, \tilde{f}_1(k) = \frac{A}{\alpha} \exp(-ikx_0) \, \tilde{f}(k/\alpha).$$

Solution 3.16

The scaling rule implies that the Fourier transform of $f(ax)$ is $\tilde{f}(k/a)/a$. The Fourier transform of $\exp(-|x|)$ is $\sqrt{\frac{2}{\pi}} \frac{1}{1+k^2}$. The Fourier transform of $f_1(x) = \exp(-a|x|)$ is therefore

$$\tilde{f}_1(k) = \frac{1}{a} \sqrt{\frac{2}{\pi}} \frac{1}{1 + (k/a)^2} = \sqrt{\frac{2}{\pi}} \frac{a}{a^2 + k^2}.$$

The Fourier transform of $f(x) = \exp(-x^2/2)$ is one of our standard examples. The function $F(x) = \exp(-ax^2)$ is in the form $F(x) = f(x/\alpha)$, where $\alpha = 1/\sqrt{2a}$. We use the formula in Example 3.5, and find $\tilde{F}(k) = \alpha \exp(-\alpha^2 k^2/2)$, so the Fourier transform of $f_2(x) = \exp(-ax^2)$ is

$$\tilde{f}_2(k) = \frac{\exp(-k^2/4a)}{\sqrt{2a}}.$$

Similarly, with $\alpha = a/2$, the other Fourier transform is

$$\tilde{f}_3(k) = \sqrt{\frac{2}{\pi}} \frac{\sin(ka/2)}{k}.$$

Solution 3.17

Note that the function $f(x) = (\sin x)/x$ is proportional to the Fourier transform of $\chi(x)$. Using the rule for inverting Fourier transforms, together with the fact that $(\sin x)/x$ is an even function, its Fourier transform is

$$\tilde{f}(k) = \sqrt{\frac{\pi}{2}}\, \chi(k).$$

Using the scaling rule, the Fourier transform of $F(x) = \sin(ax)/x = af(ax)$ is

$$\tilde{F}(k) = \sqrt{\frac{\pi}{2}}\, \chi(k/a) = \sqrt{\frac{\pi}{2}}\, \chi_a(k).$$

The final step uses the definition in equation (3.26).

Solution 3.18

The Fourier transform of $\exp(-ax^2)$ is $\exp(-k^2/4a)/\sqrt{2a}$ (found in Exercise 3.16). The function $f(x)$ is a translation of this function, and using the result of Example 3.6 we see that its Fourier transform is

$$\tilde{f}(k) = \frac{1}{\sqrt{2a}}\exp(-ikx_0)\exp(-k^2/4a).$$

The Fourier transform of the second function follows from the rule obtained in Exercise 3.14:

$$\tilde{g}(k) = \frac{1}{\sqrt{2a}}\exp\left[-\frac{(k-K)^2}{4a}\right].$$

Solution 3.19

Let $h_1 = f \otimes g$ and $h_2 = g \otimes f$. Then, using the change of variable $v = x - u$, we have

$$h_1(x) = \int_{-\infty}^{\infty} du\, f(x-u)\, g(u)$$

$$= \int_{-\infty}^{\infty} dv\, f(v)\, g(x-v) = h_2(x).$$

Solution 3.20

The Fourier transforms of the functions f_1 and f_2 introduced in Example 3.7 are

$$\tilde{f}_1(k) = \exp(-k^2/2), \quad \tilde{f}_2(k) = \frac{1}{\sqrt{\alpha}}\exp[-k^2/(2\alpha)].$$

The Fourier transform of their convolution is therefore

$$\tilde{F}(k) = \sqrt{2\pi}\,\tilde{f}_1(k)\,\tilde{f}_2(k) = \sqrt{\frac{2\pi}{\alpha}}\exp\left[-\frac{k^2}{2} - \frac{k^2}{2\alpha}\right]$$

$$= \sqrt{\frac{2\pi}{\alpha}}\exp[-(\alpha+1)k^2/2\alpha],$$

which is in agreement with Example 3.7.

Solution 3.21

We define

$$f(x) = \frac{1}{1+x^2}, \quad g(x) = \frac{1}{a^2+x^2} = \frac{1}{a^2}f(x/a).$$

The Fourier transforms of these Lorentzian functions are decreasing exponential functions of $|k|$: using the results in Table 3.2 and Example 3.5 (the scaling property), we have

$$\tilde{f}(k) = \sqrt{\frac{\pi}{2}}\exp(-|k|), \quad \tilde{g}(k) = \frac{1}{a}\sqrt{\frac{\pi}{2}}\exp(-a|k|).$$

Solutions to Exercises in Chapter 3

The Fourier transform of the convolution $F = f \otimes g$ is

$$\tilde{F}(k) = \sqrt{2\pi}\,\tilde{f}(k)\,\tilde{g}(k) = \pi\,\frac{\sqrt{2\pi}}{2a}\exp\bigl[-(a+1)|k|\bigr].$$

This is also a decreasing exponential function of $|k|$, so the convolution F is also a Lorentzian function. Using the scaling property and the same Fourier transform pair,

$$F(x) = \pi\,\frac{\sqrt{2\pi}}{2a}\sqrt{\frac{2}{\pi}}\,\frac{1}{a+1}\,\frac{1}{1+\frac{x^2}{(a+1)^2}}$$

$$= \frac{\pi}{a}\,\frac{a+1}{(a+1)^2 + x^2}.$$

Solution 3.22

In the convolution integral

$$F(x) = \int_{-\infty}^{\infty} du\,\chi(x-u)\,\chi(u),$$

the integrand is equal to unity in the overlapping section of the intervals $[-1, 1]$ and $[x-1, x+1]$, and zero elsewhere. When $0 < x < 2$ this overlap interval extends from $x-1$ to 1 and has length $2-x$, and when $-2 < x < 0$ it extends from -1 to $1+x$ and has length $2+x$. In other cases there is no overlapping region. The integral is equal to the length of the overlapping region, so

$$F(x) = \begin{cases} 2 - |x| & \text{for } |x| < 2, \\ 0 & \text{otherwise.} \end{cases}$$

The Fourier transform $\tilde{F}(k)$ of $F(x)$ is most conveniently calculated using the convolution theorem, using the Fourier transform of $\chi(x)$ quoted in Table 3.2:

$$\tilde{F}(k) = \sqrt{2\pi}\left[\sqrt{\frac{2}{\pi}}\,\frac{\sin k}{k}\right]^2 = \frac{2\sqrt{2}}{\sqrt{\pi}}\,\frac{\sin^2 k}{k^2}.$$

Solution 3.23

The Fourier transforms of the functions

$$f(x) = \frac{1}{1+x^2}, \quad g(x) = f'(x) = -\frac{2x}{(1+x^2)^2}$$

are

$$\tilde{f}(k) = \sqrt{\frac{\pi}{2}}\exp(-|k|), \quad \tilde{g}(k) = ik\,\tilde{f}(k) = i\sqrt{\frac{\pi}{2}}\,k\exp(-|k|).$$

Solution 3.24

The Fourier transform of $F(x) = x f(x)$ is

$$\tilde{F}(k) = \frac{1}{\sqrt{2\pi}}\int_{-\infty}^{\infty} dx\,x\exp(-ikx)\,f(x)$$

$$= \frac{i}{\sqrt{2\pi}}\int_{-\infty}^{\infty} dx\,\frac{\partial}{\partial k}\exp(-ikx)\,f(x)$$

$$= \frac{i}{\sqrt{2\pi}}\,\frac{d}{dk}\left[\int_{-\infty}^{\infty} dx\exp(-ikx)\,f(x)\right] = i\,\frac{d\tilde{f}(k)}{dk}.$$

Alternatively,

$$\frac{d\tilde{f}}{dk} = \frac{1}{\sqrt{2\pi}}\,\frac{d}{dk}\int_{-\infty}^{\infty} dx\exp(-ikx)\,f(x)$$

$$= \frac{-i}{\sqrt{2\pi}}\int_{-\infty}^{\infty} dx\,x\exp(-ikx)\,f(x) = -i\,\tilde{F}(k).$$

Solution 3.25

If $f(x)$ satisfies the differential equation $df/dx + xf = 0$ for all x, then the Fourier transform of the left-hand side is equal to zero for all k. The Fourier transform of df/dx is $ik\tilde{f}(k)$, where $\tilde{f}$ is the Fourier transform of f. The Fourier transform of $xf(x)$ was found in Exercise 3.24. Combining these, $\tilde{f}$ satisfies

$$ik\tilde{f} + i\frac{d\tilde{f}}{dk} = 0.$$

Dividing by i shows that f and $\tilde{f}$ satisfy the same differential equation.

The differential equation $df/dx + xf = 0$ may be rearranged to give

$$\int \frac{df}{f} + \int dx\, x = 0,$$

which may be evaluated to give

$$\ln f(x) = -\tfrac{1}{2}x^2 + \text{constant},$$

or (upon exponentiating this relation)

$$f(x) = C\exp(-x^2/2),$$

for some constant C. The Fourier transform of $f(x)$ satisfies the same differential equation, and must therefore be of the form $\tilde{f}(k) = D\exp(-k^2/2)$, for some constant D. From the general formula for the Fourier transform, we have

$$D = \tilde{f}(0) = \frac{1}{\sqrt{2\pi}}\int_{-\infty}^{\infty} dx\, f(x)$$
$$= \frac{C}{\sqrt{2\pi}}\int_{-\infty}^{\infty} dx\, \exp(-x^2/2)$$
$$= C,$$

so $D = C$. Choosing $C = 1$ gives our standard form for the Fourier transform of a Gaussian function.

Solution 3.26

When $f(x)$ is a periodic function, the choice of x_0 is arbitrary, but the degree of difficulty in finding the Fourier coefficients can depend upon making a good choice. We take $x_0 = 0$, $L = \pi$. Using the formula for Fourier coefficients (3.14), we obtain

$$c_n = \frac{1}{\pi}\int_0^{\pi} dx\, \exp(-2inx)\sin x$$
$$= \frac{1}{\pi}\int_0^{\pi} dx\, \exp(-2inx)\frac{1}{2i}[\exp(ix) - \exp(-ix)]$$
$$= \frac{1}{2\pi i}\int_0^{\pi} dx\, \left(\exp[-(2n-1)ix] - \exp[-(2n+1)ix]\right).$$

Now define integrals I_m (for integer m) by

$$I_m = \int_0^{\pi} dx\, \exp(-imx)$$
$$= \frac{i}{m}[\exp(-i\pi m) - 1]$$
$$= \frac{i[(-1)^m - 1]}{m}.$$

Note that $I_m = -2i/m$ when m is odd, and $I_m = 0$ when m is even. So

$$c_n = \frac{1}{2\pi i}(I_{2n-1} - I_{2n+1})$$
$$= \frac{1}{2\pi i}\left(\frac{-2i}{2n-1} - \frac{-2i}{2n+1}\right)$$
$$= \frac{2}{\pi(1-4n^2)}.$$

Solutions to Exercises in Chapter 3

If you repeat this exercise setting $x_0 = -\pi/2$, $L = \pi$, you will reach the same result, but the calculation will be more complicated.

Solution 3.27

We use (3.14), setting $L = 2$ and $x_0 = -1$. For $n \neq 0$,

$$c_n = \tfrac{1}{2} \int_{-1}^{1} dx\, x^2 \exp(-\pi i n x)$$

$$= -\frac{1}{2\pi i n} \left[x^2 \exp(-\pi i n x) \right]_{-1}^{1} + \frac{1}{2\pi i n} \int_{-1}^{1} dx\, 2x \exp(-\pi i n x)$$

$$= 0 + \frac{1}{\pi i n} \left[\frac{-1}{\pi i n} x \exp(-\pi i n x) \right]_{-1}^{1} + \frac{1}{(\pi i n)^2} \int_{-1}^{1} dx\, \exp(-\pi i n x)$$

$$= \frac{2(-1)^n}{\pi^2 n^2},$$

where integration by parts was used in the second and third equalities.

Also,

$$c_0 = \tfrac{1}{2} \int_{-1}^{1} dx\, x^2$$

$$= \tfrac{1}{3}.$$

Solution 3.28

The function $f(x)$ is shown in Figure 3.13.

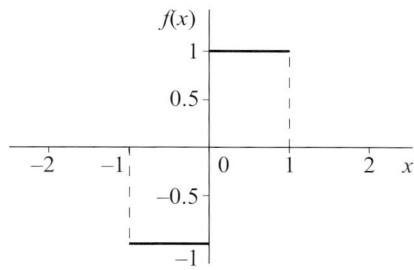

Figure 3.13 The function $f(x)$ defined in Exercise 3.28

If $\chi_a(x)$ is the characteristic function for the interval $[-a, a]$, and $\chi(x)$ is the characteristic function for the interval $[-1, 1]$, then the function $f(x)$ may be written

$$f(x) = \chi_{1/2}(x - \tfrac{1}{2}) - \chi_{1/2}(x + \tfrac{1}{2})$$

$$= \chi(2x - 1) - \chi(2x + 1).$$

The scaling rule shows that the Fourier transform of $\chi(2x)$ is

$$\sqrt{\frac{2}{\pi}} \sin(k/2)/k.$$

Applying the translation rule to each component, the Fourier transform of f is

$$\tilde{f}(k) = \sqrt{\frac{2}{\pi}} [\exp(-ik/2) - \exp(ik/2)] \frac{\sin(k/2)}{k}$$

$$= -i\sqrt{\frac{2}{\pi}} \frac{\sin^2(k/2)}{k/2}.$$

Solution 3.29

This is similar to Exercise 3.22, but it asks for the convolution of $\chi_{1/2}(x)$ with itself, rather than $\chi_1(x)$. The same reasoning shows that $F(x)$ is a scaled version of the result of that exercise. By inspection of the integral defining F, it is clear that $F = -1$ at $x = 0$, and F vanishes when $x \leq -1$ and when $x \geq 1$. The function $F(x)$ is therefore

$$F(x) = \begin{cases} -(1 - |x|) & \text{for } |x| < 1, \\ 0 & \text{for } |x| \geq 1. \end{cases}$$

The derivative of this function is equal to zero when $x < -1$ or when $x > 1$, minus one when $-1 < x < 0$, and unity when $0 < x < 1$, so $f(x) = dF/dx$ (except at $x = -1, 0, 1$, where F' is undefined).

The Fourier transform of F is obtained using the convolution theorem. The Fourier transform of $\chi_{1/2}(x) = \chi(2x)$ is $\tilde{\chi}_{1/2}(k) = \sqrt{2/\pi}\sin(k/2)/k$. The Fourier transform of F is then

$$\tilde{F}(k) = -\sqrt{2\pi}\left[\tilde{\chi}_{1/2}(k)\right]^2 = -\sqrt{2\pi}\,\frac{2}{\pi}\,\frac{\sin^2(k/2)}{k^2}.$$

Using (3.54), the Fourier transform of f is therefore

$$\tilde{f}(k) = ik\,\tilde{F}(k) = -i\sqrt{\frac{2}{\pi}}\,\frac{\sin^2(k/2)}{k/2},$$

in agreement with the result of the previous exercise.

Solution 3.30

We have

$$\tilde{F}(k_1, k_2) = \frac{1}{2\pi}\int_{-\infty}^{\infty} dx_1 \int_{-\infty}^{\infty} dx_2 \exp[-i(k_1 x_1 + k_2 x_2)]\, f(x_1)\, g(x_2)$$

$$= \frac{1}{\sqrt{2\pi}}\int_{-\infty}^{\infty} dx_1 \exp(-ik_1 x_1)\, f(x_1) \times \frac{1}{\sqrt{2\pi}}\int_{-\infty}^{\infty} dx_2 \exp(-ik_2 x_2)\, g(x_2)$$

$$= \tilde{f}(k_1)\,\tilde{g}(k_2).$$

Solution 3.31

Let $h = f \otimes g$ be the two-dimensional convolution, as defined in the text. The Fourier transform of h is

$$\tilde{h}(k_1, k_2) = \frac{1}{2\pi}\int_{-\infty}^{\infty} dx_1 \int_{-\infty}^{\infty} dx_2 \exp[-i(k_1 x_1 + k_2 x_2)]$$

$$\times \int_{-\infty}^{\infty} du_1 \int_{-\infty}^{\infty} du_2\, f(x_1 - u_1, x_2 - u_2)\, g(u_1, u_2)$$

$$= \frac{1}{2\pi}\int_{-\infty}^{\infty} du_1 \int_{-\infty}^{\infty} du_2\, g(u_1, u_2) \exp[-i(k_1 u_1 + k_2 u_2)]$$

$$\times \int_{-\infty}^{\infty} dx_1 \int_{-\infty}^{\infty} dx_2 \exp\bigl(-i[k_1(x_1 - u_1) + k_2(x_2 - u_2)]\bigr)$$

$$\times f(x_1 - u_1, x_2 - u_2)$$

$$= \frac{1}{2\pi}\int_{-\infty}^{\infty} du_1 \int_{-\infty}^{\infty} du_2 \exp[-i(k_1 u_1 + k_2 u_2)]\, g(u_1, u_2)$$

$$\times \int_{-\infty}^{\infty} dy_1 \int_{-\infty}^{\infty} dy_2 \exp[-i(k_1 y_1 + k_2 y_2)]\, f(y_1, y_2)$$

$$= 2\pi\, \tilde{f}(k_1, k_2)\, \tilde{g}(k_1, k_2).$$

The third step uses the change of variables $y_1 = x_1 - u_1$, $y_2 = x_2 - u_2$.

The generalisation to d dimensions contains the factor $(2\pi)^{d/2}$.

CHAPTER 4
Fourier methods in one dimension

4.1 Introduction

In Chapter 2, we solved the boundary- and initial-value problem for a vibrating string with fixed ends. The strategy employed in that solution consists of three main steps:

(i) separation of variables;
(ii) normal mode solutions (for given boundary conditions);
(iii) coefficients of normal modes (from given initial data).

In step (i), we replace the partial differential equation with a system of ordinary differential equations, which is then solved in step (ii) for the corresponding boundary values by separation of variables. The solution of the boundary- and initial-value problem for the partial differential equation can be expressed as a linear combination of the normal mode solutions derived in this way. The corresponding coefficients are calculated from the initial data in step (iii), using the orthogonality of normal mode solutions. Finally, all that remains to be done is to sum up the contributions of the normal mode solutions with these coefficients. For a finite string with fixed ends, the normal modes are expressed in terms of sine functions, and the expansion of the solution in terms of normal modes leads to a Fourier sine series. In this chapter, we demonstrate how this approach can be applied to solve other boundary- and initial-value problems.

We start in Section 4.2 by continuing the discussion of vibrating strings with fixed ends, and solve initial-value problems by means of Fourier sine series. The same method can be applied to solve the partial differential equation with other boundary conditions, where, for instance, the end of the string is allowed to move freely in the transverse direction, or is subjected to external forces. Examples of such boundary conditions are discussed in Section 4.3, where we show how the corresponding solutions of the wave equation are expressed in terms of Fourier series.

See Subsection 2.6.4 of Chapter 2.

In Section 4.4, we introduce a new partial differential equation that describes damped wave motion, for instance the damped vibrations of a string subject to air resistance. We demonstrate how the methods developed for the wave equation can be adapted to cope with this new situation.

Finally, in Section 4.5, we add a further complication by considering the case where an external force acts along the entire length of a string – like the wind blowing on a washing line. In particular, the phenomenon of resonance, where a small external driving force can lead to large-amplitude motion, is discussed in detail for the the case where the time-dependence of the external force is sinusoidal.

4.2 Initial-value problem for strings with fixed ends

Recapitulating the results of Chapter 2, the general solution $u(x,t)$ of the wave equation

$$u_{tt}(x,t) = c^2 \, u_{xx}(x,t), \tag{4.1}$$

with fixed boundary condition $u(0,t) = u(l,t) = 0$, can be expressed as an infinite sum of normal modes, giving

$$u(x,t) = \sum_{n=1}^{\infty} \sin\left(\frac{\pi n x}{l}\right) \left[A_n \cos\left(\frac{\pi n c t}{l}\right) + B_n \sin\left(\frac{\pi n c t}{l}\right) \right]. \tag{4.2}$$

The initial conditions $u(x,0) = a(x)$ and $u_t(x,0) = b(x)$ are expressed in terms of Fourier sine series as

$$a(x) = \sum_{n=1}^{\infty} A_n \sin\left(\frac{\pi n x}{l}\right), \quad 0 \le x \le l, \tag{4.3}$$

$$b(x) = \frac{\pi c}{l} \sum_{n=1}^{\infty} n B_n \sin\left(\frac{\pi n x}{l}\right), \quad 0 \le x \le l. \tag{4.4}$$

The coefficients A_n and B_n are obtained using the orthogonality relation

$$\frac{2}{l} \int_0^l dx \, \sin\left(\frac{\pi m x}{l}\right) \sin\left(\frac{\pi n x}{l}\right) = \delta_{mn}. \tag{4.5}$$

This gives the results

$$A_n = \frac{2}{l} \int_0^l dx \, a(x) \sin\left(\frac{\pi n x}{l}\right), \tag{4.6}$$

$$B_n = \frac{2}{\pi n c} \int_0^l dx \, b(x) \sin\left(\frac{\pi n x}{l}\right); \tag{4.7}$$

compare equations (2.62), (2.64) and (2.65) of Chapter 2.

In the following example, we consider the solution for an initial condition that simulates a plucked string.

Example 4.1

Consider a string of length l, fixed at its end points at $x = 0$ and $x = l$, which at time $t = 0$ is released from rest, so $b(x) = 0$, with the triangular initial shape

$$a(x) = \begin{cases} x & \text{for } 0 \le x \le l/2, \\ l - x & \text{for } l/2 < x \le l, \end{cases} \tag{4.8}$$

as shown in Figure 4.1. Calculate the corresponding solution $u(x,t)$ of the wave equation.

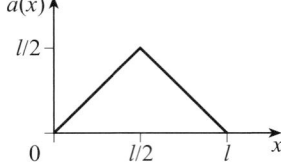

Figure 4.1 The triangular wave $a(x)$

4.2 Initial-value problem for strings with fixed ends

Solution

The solution is given by equation (4.2) with the appropriate values of the Fourier coefficients A_n and B_n. The string is released from rest, so $b(x) = 0$, and equation (4.7) gives $B_n = 0$ for all n. For the Fourier coefficients A_n, we obtain from equation (4.6)

$$A_n = \frac{2}{l} \int_0^{l/2} dx \, x \sin\left(\frac{\pi n x}{l}\right) + \frac{2}{l} \int_{l/2}^{l} dx \, (l-x) \sin\left(\frac{\pi n x}{l}\right). \tag{4.9}$$

Substituting $y = l - x$ in the second integral gives

$$\begin{aligned} \int_{l/2}^{l} dx \, (l-x) \sin\left(\frac{\pi n x}{l}\right) &= -\int_{l/2}^{0} dy \, y \sin\left(\frac{\pi n (l-y)}{l}\right) \\ &= \int_0^{l/2} dy \, y \sin\left(\pi n - \frac{\pi n y}{l}\right) \\ &= -(-1)^n \int_0^{l/2} dy \, y \sin\left(\frac{\pi n y}{l}\right). \end{aligned} \tag{4.10}$$

The last relation follows because

$$\begin{aligned} \sin(\pi n - x) &= \sin(\pi n) \cos x - \cos(\pi n) \sin x \\ &= -\cos(\pi n) \sin x \\ &= -(-1)^n \sin x \end{aligned}$$

for integer n. Thus the second integral in equation (4.9) equals $-(-1)^n$ times the first, and we obtain

$$A_n = [1 - (-1)^n] \frac{2}{l} \int_0^{l/2} dx \, x \sin\left(\frac{\pi n x}{l}\right). \tag{4.11}$$

In particular, this shows that all the even coefficients vanish; that is, $A_n = 0$ for $n = 2m$.

Symmetry arguments show that this is true for all initial profiles $a(x)$ which are symmetric about $x = l/2$, i.e. $a(x) = a(l-x)$.

We still need to calculate the integral in equation (4.11). You are asked to show in Exercise 4.1 that

$$\int_0^z dx \, x \sin(kx) = \frac{\sin(kz) - kz \cos(kz)}{k^2}. \tag{4.12}$$

With $k = \pi n / l$ and $z = l/2$, we obtain $kz = \pi n / 2$. For odd n, writing $n = 2m - 1$, this gives $\sin(kz) = (-1)^{m-1}$ and $\cos(kz) = 0$. Thus we obtain

$$A_{2m-1} = \frac{4l(-1)^{m-1}}{\pi^2 (2m-1)^2}, \quad A_{2m} = 0, \quad m = 1, 2, 3, \ldots. \tag{4.13}$$

Therefore

$$u(x,t) = \frac{4l}{\pi^2} \sum_{m=1}^{\infty} \frac{(-1)^{m-1}}{(2m-1)^2} \sin\left(\frac{\pi(2m-1)x}{l}\right) \cos\left(\frac{\pi(2m-1)ct}{l}\right) \tag{4.14}$$

is the appropriate solution for the given initial condition. ∎

Note that in this example, the time-dependent part involves only the cosine function. So, at time $t = 0$, each mode contributes with its maximal displacement, which makes sense if the string is released from rest.

Exercise 4.1

Verify equation (4.12), using integration by parts.

What does this solution look like, and how useful is it as an infinite series? It depends very much on the convergence properties of the series. Consider the truncated series

$$\frac{4l}{\pi^2} \sum_{m=1}^{M} \frac{(-1)^{m-1}}{(2m-1)^2} \sin\left(\frac{\pi(2m-1)x}{l}\right) \cos\left(\frac{\pi(2m-1)ct}{l}\right), \quad (4.15)$$

where we sum only the first M terms. If, for a reasonably small value of M, this gives a good approximation to the true solution with $M = \infty$, then this approach is of practical use, because the finite sum can easily be calculated on a computer. For the example at hand, the behaviour is shown in Figure 4.2, which shows snapshots of the form of the string as a function of time over a whole period $\tau = 2l/c$ of the motion. The snapshots are taken at times t which are multiples of $\tau/16$.

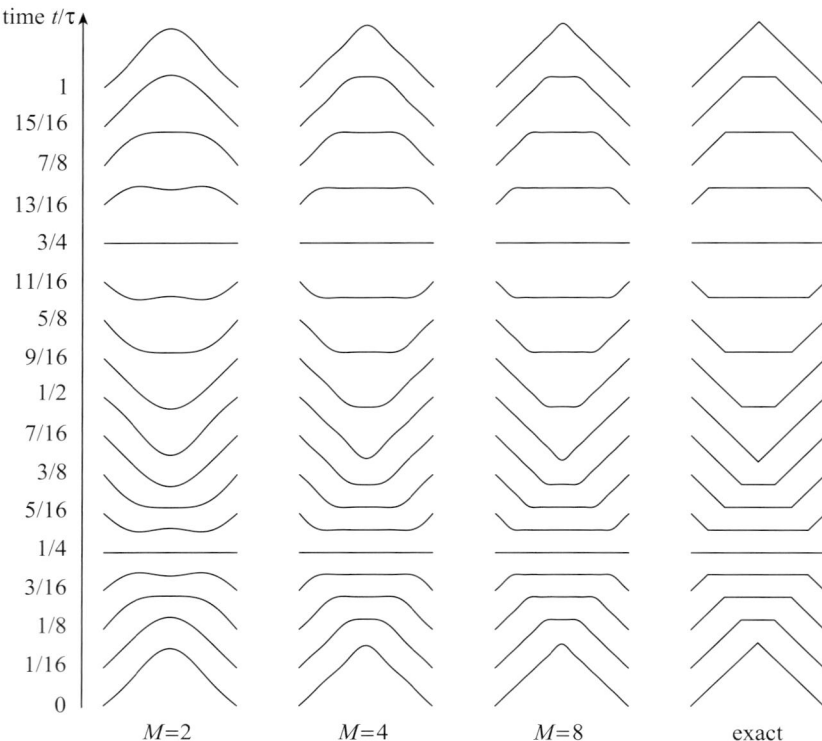

Figure 4.2 Snapshots of the exact solution (4.14) of the initial-value problem (extreme right) and three approximations by truncated Fourier series (4.15)

From Figure 4.2 it is clear, even for such small numbers of terms in the series (4.15), that we have obtained a reasonable approximation to the true solution, and the accuracy increases with increasing M. This is particularly apparent close to the 'kinks' in this solution, where the derivative changes discontinuously.

> Note that the exact solution shown in Figure 4.2 was, in fact, not obtained by means of the infinite series (4.14), but by the methods discussed in Chapter 2, using d'Alembert's solution and reflection at the boundary. According to this approach, the solution is given by
>
> $$u(x,t) = \tfrac{1}{2}\left[a(x-ct) + a(x+ct)\right], \quad (4.16)$$
>
> where $a(x)$, defined for all real x, is now the *odd periodic extension* of the initial function on $0 \leq x \leq l$ defined in equation (4.8), with period $L = 2l$, which means that it satisfies $a(-x) = -a(x)$ and $a(x+2l) = a(x)$ for all x.

4.2 Initial-value problem for strings with fixed ends

This odd periodic extension, depicted in Figure 4.3, agrees with the Fourier sine series deduced above.

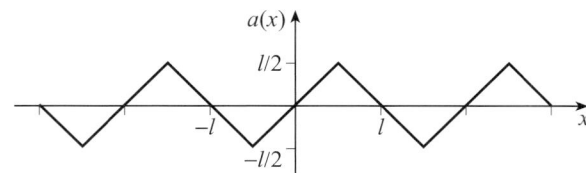

Figure 4.3 The odd periodic extension of the triangular wave $a(x)$

The Fourier sine series of a function defined on an interval $[0, l]$ is always an odd periodic function of period $L = 2l$.

Exercise 4.2

Consider the initial-value problem of a string released from rest, but now with an initial displacement

$$a(x) = \begin{cases} x/10 & \text{for } 0 \leq x \leq l/2, \\ (l-x)/10 & \text{for } l/2 < x \leq l. \end{cases}$$

(a) What is the solution in this case? [Hint: Use the results of Example 4.1.]

(b) With regard to the assumptions that entered the derivation of the wave equation, is there any difference between the solutions for these initial conditions and those of equation (4.8), with respect to their applicability to the motion of a real string?

We now consider an example where the string is initially at its equilibrium position, but with a given initial velocity. So the initial conditions are $a(x) = u(x,0) = 0$ and $b(x) = u_t(x,0)$. You may think of a string in equilibrium being given an initial 'kick' at time $t = 0$, described by a function $b(x)$.

Example 4.2

Consider a string of length l, fixed at its end points at $x = 0$ and $x = l$, which at time $t = 0$ is in its equilibrium position with an initial velocity

$$b(x) = \begin{cases} x & \text{for } 0 \leq x \leq l/2, \\ l - x & \text{for } l/2 < x \leq l, \end{cases} \qquad (4.17)$$

which is the same as the function shown in Figure 4.1. Calculate the corresponding solution $u(x,t)$ of the wave equation.

Solution

The Fourier coefficients B_n are given by equation (4.7). Since the function $b(x)$ is identical to the function $a(x)$ of equation (4.8), the coefficients B_n must be given by $B_n = lA_n/(\pi n c)$ with A_n from equations (4.13). Hence, for $m = 1, 2, 3, \ldots$, we obtain

$$B_{2m-1} = \frac{l}{\pi(2m-1)c} \frac{4l(-1)^{m-1}}{\pi^2(2m-1)^2} = \frac{4l^2(-1)^{m-1}}{\pi^3(2m-1)^3 c}, \quad B_{2m} = 0. \qquad (4.18)$$

The string is initially in its equilibrium position, so $a(x) = u(x,0) = 0$, and the coefficients A_n vanish for all n. So

$$u(x,t) = \sum_{m=1}^{\infty} \frac{4l^2(-1)^{m-1}}{\pi^3(2m-1)^3 c} \sin\left(\frac{\pi(2m-1)x}{l}\right) \sin\left(\frac{\pi(2m-1)ct}{l}\right) \qquad (4.19)$$

is the appropriate solution of the wave equation. ∎

In Figure 4.4, the functions obtained by truncating this series to $m \leq M$ are shown for $M = 2$, $M = 4$ and $M = 8$. Comparing this with Figure 4.2, it is apparent that this series converges faster; in fact, it is hard to spot any difference between the curves of Figure 4.4 for different M.

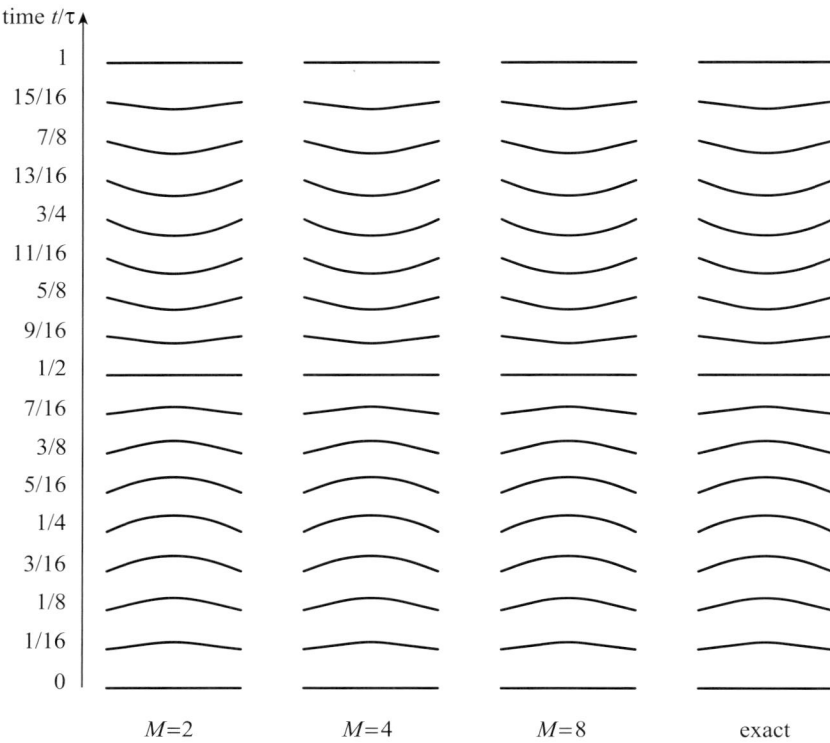

Figure 4.4 Snapshots of the exact solution of the initial-value problem (extreme right) and approximations obtained by truncating the Fourier series after M terms

The reason for the faster convergence is that the coefficients B_n are proportional to n^{-3}, whereas the coefficients A_n of equations (4.13) are proportional to n^{-2}. This additional factor of n^{-1} comes from taking the derivative $u_t(x, 0)$ in the initial condition.

Exercise 4.3

Verify by explicit calculation that the solution $u(x, t)$ given in equation (4.19) satisfies the boundary conditions $u(0, t) = u(l, t) = 0$ and the initial conditions $u(x, 0) = 0$ and $u_t(x, 0) = b(x)$, with $b(x)$ as given in equation (4.17).

In the following exercise, you are asked to deal with a case where neither the initial displacement $a(x) = u(x, 0)$ nor the velocity $b(x) = u_t(x, 0)$ vanishes.

Exercise 4.4

Consider the motion of a string of length l, fixed at $x = 0$ and $x = l$, with initial conditions

$$a(x) = \begin{cases} x/10 & \text{for } 0 \leq x \leq l/2, \\ (l-x)/10 & \text{for } l/2 < x \leq l, \end{cases} \quad b(x) = \begin{cases} -x/20 & \text{for } 0 \leq x \leq l/2, \\ -(l-x)/20 & \text{for } l/2 < x \leq l. \end{cases}$$

Determine the solution of the initial-value problem.

4.3 Strings with moving ends

So far, we have concentrated on the example of a vibrating string which is fixed at its end points, such as a guitar string. However, there are also physical systems which correspond to different boundary conditions. For instance, an organ pipe or a wind instrument is usually open at one end. For a vibrating string, an analogy of this situation is a free end, say at $x = 0$, which means that $u(0, t)$ is not constrained to be zero. However, we still assume that the motion is transverse, so the end is always at $x = 0$. As sketched in Figure 4.5, this may be realised by attaching a ring to the end of the string such that the ring slides without friction along a rod fixed at position $x = 0$. Later, we shall also consider the case where friction is present at the end, or where the end is attached to the equilibrium position $u(0, t) = 0$ by a spring.

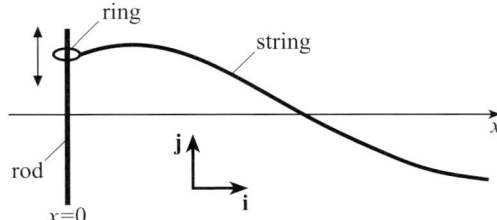

Figure 4.5 A string with a ring at its end at $x = 0$

4.3.1 A free end

Consider a string of length l, extending from $x = 0$ to $x = l$. Suppose that it performs transverse motion, with a free end, as depicted in Figure 4.5, at $x = 0$. Note that we do not allow longitudinal motion – a marked point at position $x\mathbf{i}$ along the equilibrium string can move only transversely (perpendicular to the string direction), so at time t, its position is $x\mathbf{i} + u(x, t)\mathbf{j}$.

The free end at $x = 0$ is allowed to move, so $u(0, t) \neq 0$ in general. In order to find the appropriate boundary condition for this case, let us consider the forces acting on the ring; see Figure 4.6. We assume that the ring is small, so we can treat it as a particle moving according to Newton's second law. The ring has a small mass, M, and eventually we shall take the limit as $M \to 0$. We are not interested in the effect of gravity; you may think of the motion as taking place in a horizontal plane, so the gravitational force on the ring does not affect the motion. As in Chapter 1, we assume that we can neglect the change in length of the string, and hence the change in tension.

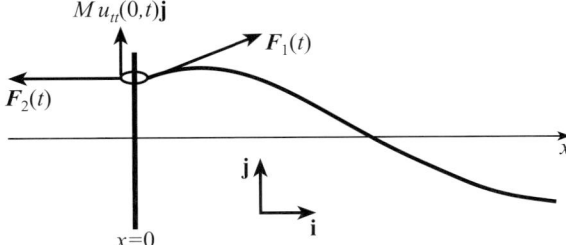

Figure 4.6 Forces acting on a ring of mass M attached to the end of a string

At time t, the positive string tension $T > 0$ yields a force $\boldsymbol{F}_1(t)$ on the ring. As discussed in Chapter 1, the force is tangential to the string, and its magnitude is equal to the tension T, which, within the same assumptions and approximations as in Chapter 1, is constant for small deviations from equilibrium. For small $u_x(0,t)$, the force is

$$\boldsymbol{F}_1(t) = T[\mathbf{i} + u_x(0,t)\mathbf{j}], \tag{4.20}$$

which follows from equation (1.16) on page 21 with $\phi_1 = u_x(0,t)$, noting that there is a change of sign due to the definitions of the forces.

The force $\boldsymbol{F}_1$ is partly compensated by a force $\boldsymbol{F}_2$, exerted by the solid rod on the ring to keep it at $x = 0$; see Figure 4.6. As the solid rod resists motion only in the $\mathbf{i}$ direction, the component of $\boldsymbol{F}_2$ in the $\mathbf{j}$ direction vanishes, so $\boldsymbol{F}_2(t) = -F_2(t)\mathbf{i}$.

Newton's second law for the motion of the ring gives

$$M\, u_{tt}(0,t)\,\mathbf{j} = \boldsymbol{F}_1(t) + \boldsymbol{F}_2(t) = [T - F_2(t)]\,\mathbf{i} + T\, u_x(0,t)\,\mathbf{j}$$
$$= T\, u_x(0,t)\,\mathbf{j}, \tag{4.21}$$

because the component of the resultant force in the $\mathbf{i}$ direction must vanish, so $F_2(t) = T$ within our approximation. The boundary condition for the string with the ring of mass M attached to its end is thus

$$M\, u_{tt}(0,t) = T\, u_x(0,t). \tag{4.22}$$

For vanishing mass, $M \to 0$, we must have $u_x(0,t) = 0$, as otherwise the acceleration $u_{tt}(0,t)$ would become arbitrarily large, and the end of the string would move infinitely fast.

We hence obtain the boundary condition for a free end

Boundary condition for a free end
$$u_x(0,t) = 0, \tag{4.23}$$

corresponding to the limiting case of a massless ring. This boundary condition means that at $x = 0$ the string is horizontal, as depicted in Figure 4.7.

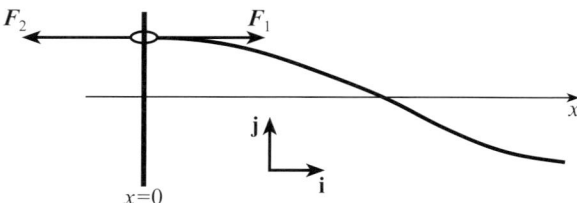

Figure 4.7 A string attached to a massless ring at $x = 0$

The boundary condition (4.23) is an example of a *Neumann condition*, named after Carl Gottfried Neumann (1832–1925), where the value of the partial derivative of the function u is specified on the boundary.

4.3.2 Damped and sprung ends

In the above example of a string with free ends, the ring at the end of the string was supposed to move freely, which means without any friction, along the transverse rod. In any real experiment, some friction will be present, which means that there will be a force that damps the motion of the end along the rod; see Figure 4.8.

4.3 Strings with moving ends

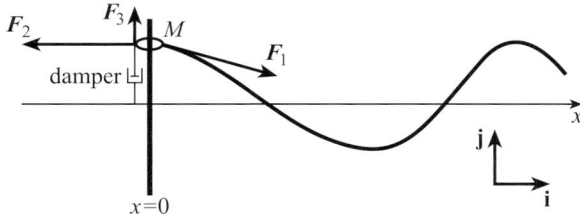

Figure 4.8 A string with a damped end at $x = 0$. At the instant depicted here, the ring is moving downwards, and the damping force acts upwards.

Now, in addition to the forces $\boldsymbol{F}_1 = T(\mathbf{i} + u_x(0,t)\mathbf{j})$ and $\boldsymbol{F}_2 = -T\mathbf{i}$, we have an additional force $\boldsymbol{F}_3$ which acts in the $\mathbf{j}$ direction. For the damped end, we consider a force $\boldsymbol{F}_3$ which is proportional to velocity in magnitude and opposed in direction, so it tends to slow down any motion. This means that

$$\boldsymbol{F}_3(t) = -r\, u_t(0,t)\,\mathbf{j}, \tag{4.24}$$

where r is called the *damping constant*. For small velocities, this simple force provides a good description of damping in many physical systems, for instance for air resistance.

Again, the component of the resultant force $\boldsymbol{F} = \boldsymbol{F}_1 + \boldsymbol{F}_2 + \boldsymbol{F}_3$ in the $\mathbf{i}$ direction vanishes, so

$$\boldsymbol{F}(t) = [T\, u_x(0,t) - r\, u_t(0,t)]\,\mathbf{j}. \tag{4.25}$$

The equation of motion of the ring in the $\mathbf{j}$ direction is then given by Newton's second law as

$$M\, u_{tt}(0,t) = T\, u_x(0,t) - r\, u_t(0,t). \tag{4.26}$$

In the limit as $M \to 0$, we obtain the boundary condition for a damped end

Boundary condition for a damped end
$$u_x(0,t) - R\, u_t(0,t) = 0, \tag{4.27}$$

where $R = r/T$.

Another interesting case occurs if the end is connected to a spring such that the spring force acts on the end towards an equilibrium position, as depicted in Figure 4.9, which we refer to as a 'sprung end'.

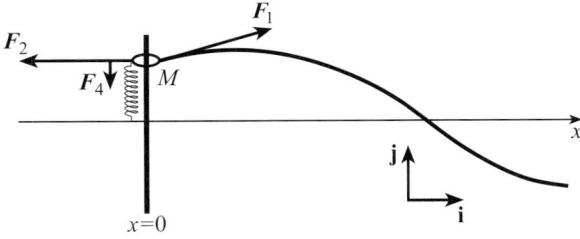

Figure 4.9 A string with a sprung end at $x = 0$

For a sprung end, characterised by a spring constant k and equilibrium position at $u = 0$, the additional force is

$$\boldsymbol{F}_4 = -k\, u(0,t)\,\mathbf{j}. \tag{4.28}$$

You are asked in Exercise 4.5 to verify that this leads to the boundary condition for a sprung end

Boundary condition for a sprung end
$$u_x(0,t) - \kappa\, u(0,t) = 0 \tag{4.29}$$

κ is the Greek letter kappa.

with $\kappa = k/T$.

Note that the letter k occurs with two different meanings, as a spring constant and as a wave number in the solution. However, this notation is conventional, and you will find it in most textbooks. In what follows, we try to avoid any confusion by using $\kappa = k/T$ to characterise a sprung end.

Exercise 4.5

Show that the boundary condition for a sprung end is given by equation (4.29).

If we consider damped or sprung ends at the other end of the string, we obtain the same boundary conditions, apart from a change in sign of the term $u_x(0,t)$. This arises because, at $x = l$, the tension force $\boldsymbol{F}_1$, which is tangential to the string, has a *negative* component in the $\mathbf{i}$ direction, so $\boldsymbol{F}_1(t) = T(-\mathbf{i} - u_x(l,t)\mathbf{j}) = -T(\mathbf{i} + u_x(l,t)\mathbf{j})$. Correspondingly, the force $\boldsymbol{F}_2$ that keeps the end at $x = l$ is then $\boldsymbol{F}_2(t) = T\mathbf{i}$.

This is equation (1.16) with $\phi_1 = u_x(l,t)$.

For a damped and sprung end at $x = l$, the appropriate boundary conditions are

$$u_x(l,t) + R\,u_t(l,t) = 0 \tag{4.30}$$

and

$$u_x(l,t) + \kappa\,u(l,t) = 0, \tag{4.31}$$

respectively. You are asked to verify the latter equation in the following exercise.

Exercise 4.6

Consider a string of tension T with a mass M attached to its end. The end is held at $x = l$ by a spring with spring constant κT and equilibrium position $u(l,t) = 0$.

(a) Use Newton's second law to derive the equation of motion of the mass M, assuming that displacements are small, and that gravity can be neglected.

(b) Show that as $M \to 0$, the condition (4.31) is recovered.

The boundary conditions considered here are such that they lead to linear equations in the function u and its partial derivatives at the ends; see equations (4.23), (4.27) and (4.29)–(4.31). This property makes it possible to solve the wave equation with these boundary conditions along the same lines as for the case of fixed ends.

4.3.3 Normal modes for strings with moving ends

We now follow the same approach as before to solve the wave equation with one or two moving ends of the type introduced above. We first derive the normal mode solution for the case where one end is free and the other end is fixed.

One fixed and one free end

For the string with a free end at $x = 0$ and a fixed end at $x = l$, the boundary conditions are

$$u_x(0,t) = u(l,t) = 0. \tag{4.32}$$

4.3 Strings with moving ends

The wave equation itself is unchanged, so separation of variables $u(x,t) = f(x)g(t)$ is applicable, and the resulting ordinary differential equations are the same as for fixed boundary conditions:

$$f''(x) - C f(x) = 0, \tag{4.33}$$
$$g''(t) - c^2 C g(t) = 0; \tag{4.34}$$

compare equations (2.50) and (2.51).

In the second step, we have to solve the ordinary differential equations with the new boundary conditions, i.e. $f'(0) = f(l) = 0$. As in the case of fixed ends, only negative values of the separation constant $C = -k^2 < 0$ lead to non-trivial solutions. This can be shown in the same way as in Subsection 2.6.1; compare Exercise 2.23. The general solution for $f(x)$ is

$$f(x) = a\cos(kx) + b\sin(kx), \quad a, b \in \mathbb{R}, \tag{4.35}$$

as derived in equation (2.52) for the string with fixed ends. So far, everything is exactly as for fixed ends, but the boundary condition at $x = 0$ now yields $f'(0) = bk = 0$, which implies $b = 0$, so $f(x) = a\cos(kx)$. At $x = l$, we obtain

$$f(l) = a\cos(kl) = 0, \tag{4.36}$$

so we arrive at the condition $\cos(kl) = 0$ for the possible values of k leading to non-trivial solutions. This means that the eigenvalues are given by $k_n l = n\pi - \pi/2 = \pi(2n-1)/2$ with integer n, so $k_n = \pi(2n-1)/(2l)$. The corresponding solutions or eigenmodes are

$$f_n(x) = a_n \cos\left(\frac{\pi(2n-1)x}{2l}\right), \quad n = 1, 2, 3, \ldots. \tag{4.37}$$

We need to consider only positive values of n. This can be seen by considering $n = 1 - m$, which gives

$$\cos[(2(1-m) - 1)\pi x/(2l)] = \cos[-(2m-1)\pi x/(2l)]$$
$$= \cos[(2m-1)\pi x/(2l)]$$

as the cosine function is symmetric about 0. Hence $f_n(x) = f_{1-n}(x)$ if we choose $a_n = a_{1-n}$.

The differential equation (4.34) for the function $g(t)$ is the same as equation (2.51), so its solution is

$$g_n(t) = A_n \cos(k_n ct) + B_n \sin(k_n ct); \tag{4.38}$$

compare equation (2.57). Hence we have derived the normal modes of the string with one free end

Normal modes of the string with one free end

$$u_n(x,t) = \cos\left(\frac{\pi(2n-1)x}{2l}\right)$$
$$\times \left[A_n \cos\left(\frac{\pi(2n-1)ct}{2l}\right) + B_n \sin\left(\frac{\pi(2n-1)ct}{2l}\right)\right], \tag{4.39}$$

with arbitrary real constants A_n and B_n, for $n = 1, 2, 3, \ldots$. The difference from the case of two fixed ends considered in Chapter 2 is the cosine function for the position-dependent part in place of the sine function, and the different values of k_n. The presence of a cosine function means that the end point at $x = 0$, rather than being a node, now is a point where the string moves the furthest from its equilibrium position during a normal mode motion. This may seem surprising, but indeed it is a direct consequence of the boundary conditions, as the equation $f'(0) = 0$ means that f must be extremal at $x = 0$.

The general solution of the boundary-value problem is, as before, a linear combination of all normal mode solutions:

$$u(x,t) = \sum_{n=1}^{\infty} \cos\left(\frac{\pi(2n-1)x}{2l}\right)$$
$$\times \left[A_n \cos\left(\frac{\pi(2n-1)ct}{2l}\right) + B_n \sin\left(\frac{\pi(2n-1)ct}{2l}\right)\right]. \quad (4.40)$$

The third step, the computation of the coefficients A_n and B_n from given initial conditions $u_x(x,0) = a(x)$ and $u_t(x,0) = b(x)$, is deferred to Exercise 4.21 at the end of this chapter.

Two free ends: Fourier cosine series

For two free ends, the boundary conditions are

$$u_x(0,t) = u_x(l,t) = 0. \quad (4.41)$$

Again, the wave equation is the same as before. With $u(x,t) = f(x)g(t)$, we again obtain the ordinary differential equations (4.33) and (4.34), with boundary conditions

$$f'(0) = f'(l) = 0. \quad (4.42)$$

There is one important change compared to the previous cases. For the case of two free ends, it is no longer true that only separation constants $C < 0$ lead to non-trivial solutions. Here, we also need to take into account the case $C = 0$, which leads to a solution

$$u_0(x,t) = \tfrac{1}{2}(A_0 + B_0 t); \quad (4.43)$$

The factor $\tfrac{1}{2}$ is introduced for later convenience.

compare equations (2.50) and (2.51) and Exercise 2.23 in Chapter 2. The interpretation of this solution is that the string is allowed to move with a constant velocity $B_0/2$ up (or down) the end rods, starting from a constant equilibrium position $A_0/2$ at $t = 0$. With one or two fixed ends, this is not possible. Although we may not be interested in this solution, we need to take it into account in order to obtain solutions for any given initial conditions.

Consider now solutions with $C = -k^2 < 0$. The general solution for $f(x)$ is once more given by equation (4.35). Inserting the boundary conditions (4.42) yields

$$f'(0) = bk = 0, \quad (4.44)$$

which implies $b = 0$, and

$$f'(l) = -ak\sin(kl) = 0, \quad (4.45)$$

so we arrive at the condition $\sin(kl) = 0$. This is the same condition as for the string with two fixed ends, thus $k_n = \pi n/l$. The corresponding eigenmodes are

$$f_n(x) = a_n \cos\left(\frac{\pi n x}{l}\right), \quad n = 1, 2, 3, \ldots. \quad (4.46)$$

Because the cosine function is even, positive and negative values of n give the same function.

Once again the corresponding solutions for $g(t)$ are given by equation (4.38). We have now derived the normal modes of the string with two free ends

Normal modes of the string with two free ends

$$u_n(x,t) = \cos\left(\frac{\pi n x}{l}\right)\left[A_n \cos\left(\frac{\pi n c t}{l}\right) + B_n \sin\left(\frac{\pi n c t}{l}\right)\right], \quad (4.47)$$

with arbitrary real constants A_n and B_n, for $n = 1, 2, 3, \ldots$. Together with the solution u_0 given in equation (4.43), this forms a complete set in the sense that any solution can be expressed as a linear combination of these.

4.3 Strings with moving ends

The general solution of the boundary-value problem is now expressed in terms of the solution (4.43) and the normal modes of equation (4.47), giving

$$u(x,t) = \tfrac{1}{2}(A_0 + B_0 t)$$
$$+ \sum_{n=1}^{\infty} \cos\left(\frac{\pi n x}{l}\right)\left[A_n \cos\left(\frac{\pi n c t}{l}\right) + B_n \sin\left(\frac{\pi n c t}{l}\right)\right]. \quad (4.48)$$

In the third and final step of our procedure, we need to compute the coefficients A_n and B_n, for $n = 0, 1, 2, \ldots$, from given initial displacement $u(x,0) = a(x)$ and initial velocity $u_t(x,0) = b(x)$. We do this by following the same steps as for the string with fixed ends. The difference is that we now deal with cosine rather than sine functions of the position coordinate x. We thus expect a result expressed in terms of a Fourier series containing only cosine functions.

Inserting the initial conditions in equation (4.48) gives

Fourier cosine series

$$a(x) = \tfrac{1}{2} A_0 + \sum_{n=1}^{\infty} A_n \cos\left(\frac{\pi n x}{l}\right) \quad (4.49)$$

and

$$b(x) = \tfrac{1}{2} B_0 + \frac{\pi c}{l} \sum_{n=1}^{\infty} n B_n \cos\left(\frac{\pi n x}{l}\right). \quad (4.50)$$

Apart from replacing the sine functions by cosine functions, these have forms similar to the equations for the string with fixed ends; compare equations (4.3) and (4.4). The coefficients A_n and B_n are thus essentially the Fourier coefficients of a *Fourier cosine series*.

As for the sine functions, there exists an orthogonality relation

$$\frac{2}{l}\int_0^l dx \cos\left(\frac{\pi m x}{l}\right)\cos\left(\frac{\pi n x}{l}\right) = \delta_{mn}, \quad m,n \geq 1. \quad (4.51)$$

In Exercise 4.7, you are asked to verify this relation and to derive the Fourier coefficients of the Fourier cosine series

Fourier coefficients of the Fourier cosine series

$$A_n = \frac{2}{l}\int_0^l dx\, a(x) \cos\left(\frac{\pi n x}{l}\right), \quad n \geq 0. \quad (4.52)$$

Exercise 4.7

(a) Show that the orthogonality relation (4.51) holds. [Hint: Use the relation $2\cos\alpha\cos\beta = \cos(\alpha-\beta) + \cos(\alpha+\beta)$, and consider separately the cases $m \neq n$ and $m = n$ ($m, n \geq 1$).]

(b) Show that equation (4.49) and this orthogonality relation imply formula (4.52) for the Fourier coefficients.

The coefficients A_n and B_n of equations (4.49) and (4.50), for $n = 1, 2, 3, \ldots$, are given by

$$A_n = \frac{2}{l}\int_0^l dx\, a(x) \cos\left(\frac{\pi n x}{l}\right), \quad B_n = \frac{2}{\pi n c}\int_0^l dx\, b(x) \cos\left(\frac{\pi n x}{l}\right), \quad (4.53)$$

and for $n = 0$ we obtain

$$A_0 = \frac{2}{l}\int_0^l dx\, a(x), \quad B_0 = \frac{2}{l}\int_0^l dx\, b(x), \quad (4.54)$$

which means that the coefficients $A_0/2$ and $B_0/2$ that enter in equation (4.48) are the mean values of the functions $a(x)$ and $b(x)$ on the interval $[0, l]$, respectively.

As an explicit example, we consider the case of a linear initial displacement given by

$$a(x) = \frac{l}{2} - x, \quad 0 \leq x \leq l; \tag{4.55}$$

this function is shown in Figure 4.10.

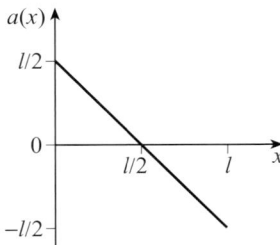

Figure 4.10 The function $a(x)$ of equation (4.55)

Because cosine is an *even* function, the corresponding Fourier cosine series of $a(x)$ is an *even periodic extension* of the function $a(x)$, which is extended to a function on the whole real line such that $a(x) = a(-x)$ with period $L = 2l$; this function is shown in Figure 4.11.

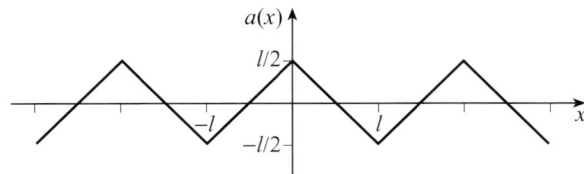

Figure 4.11 An even periodic extension of the function $a(x)$ of equation (4.55)

In fact, this function is the same as the triangular wave considered in Section 4.2, apart from a shift of $l/2$ in the x direction; compare Figure 4.3. So it is to be expected that the resulting Fourier coefficients are intimately related. This is indeed the case, as you are asked to show in Exercise 4.8, where the corresponding initial-value problem for the string with free ends is to be solved. A second example is considered in Exercise 4.20 at the end of this chapter.

Exercise 4.8

(a) Calculate the integral $\int_0^z dx\, x \cos(kx)$ using integration by parts.

(b) Calculate the Fourier cosine coefficients A_n of the function $a(x)$ of equation (4.55) from equations (4.52).

(c) Compare the result of part (b) with the coefficients A_n of the corresponding Fourier sine series given in equations (4.13). How are the coefficients related?

(d) Assume that $b(x) = 0$. Calculate the solution of the corresponding initial-value problem for the string with two free ends at $x = 0$ and $x = l$, and compare the result to equation (4.14).

4.3 Strings with moving ends

Other boundary conditions

In principle, we can calculate the normal modes for other linear boundary conditions analogously. However, for the sprung or damped ends the boundary condition involves two terms, so the result will be more complicated. An example is considered in Exercise 4.9.

Exercise 4.9

Consider a string of length l with constant tension $T > 0$ with a free end at $x = 0$ and a sprung end at $x = l$. The sprung end is characterised by a spring constant κT.

(a) Derive the condition $\tan(kl) = \kappa/k$ for the wave numbers k of the normal modes.

(b) Assume that $\kappa l = 1$. Sketch the two functions $\tan z$ and $\kappa l/z = 1/z$ for $z > 0$, where $z = kl$. What can you say about the solutions z_n of $\tan z = 1/z$?

(c) Starting from the condition derived in part (a), determine the limiting values of k when the spring at the end is very weak, that is, in the limit as $\kappa \to 0$.

(d) Similarly, what are the limiting values for k for a very stiff spring at the end, that is, for $\kappa \to \infty$?

(e) If you compare this with the wave numbers k_n for finite κ, what would you conclude about the effect of the sprung end on the corresponding normal modes? Is it more pronounced for low- or for high-frequency modes?

The above discussion shows how to solve the initial-value problem for strings with free boundary conditions within the framework of Fourier series. The damped and sprung boundaries are more involved, because the values of k_n are not simply proportional to n and, as evinced by the example of Exercise 4.9, cannot be simply evaluated. This means that the resulting expression will not be a Fourier series in the usual sense. However, it is possible to generalise the notion of Fourier series so that an analogous treatment extends to this case. We shall encounter generalised Fourier series again later in the course.

4.3.4 Inhomogeneous boundary conditions

The boundary conditions considered so far have the form

$$\alpha\, u(x_{\mathrm{b}}, t) + \beta\, u_t(x_{\mathrm{b}}, t) + \gamma\, u_x(x_{\mathrm{b}}, t) = 0$$

at the boundary point x_{b}, with real constants α, β and γ. These are linear homogeneous boundary conditions, which involve the function u and its first partial derivatives. The word *linear* refers to the fact the u and its derivatives enter linearly, and *homogeneous* means that all single terms depend on the dependent variable u or its derivatives, such that they vanish for $u = 0$. We call boundary conditions *inhomogeneous* if that is not the case.

As an example, consider a string with free ends. We grip the end at $x = l$ and move it up and down, according to some given function $p(t)$; so the boundary condition at $x = l$ becomes $u(l, t) = p(t)$. For example, you might think of a musical wind instrument, where the motion of the reed excites a wave motion of the air in the instrument; or a loudspeaker, where the membrane undergoes forced vibrations which are produced by moving one point of the membrane.

Two-dimensional wave motion will be considered in Chapter 5.

Let us consider a specific example. A string of length l is fixed at $x = 0$. The other end, at $x = l$, is forced to move sinusoidally with angular frequency ω_{ex}, so the boundary conditions are

The subscript 'ex' stands for 'external'.

$$u(0, t) = 0, \quad u(l, t) = \sin(\omega_{\text{ex}} t). \tag{4.56}$$

We consider the case where the string is initially at rest in the equilibrium position, so the initial conditions are

$$u(x, 0) = u_t(x, 0) = 0, \quad 0 \leq x < l. \tag{4.57}$$

In this example, the end of the string at $x = l$ performs a harmonic oscillation with angular frequency ω_{ex}, an external driving frequency. We assume that ω_{ex} does not coincide with one of the eigenfrequencies ω_n of the system.

General solution

It can be shown that the general solution for the inhomogeneous boundary condition can be obtained as the sum of a particular solution (or particular integral) $u^{(\text{par})}(x, t)$ of the wave equation that satisfies the inhomogeneous boundary conditions, and the complementary function, which is the general solution $u^{(\text{hom})}(x, t)$ of the associated homogeneous boundary problem, in our case $u(l, t) = 0$. So

This is similar to the solution of an inhomogeneous ordinary differential equation in terms of a particular integral and the complementary function; see Block 0, Section 1.3.

$$u(x, t) = u^{(\text{par})}(x, t) + u^{(\text{hom})}(x, t), \tag{4.58}$$

where the second term has to satisfy the wave equation with homogeneous boundary conditions $u^{(\text{hom})}(0, t) = u^{(\text{hom})}(l, t) = 0$, which means we are dealing with fixed ends. According to equation (4.2), the general solution of the homogeneous case is

$$u^{(\text{hom})}(x, t) = \sum_{n=1}^{\infty} \sin(k_n x)[A_n \cos(\omega_n t) + B_n \sin(\omega_n t)] \tag{4.59}$$

with $\omega_n = c k_n = \pi n c / l$.

We now proceed as follows. First, we find a particular integral $u^{(\text{par})}(x, t)$ that obeys the inhomogeneous boundary conditions. Then we need to solve an initial-value problem for the homogeneous system, where the initial conditions are obtained from equation (4.57) after inserting the chosen particular integral. This determines the Fourier coefficients A_n and B_n in equation (4.59), and hence the homogeneous part $u^{(\text{hom})}(x, t)$. The desired solution is then obtained by adding these two contributions according to equation (4.58).

A particular solution

A particular solution of the inhomogeneous boundary problem is a function $u^{(\text{par})}(x, t)$ that satisfies the wave equation $u_{tt}^{(\text{par})} = c^2 u_{xx}^{(\text{par})}$ and the boundary conditions

$$u^{(\text{par})}(0, t) = 0, \quad u^{(\text{par})}(l, t) = \sin(\omega_{\text{ex}} t). \tag{4.60}$$

We try

$$u^{(\text{par})}(x, t) = U(x) \sin(\omega_{\text{ex}} t), \tag{4.61}$$

which corresponds to a string that follows the externally imposed oscillatory motion and has a shape given by a function $U(x)$. Inserting this into the wave equation gives

4.3 Strings with moving ends

$$U''(x) + \frac{\omega_{\text{ex}}^2}{c^2} U(x) = 0, \tag{4.62}$$

and the boundary conditions (4.60) imply $U(0) = 0$ and $U(l) = 1$. The general solution of the differential equation (4.62) is $U(x) = a\cos(k_{\text{ex}}x) + b\sin(k_{\text{ex}}x)$ with $k_{\text{ex}} = \omega_{\text{ex}}/c$. The boundary conditions require $a = 0$ and $1 = U(l) = b\sin(k_{\text{ex}}l)$, so $b = 1/\sin(k_{\text{ex}}l)$. Therefore a suitable particular solution is

$$u^{(\text{par})}(x,t) = \frac{\sin(k_{\text{ex}}x)}{\sin(k_{\text{ex}}l)} \sin(\omega_{\text{ex}}t). \tag{4.63}$$

Note that $k_{\text{ex}}l \neq \pi n$ according to our assumption that $\omega_{\text{ex}} \neq \omega_n$. The general solution for the boundary conditions (4.56) is thus

$$u(x,t) = \frac{\sin(k_{\text{ex}}x)}{\sin(k_{\text{ex}}l)} \sin(\omega_{\text{ex}}t) + \sum_{n=1}^{\infty} \sin(k_n x)[A_n \cos(\omega_n t) + B_n \sin(\omega_n t)]. \tag{4.64}$$

Initial conditions

We now impose the initial conditions (4.57), which give

$$0 = u(x,0) = u^{(\text{par})}(x,0) + u^{(\text{hom})}(x,0), \tag{4.65}$$
$$0 = u_t(x,0) = u_t^{(\text{par})}(x,0) + u_t^{(\text{hom})}(x,0). \tag{4.66}$$

Inserting the terms for the particular solution (4.63) yields

$$u^{(\text{hom})}(x,0) = 0, \quad u_t^{(\text{hom})}(x,0) = -\frac{c k_{\text{ex}} \sin(k_{\text{ex}}x)}{\sin(k_{\text{ex}}l)}, \tag{4.67}$$

which are initial conditions for the homogeneous solution $u^{(\text{hom})}(x,0)$. So this is now an initial-value problem for the string with fixed ends, which we know how to solve.

Calculation of Fourier coefficients

The first condition, $u^{(\text{hom})}(x,0) = a(x) = 0$, implies Fourier coefficients $A_n = 0$ in equation (4.64); compare equation (4.6). The coefficients B_n are given by equation (4.6) with $b(x) = -c k_{\text{ex}} \sin(k_{\text{ex}}x)/\sin(k_{\text{ex}}l)$, so

$$B_n = \frac{2}{\pi n c} \int_0^l dx\, b(x) \sin(k_n x)$$
$$= -\frac{2 k_{\text{ex}}}{\pi n \sin(k_{\text{ex}}l)} \int_0^l dx\, \sin(k_{\text{ex}}x) \sin(k_n x). \tag{4.68}$$

We can evaluate the integral by making use of the relation $2\sin\alpha\sin\beta = \cos(\alpha - \beta) - \cos(\alpha + \beta)$, to obtain

$$B_n = -\frac{k_{\text{ex}}}{\pi n \sin(k_{\text{ex}}l)} \int_0^l dx\, \left(\cos[(k_{\text{ex}} - k_n)x] - \cos[(k_{\text{ex}} + k_n)x]\right)$$
$$= -\frac{k_{\text{ex}}}{\pi n \sin(k_{\text{ex}}l)} \left[\frac{\sin[(k_{\text{ex}} - k_n)x]}{k_{\text{ex}} - k_n} - \frac{\sin[(k_{\text{ex}} + k_n)x]}{k_{\text{ex}} + k_n}\right]_0^l$$
$$= -\frac{k_{\text{ex}}}{\pi n \sin(k_{\text{ex}}l)} \left(\frac{\sin[(k_{\text{ex}} - k_n)l]}{k_{\text{ex}} - k_n} - \frac{\sin[(k_{\text{ex}} + k_n)l]}{k_{\text{ex}} + k_n}\right). \tag{4.69}$$

With $k_n = n\pi/l$ for integer n, we have

$$\sin[(k_{\text{ex}} \pm k_n)l] = \sin(k_{\text{ex}}l \pm \pi n) = (-1)^n \sin(k_{\text{ex}}l),$$

so the factor $\sin(k_{\text{ex}}l)$ cancels and we obtain

$$\begin{aligned}
B_n &= -\frac{(-1)^n k_{\text{ex}}}{\pi n}\left(\frac{1}{k_{\text{ex}}-k_n} - \frac{1}{k_{\text{ex}}+k_n}\right) \\
&= -\frac{2(-1)^n k_{\text{ex}}}{\pi n}\frac{k_n}{k_{\text{ex}}^2 - k_n^2} \\
&= -\frac{(-1)^n}{l}\frac{2k_{\text{ex}}}{k_{\text{ex}}^2 - k_n^2} \\
&= \frac{(-1)^{n-1}}{l}\frac{2c\omega_{\text{ex}}}{\omega_{\text{ex}}^2 - \omega_n^2}.
\end{aligned} \qquad (4.70)$$

Note that the denominator of B_n is proportional to $\omega_{\text{ex}}^2 - \omega_n^2$, which vanishes for $\omega_{\text{ex}} = \omega_n$. This means that whenever the driving frequency approaches one of the eigenfrequencies of the system, the corresponding Fourier coefficient grows indefinitely. This phenomenon is called *resonance*. In this case, our solution ceases to be valid near the resonance, because we rely on assumptions about small deviations from equilibrium, and neglected physical effects (such as friction or non-linear effects) which will become important and remove the divergence for $\omega_{\text{ex}} = \omega_n$. One way to achieve that is to add an arbitrarily small amount of friction, which leads to a damped motion. In fact, as friction cannot be completely avoided in any mechanical system, in reality the motion will always be damped, and the amplitude at resonance will stay finite. Still, it may become large, so resonance effects need to be taken into account in the structural analysis of bridges and buildings.

Note that the eigenfrequencies are those of the homogeneous system, which means for fixed boundary conditions.

Solution for the example

Here, we assume that we are not at resonance, so $\omega_{\text{ex}} \neq \omega_n$ for any string mode $n = 1, 2, 3, \ldots$. Then the solution for the example is

$$\begin{aligned}
u(x,t) &= \frac{\sin(k_{\text{ex}}x)}{\sin(k_{\text{ex}}l)}\sin(\omega_{\text{ex}}t) \\
&\quad + \frac{2c\omega_{\text{ex}}}{l}\sum_{n=1}^{\infty}\frac{(-1)^{n-1}}{\omega_{\text{ex}}^2 - \omega_n^2}\sin(k_n x)\sin(\omega_n t),
\end{aligned} \qquad (4.71)$$

where $\omega_n = ck_n = \pi nc/l$.

Exercise 4.10

Consider a string of length $l = 10$ and wave speed $c = 200$ with a free boundary condition at $x = 0$. The other end, at $x = l$, is forced to move according to $u(l,t) = 2\sin(25\pi t)$. The string is initially at rest in the equilibrium position. Find a particular solution that satisfies the boundary conditions. [Hint: The different boundary condition at $x = 0$ suggests using a cosine function rather than a sine function as a trial solution for $U(x)$ in equation (4.63).]

4.4 Damped waves

Previously, we considered damped ends, but we did not take into account any friction acting along the string. In a real mechanical system, however, there is always some friction present, such as friction in the string or air resistance. Let us assume that the corresponding force is proportional to

4.4 Damped waves

velocity, as in equation (4.24), but it now acts uniformly along the entire string. As it acts in the opposite direction to the velocity, it causes the wave motion to decrease with time. If we repeat the derivation of the wave equation in Chapter 1 with this additional force, we arrive at the damped wave equation

Damped wave equation
$$u_{tt}(x,t) + 2\mu\, u_t(x,t) - c^2\, u_{xx}(x,t) = 0, \tag{4.72}$$

which involves a term proportional to the first partial derivative with respect to time t, with a proportionality constant $\mu \geq 0$. Here, we are interested not in the details of the derivation of the damping term, characterised by a damping constant μ, from the friction force, which can be done along the lines of Chapter 1, but in its solution using the methods developed above. For simplicity, we concentrate on the case of fixed boundary conditions.

4.4.1 Separation of variables

The first step in solving this equation is again separation of variables. Let us see what changes if we use a product form $u(x,t) = f(x)g(t)$, as we did in equation (2.45) of Chapter 2 for the (undamped) wave equation. Inserting the product into equation (4.72) and rearranging terms gives

$$\frac{1}{c^2} \frac{g''(t) + 2\mu g'(t)}{g(t)} = \frac{f''(x)}{f(x)} = C, \tag{4.73}$$

where C is an arbitrary constant. So separation of variables still works. The position-dependent part does not change, and is still given by equation (4.33). The equation for the time-dependent part becomes

$$g''(t) + 2\mu\, g'(t) - c^2 C\, g(t) = 0, \tag{4.74}$$

which differs from the corresponding equation (4.34) by the presence of the damping term involving the first derivative of the function $g(t)$. For $\mu = 0$ the damping term vanishes, and we recover the equations of undamped wave motion.

As before, we proceed by solving the two equations for the functions $f(x)$ and $g(t)$. Since the position-dependent part does not change, the arguments remain the same. As for the undamped wave equation, only negative values of the constant C lead to non-trivial solutions for fixed boundary conditions, so we can again choose $-C = k^2 = \omega^2/c^2 > 0$ with $\omega \geq 0$ and $k \geq 0$. The general solution of the linear second-order differential equation for f is $f(x) = a\cos(kx) + b\sin(kx)$ with arbitrary real coefficients a and b; compare equation (4.35).

Solution of the time-dependent part

Equation (4.74) for $g(t)$ is the same as the equation of a *damped harmonic oscillator*, which you may have come across previously (for instance in MST207 or MST209).

We can solve it by first solving the auxiliary equation

See Block 0, Section 1.3.

$$\lambda^2 + 2\mu\lambda + \omega^2 = 0, \tag{4.75}$$

where $\omega^2 = -c^2 C$. In general, this quadratic equation has two solutions, given by

$$\lambda_{1,2} = -\mu \pm \sqrt{\mu^2 - \omega^2}. \tag{4.76}$$

We now have to distinguish three cases, depending on the sign of $\mu - \omega$.

First, for $\mu > \omega$, we are in the *strong damping* case, where equation (4.76) gives two real solutions $\lambda_{1,2} = -\mu \pm \sqrt{\mu^2 - \omega^2}$. Second, $\mu = \omega$ is the case of *critical damping*, where the square root in equation (4.76) vanishes, and we are left with a *single* solution $\lambda = -\mu$. Third, for $\mu < \omega$, we are in the *weak damping* regime. In this situation, equation (4.76) gives two complex conjugate solutions $\lambda_{1,2} = -\mu \pm i\sqrt{\omega^2 - \mu^2}$.

The corresponding general solutions are as follows. For the strongly damped case, the general solution can be expressed as the sum of two exponentials

$$g(t) = ae^{\lambda_1 t} + be^{\lambda_2 t} \quad (\mu > \omega), \tag{4.77}$$

with negative $\lambda_{1,2} = -\mu \pm \sqrt{\mu^2 - \omega^2} < 0$ and arbitrary real coefficients a and b. At critical damping, we have only one exponential function, multiplied by a linear function in t,

$$g(t) = (a + bt)e^{-\mu t} \quad (\mu = \omega), \tag{4.78}$$

where again a and b are arbitrary real constants. Finally, for weak damping, we obtain

$$g(t) = [a\cos(\Omega t) + b\sin(\Omega t)]e^{-\mu t} \quad (\mu < \omega), \tag{4.79}$$

where the angular frequency $\Omega > 0$ is given by

$$\Omega = \sqrt{\omega^2 - \mu^2}. \tag{4.80}$$

As for strong and critical damping, the function $g(t)$ in equation (4.79) decreases exponentially, but the trigonometric factor means that it oscillates with an angular frequency Ω. This function $g(t)$ therefore corresponds to an oscillatory motion with an amplitude $A\exp(-\mu t)$ that decreases exponentially with time.

Example solutions for strong, critical and weak damping are shown in Figure 4.12.

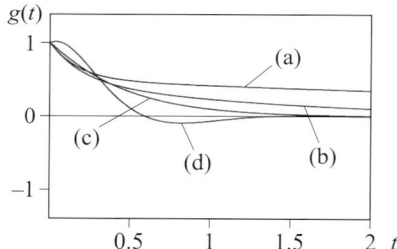 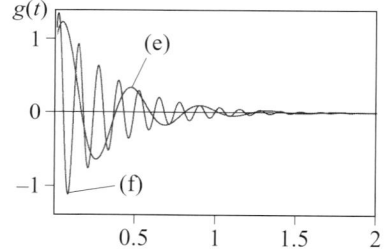

Figure 4.12 Plot of solutions $g(t)$ of equation (4.74) with $\mu = 3$. Curves (a) and (b) correspond to strongly damped solutions (4.77) for $\omega = 1$ and $\omega = 2$, respectively. Curve (c) is a critically damped solution (4.78) for $\omega = 3$, and curves (d)–(f) are weakly damped solutions (4.79) for $\omega = 5$, $\omega = 15$ and $\omega = 50$, respectively. The coefficients a and b in equations (4.77)–(4.79) are chosen as $a = b = \frac{1}{2}$ for curves (a) and (b), and $a = b = 1$ for curves (c)–(f).

For our example of the string, strong and critical damping mean that a plucked string returns to equilibrium without any oscillations – which might happen if you considered a string immersed in oil or a similar environment that produces sufficiently strong damping. However, for a guitar string that does not usually happen, as you can hear the sound of a certain pitch that corresponds to an oscillatory motion of angular frequency Ω. Still, the motion of a real guitar string is clearly damped – if you listen to the sound after you pluck the string, its volume decreases with time, until after a while it cannot be heard. At the same time, you can also see that the visible vibration of the string diminishes and eventually fades.

Exercise 4.11

If we consider the motion of a damped string with two free ends, we can again have a mode $u_0(x,t)$ with $\omega_0 = 0$.

(a) According to our characterisation, is this mode weakly, critically or strongly damped for $\mu > 0$?

(b) Accordingly, what solution would you obtain from equations (4.77)–(4.79)?

(c) Now solve the differential equation (4.74) for $g(t)$ with $C = 0$ by integrating it directly, and determine the general solution.

(d) Write down the corresponding solution $u_0(x,t)$ of the wave equation for a string with free ends.

4.4.2 Solution of the damped wave equation

In this section, we concentrate on the weakly damped case. If we consider a string of length l, fixed at its end points at $x = 0$ and $x = l$, we need to impose the boundary conditions $u(0,t) = u(l,t) = 0$. Because $u(x,t) = f(x)g(t)$, this implies $f(0) = f(l) = 0$, exactly as for the undamped wave equation. Thus in the function $f(x)$ of equation (4.35) we have $a = 0$ and $k = \pi n/l$, for $n = 1, 2, 3, \ldots$. This gives

$$f_n(x) = b_n \sin\left(\frac{\pi n x}{l}\right), \quad n = 1, 2, 3, \ldots, \tag{4.81}$$

as for the undamped wave equation. The corresponding functions $g_n(t)$ then become

$$g_n(t) = e^{-\mu t}[A_n \cos(\Omega_n t) + B_n \sin(\Omega_n t)] \tag{4.82}$$

where

$$\Omega_n = \sqrt{\omega_n^2 - \mu^2}, \quad \omega_n = ck_n = \frac{\pi n c}{l}, \quad n = 1, 2, 3, \ldots. \tag{4.83}$$

The general solution of the weakly damped wave equation (4.72) with boundary conditions $u(0,t) = u(l,t) = 0$ is thus

$$u(x,t) = \sum_{n=1}^{\infty} f_n(x) g_n(t)$$

$$= e^{-\mu t} \sum_{n=1}^{\infty} \sin(k_n x) [A_n \cos(\Omega_n t) + B_n \sin(\Omega_n t)], \tag{4.84}$$

choosing $b_n = 1$ without loss of generality.

This solution holds true if the damping is so weak that all modes are weakly damped, which means that $\mu < \omega_n$ for all n, or

$$\mu < \omega_1 = \frac{\pi c}{l}. \tag{4.85}$$

Otherwise, the solution contains a mixture of strongly and weakly damped modes, and occasionally a single critically damped mode. For $\mu = 0$, we have $\Omega_n = \omega_n$ and thus recover the equations for the undamped case.

Note that the normal modes governing the position-dependent part of the solutions are unchanged. In particular, if the motion is started in a particular normal mode, then the system will stay in this mode throughout its motion. The damping means only that the frequency of the mode is slightly shifted, according to equations (4.83), and that the amplitude of the oscillations decays exponentially. If the damping is sufficiently strong, then the lower-eigenvalue modes may be strongly or critically damped and thus not lead to oscillatory motion.

Exercise 4.12

Consider a string of length $l = 1$ with fixed ends, wave speed $c = 10$ and damping constant $\mu = 30\pi$. Are these normal modes strongly, critically or weakly damped?

(a) $n = 1$ (b) $n = 2$ (c) $n = 3$ (d) $n = 4$ (e) $n = 5$

Exercise 4.13

What condition has the damping constant μ to satisfy for a mode n of a string of length l with fixed ends, wave speed c and damping constant μ to be critically damped or strongly damped? Write down the general form of the contribution to the general solution from each of the following.

(a) A critically damped mode
(b) A strongly damped mode

Exercise 4.14

Write down the general solution of the damped wave equation for a string of length $l = 1$ with fixed ends, wave speed $c = 10$ and damping constant $\mu = 30\pi$.

Solution of the initial-value problem

The final step in the solution of the damped wave equation concerns again the initial-value problem for fixed ends. We know what we have to do – we write down the initial conditions $u(x,0) = a(x)$ and $u_t(x,0) = b(x)$ in terms of our general solution (4.84), and determine from this the arbitrary coefficients A_n and B_n. For simplicity, we again concentrate on the weakly damped case; the other cases can be treated analogously.

At time $t = 0$, the solution (4.84) gives

$$a(x) = \sum_{n=1}^{\infty} A_n \sin\left(\frac{\pi n x}{l}\right), \tag{4.86}$$

$$b(x) = \sum_{n=1}^{\infty} (\Omega_n B_n - \mu A_n) \sin\left(\frac{\pi n x}{l}\right). \tag{4.87}$$

Exercise 4.15

Verify equation (4.87) for $b(x) = u_t(x,0)$, with u given by equation (4.84).

If we compare this with the result for the undamped wave equation, we find that the differences are the term $-\mu A_n$ appearing in the sum for $b(x)$, and the presence of Ω_n rather than ω_n. Nevertheless, we still obtain the coefficients A_n and B_n from the Fourier sine series of the functions $a(x)$ and $b(x)$ that specify the initial conditions. This gives

See equations (4.3) and (4.4).

$$A_n = \frac{2}{l} \int_0^l dx\, a(x) \sin\left(\frac{\pi n x}{l}\right), \tag{4.88}$$

which is the same as in equation (4.6) for the undamped case, and

$$B_n = \frac{1}{\Omega_n}\left(\mu A_n + \frac{2}{l} \int_0^l dx\, b(x) \sin\left(\frac{\pi n x}{l}\right)\right), \tag{4.89}$$

which differs from the corresponding equation (4.7).

Exercise 4.16

Consider a string of length $l = 10$ with fixed ends, wave speed $c = 200$ and damping constant $\mu = 2\pi$. It is initially released from rest with

$$u(x,0) = a(x) = \begin{cases} x/10 & \text{for } 0 \leq x \leq 5, \\ (10-x)/10 & \text{for } 5 < x \leq 10. \end{cases}$$

Determine the solution of the initial-value problem.

4.5 Inhomogeneous damped wave equation

So far, we have considered *homogeneous* partial differential equations. Neither the undamped wave equation (4.1) nor the damped wave equation (4.72) involves functions of position or time other than the function $u(x,t)$ or its derivatives. In what follows, we are going to apply the methods developed so far in a more general set-up. We shall consider *inhomogeneous* equations, such as, for instance, the wave equation for a string that is subject to an external force along its length. As an example, you may think of a string which is subject to a continuous external driving force, like a washing line or a power line blowing in the wind. Other examples are gravity acting on a string, or a tuning fork exciting the vibration of a string.

To describe such systems, we consider the inhomogeneous damped wave equation

Inhomogeneous damped wave equation
$$u_{tt}(x,t) + 2\mu\, u_t(x,t) - c^2\, u_{xx}(x,t) = p(x,t), \qquad (4.90)$$

which includes a term that corresponds to an external force (per unit length) of magnitude $F(x,t) = \rho\, p(x,t)$ acting at position x along the string at time t, where ρ is the density (1.11). We assume that the function p is given, and we wish to solve equation (4.90) for the function u.

In principle, we can again consider various boundary conditions. For definiteness, we concentrate again on our main example, the finite string with fixed boundaries, so the boundary conditions are $u(0,t) = u(l,t) = 0$. To conform with the fixed end boundary conditions, we require that $p(0,t) = p(l,t) = 0$ for all t. Initial conditions are specified by $u(x,0) = a(x)$ and $u_t(x,0) = b(x)$.

The homogeneous equation, which corresponds to taking $p(x,t) = 0$ in equation (4.90), is the damped wave equation discussed before. We know that the normal modes $f_n(x)$ of the homogeneous equation are given by equations (4.81), which takes care of the fixed boundary conditions. We can express the solution of the initial-value problem in terms of normal modes as $u(x,t) = \sum_{n=1}^{\infty} f_n(x) g_n(t)$. This is the approach that we shall take for the driven system as well.

As a first step, we express the function $p(x,t)$ in terms of the normal modes $f_n(x)$ of the homogeneous equation which were given in equations (4.81).

This means that we write $p(x,t)$ as a Fourier sine series, so

$$p(x,t) = \sum_{n=1}^{\infty} p_n(t) \sin(k_n x), \quad k_n = \frac{\pi n}{l}, \tag{4.91}$$

where the time-dependence is contained in the coefficients

$$p_n(t) = \frac{2}{l} \int_0^l dx \, p(x,t) \sin(k_n x). \tag{4.92}$$

The expansion of the solution of equation (4.90) in terms of the normal modes is

$$u(x,t) = \sum_{n=1}^{\infty} f_n(x) g_n(t) = \sum_{n=1}^{\infty} g_n(t) \sin(k_n x), \tag{4.93}$$

where $g_n(t)$ are functions of time t. Inserting equations (4.91) and (4.93) into the inhomogeneous wave equation (4.90) gives

$$\sum_{n=1}^{\infty} \left[g_n''(t) + 2\mu g_n'(t) + c^2 k_n^2 g_n(t) - p_n(t) \right] \sin(k_n x) = 0. \tag{4.94}$$

The left-hand side is therefore a Fourier series for the constant function 0, thus all Fourier coefficients must vanish. This yields

$$g_n''(t) + 2\mu g_n'(t) + c^2 k_n^2 g_n(t) = p_n(t), \quad n = 1, 2, 3, \ldots, \tag{4.95}$$

which is a set of ordinary differential equations for the coefficients $g_n(t)$. For a given function $p(x,t)$, we can determine the Fourier coefficients $p_n(t)$, and then solve this set of equations.

For the initial conditions $a(x) = u(x,0)$ and $b(x) = u_t(x,0)$, equation (4.93) gives

$$a(x) = \sum_{n=1}^{\infty} g_n(0) \sin(k_n x), \quad b(x) = \sum_{n=1}^{\infty} g_n'(0) \sin(k_n x). \tag{4.96}$$

The initial conditions $g_n(0)$ and $g_n'(0)$ for the differential equations (4.95) are thus given by the Fourier coefficients of the functions $a(x)$ and $b(x)$; compare equations (4.3) and (4.4). Therefore we obtain

$$g_n(0) = \frac{2}{l} \int_0^l dx \, a(x) \sin(k_n x), \quad g_n'(0) = \frac{2}{l} \int_0^l dx \, b(x) \sin(k_n x). \tag{4.97}$$

The second equation looks somewhat different from its counterpart in equation (4.4), because we also used a Fourier decomposition for the time-dependent part in that case, and actually performed the time derivative.

4.5.1 Harmonic driving

As an example, we consider a case where the Fourier expansion (4.91) of $p(x,t)$ consists of a single term

$$p(x,t) = p_m(t) \sin(k_m x), \tag{4.98}$$

and the system is initially at rest at its equilibrium position, so

$$a(x) = b(x) = 0. \tag{4.99}$$

In this case, the set of differential equations (4.95) becomes

$$g_n''(t) + 2\mu g_n'(t) + c^2 k_n^2 g_n(t) = p_m(t) \delta_{nm}$$

$$= \begin{cases} p_m(t) & \text{for } n = m, \\ 0 & \text{for } n \neq m, \end{cases} \tag{4.100}$$

4.5 Inhomogeneous damped wave equation

and the initial conditions (4.99) amount to

$$g_n(0) = g'_n(0) = 0, \quad n = 1, 2, 3, \ldots; \tag{4.101}$$

compare equations (4.96). For $n \neq m$, the solution of the second-order differential equation (4.100) with $g_n(0) = g'_n(0) = 0$ is $g_n(t) = 0$. This means that the Fourier series of the solution $u(x,t)$ in equation (4.93) consists of the single normal mode $n = m$ only, irrespective of the time-dependence of $p_m(t)$, so

$$u(x,t) = g_m(t) \sin(k_m x). \tag{4.102}$$

Now we have to solve the equation for $n = m$. This is the differential equation for a damped harmonic oscillator with external driving $p_m(t)$. We shall solve this equation only for the special case where p varies sinusoidally with time.

Single-frequency driving

Consider a harmonic driving with

$$p_m(t) = p_m \cos(\omega_{\text{ex}} t) \tag{4.103}$$

in equation (4.98). The string is subject to a periodic time-dependent force (with period $\tau = 2\pi/\omega_{\text{ex}}$) which varies sinusoidally along the string.

The motion of the string involves only the mode $n = m$, so the solution $u(x,t)$ has the form of equation (4.102). To calculate the function $g_m(t)$, we need to solve equation (4.100) for $n = m$, which becomes

$$g''_m(t) + 2\mu g'_m(t) + \omega_m^2 g_m(t) = p_m \cos(\omega_{\text{ex}} t). \tag{4.104}$$

It turns out that it is advantageous to solve this equation in terms of complex exponential functions rather than trigonometric functions. To this end, we note that

$$p_m(t) = p_m \cos(\omega_{\text{ex}} t) = p_m \operatorname{Re}(e^{i\omega_{\text{ex}} t}), \tag{4.105}$$

so we consider the complex equation

$$g''_m(t) + 2\mu g'_m(t) + \omega_m^2 g_m(t) = p_m e^{i\omega_{\text{ex}} t}, \tag{4.106}$$

keeping in mind that it is the real part of its solution $g_m(t)$ that actually describes the motion.

The general solution of the inhomogeneous equation is the sum of the general solution of the homogeneous equation, also called the complementary function, and a particular integral of the inhomogeneous equation. Again, the homogeneous part consists of a damped motion, which eventually dies out. This part of the solution is referred to as the *transient*. Particular integrals correspond to the forced motion. A natural assumption is that the system oscillates with the frequency ω_{ex} of the driving force. We check if this works by inserting

$$g_m^{(\text{par})} = A_m e^{i\omega_{\text{ex}} t}, \tag{4.107}$$

where the coefficient A_m may, in general, be complex. This yields

$$A_m e^{i\omega_{\text{ex}} t} \left[(i\omega_{\text{ex}})^2 + 2\mu i\omega_{\text{ex}} + \omega_m^2\right] = p_m e^{i\omega_{\text{ex}} t}. \tag{4.108}$$

Cancelling $\exp(i\omega_{\text{ex}} t)$ on both sides leaves us with

$$A_m = \frac{p_m}{\omega_m^2 - \omega_{\text{ex}}^2 + 2i\mu\omega_{\text{ex}}}, \tag{4.109}$$

which is complex for non-vanishing damping constant μ.

The real and imaginary parts of $A_m = A_R + iA_I$ of equation (4.109) are given by

$$A_R + iA_I = \frac{p_m}{\omega_m^2 - \omega_{ex}^2 + 2i\mu\omega_{ex}} \times \frac{\omega_m^2 - \omega_{ex}^2 - 2i\mu\omega_{ex}}{\omega_m^2 - \omega_{ex}^2 - 2i\mu\omega_{ex}}$$

$$= \frac{p_m(\omega_m^2 - \omega_{ex}^2)}{(\omega_m^2 - \omega_{ex}^2)^2 + 4\mu^2\omega_{ex}^2} - i\frac{2p_m\mu\omega_{ex}}{(\omega_m^2 - \omega_{ex}^2)^2 + 4\mu^2\omega_{ex}^2}. \quad (4.110)$$

Alternatively, we may express A_m in terms of its absolute value $|A_m| = \sqrt{A_R^2 + A_I^2}$ and its argument ϕ_m, giving

$$A_m = |A_m|e^{i\phi_m} = |A_m|(\cos\phi_m + i\sin\phi_m), \quad (4.111)$$

with

$$|A_m| = \frac{|p_m|}{\sqrt{(\omega_m^2 - \omega_{ex}^2)^2 + 4\mu^2\omega_{ex}^2}}, \quad (4.112)$$

and a phase ϕ_m which is determined, up to multiples of 2π, by

$$\cos\phi_m = \frac{A_R}{|A_m|} = \frac{p_m}{|p_m|}\frac{\omega_m^2 - \omega_{ex}^2}{\sqrt{(\omega_m^2 - \omega_{ex}^2)^2 + 4\mu^2\omega_{ex}^2}}, \quad (4.113)$$

$$\sin\phi_m = \frac{A_I}{|A_m|} = -\frac{p_m}{|p_m|}\frac{2\mu\omega_{ex}}{\sqrt{(\omega_m^2 - \omega_{ex}^2)^2 + 4\mu^2\omega_{ex}^2}}. \quad (4.114)$$

Taking the ratio of these two equations gives

$$\tan\phi_m = \frac{\sin\phi_m}{\cos\phi_m} = -\frac{2\mu\omega_{ex}}{\omega_m^2 - \omega_{ex}^2}, \quad (4.115)$$

which, however, determines ϕ_m only up to multiples of π.

Exercise 4.17

Assume that $p_m > 0$.
(a) Calculate $|A_m|$ for the driving frequency $\omega_{ex} = \omega_m$.
(b) Calculate the phase ϕ_m.

Now the real part of the particular solution is

$$\text{Re}[g_m^{(\text{par})}] = \text{Re}[A_m e^{i\omega_{ex}t}]$$
$$= \text{Re}[\,|A_m|e^{i(\omega_{ex}t + \phi_m)}]$$
$$= |A_m|\cos(\omega_{ex}t + \phi_m). \quad (4.116)$$

The transient part is given by the general solution of the associated homogeneous equation (4.74). Here we shall assume that we are in the weakly damped case, so the general solution is

$$g_m^{(\text{hom})}(t) = e^{-\mu t}[a_m \cos(\Omega_m t) + b_m \sin(\Omega_m t)], \quad (4.117)$$

with

$$\Omega_m = \sqrt{\omega_m^2 - \mu^2}; \quad (4.118)$$

compare equations (4.79) and (4.80). Hence the general solution of equation (4.104) is

$$g_m(t) = |A_m|\cos(\omega_{ex}t + \phi_m)$$
$$+ e^{-\mu t}[a_m \cos(\Omega_m t) + b_m \sin(\Omega_m t)]. \quad (4.119)$$

An example is shown in Figure 4.13.

4.5 Inhomogeneous damped wave equation

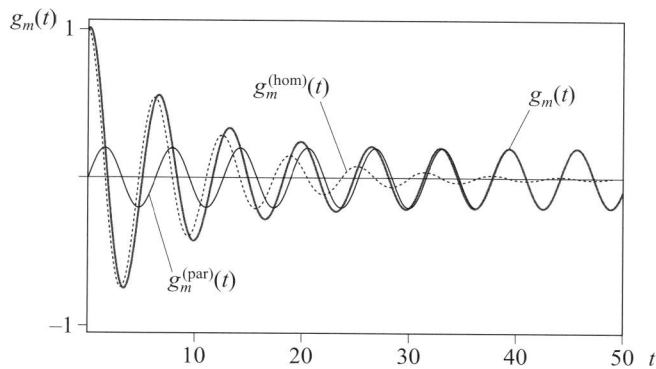

Figure 4.13 Plot of the time-dependent part $g_m(t) = g_m^{(\mathrm{par})}(t) + g_m^{(\mathrm{hom})}(t)$ of the forced wave motion, for $\mu = \omega_m/10$, $\omega_{\mathrm{ex}} = \omega_m = 1$, $p_m = 1/5$ and coefficients $a = 1$ and $b = 0$ in equation (4.119)

The constants a and b are determined by the initial conditions. For a string that is initially at rest in equilibrium, the initial conditions (4.101) give

$$a_m + |A_m|\cos\phi_m = 0, \quad b_m\Omega_m - a_m\mu - |A_m|\omega_{\mathrm{ex}}\sin\phi_m = 0. \tag{4.120}$$

This yields

$$a_m = -|A_m|\cos\phi_m, \quad b_m = -\frac{|A_m|}{\Omega_m}(\mu\cos\phi_m - \omega_{\mathrm{ex}}\sin\phi_m). \tag{4.121}$$

So the complete, admittedly rather complicated, solution turns out to be

$$u(x,t) = \sin(k_m x)\left(|A_m|\cos(\omega_{\mathrm{ex}}t + \phi_m)\right. \\ \left. + e^{-\mu t}[a_m\cos(\Omega_m t) + b_m\sin(\Omega_m t)]\right); \tag{4.122}$$

see equation (4.93). It consists of a transient motion of frequency Ω_m from equation (4.118), and an undamped harmonic motion with the driving frequency ω_{ex}; see Figure 4.13. The response of the system is shifted with respect to the driving force by a phase shift ϕ_m.

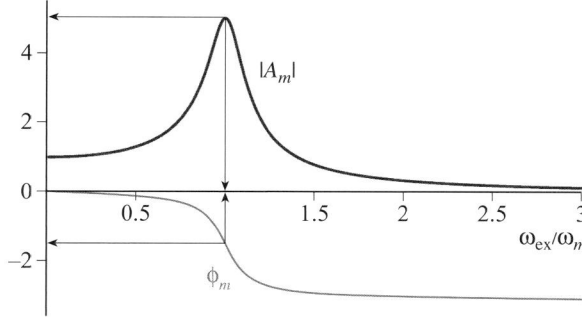

Figure 4.14 Sketch of the amplitude $|A_m|$ and the phase shift ϕ_m for weak damping, $\mu = \omega_m/10$ and $p_m/\omega_m^2 = 1$, as a function of the external driving frequency ω_{ex}. At $\omega_{\mathrm{ex}} = \omega_m$, $\phi_m = -\pi/2$, whereas the maximal amplitude occurs at a frequency that is slightly smaller than ω_m; compare Exercise 4.18 below.

As the driving frequency ω_{ex} is varied, the amplitude $|A_m|$ goes through a maximum; see Figure 4.14. This is the phenomenon of *resonance*, which we met previously in Section 4.3 when discussing inhomogeneous boundary conditions. You may also be familiar with resonance occurring in forced oscillations in mechanics. It occurs when the damping is weak and the driving frequency ω_{ex} is close to the natural frequency ω_m of the normal mode. For vanishing damping $\mu \to 0$, the amplitude becomes infinitely large at the resonance, which is a reason why damping should be included in the discussion of forced motion. Roughly speaking, the closer the external driving frequency matches a natural frequency of a system, the easier it is to force the system to oscillate.

Resonance is important in many applications. In some cases, like in musical instruments, resonance is desired. However, in other cases it can have destructive effects, and needs to be avoided. This is the reason why, for instance, large groups of soldiers should not cross a bridge in step.

Exercise 4.18

Consider the amplitude $|A_m|$ of the forced motion. We are interested in the driving frequency ω_{ex} where the amplitude is largest.

(a) Argue that the amplitude is maximal for the driving frequency ω_{ex} that minimises the expression $(\omega_m^2 - \omega_{\text{ex}}^2)^2 + 4\mu^2\omega_{\text{ex}}^2$.

(b) Calculate the derivative of this expression with respect to ω_{ex}, and determine the values of ω_{ex} where this derivative vanishes.

(c) What condition must ω_m and μ satisfy to give a positive value for ω_{ex} where the derivative vanishes?

(d) Assuming that there is a positive real solution, what is $\tan\phi_m$ for this value of ω_{ex}? What happens to the phase ϕ_m as the damping constant μ becomes small?

4.6 Summary and Outcomes

In this chapter, Fourier methods were used to solve the one-dimensional wave equation. Solving the initial-value problem relies upon calculating the Fourier series of the functions specifying the initial conditions.

A variety of boundary conditions have been considered, including free, damped and sprung ends, as well as inhomogeneous boundary conditions which correspond to an external driving of the motion. The phenomenon of resonance occurs when the external driving frequency coincides with an eigenfrequency of the system. Without damping, the amplitude of the solution is unbounded.

The damped wave equation, with a damping term that is proportional to velocity, can be solved with the same methods. If there is an external driving force, the damped wave equation becomes inhomogeneous, and the general solution is the sum of a particular solution of the inhomogeneous equation and the general solution of the homogeneous equation. For the damped wave equation, the latter always decays exponentially, and is called a transient. After a long time has elapsed, the motion is given completely by the particular solution which, for a harmonic driving force, has the same frequency as the external driving force.

After working through this chapter, you should:
- know how to treat fixed and free boundary conditions;
- be able to calculate normal modes for free, sprung or damped boundary conditions;
- know how to solve initial-value problems for these boundary conditions;
- be able to apply Fourier sine and cosine series;
- know how to account for inhomogeneous boundary conditions;
- be able to solve the damped wave equation;
- know the difference between strong, critical and weak damping;
- understand what resonance means and how it arises;
- be able to solve inhomogeneous wave equations for simple driving forces.

4.7 Further Exercises

Exercise 4.19

Consider the motion of a stretched string of length $l = 3$, with fixed ends and the initial conditions

$$a(x) = \begin{cases} 0 & \text{for } 0 \leq x < 1, \\ 1/10 & \text{for } 1 \leq x \leq 2, \\ 0 & \text{for } 2 < x \leq 3, \end{cases} \quad b(x) = 0;$$

i.e. the string is released at rest with a square-shaped pulse of one third of its length at the centre.

This pulse form with discontinuities cannot be applied to a real string, but we may think of a square pulse with slightly rounded edges that we can approximate in this way.

(a) Draw a sketch of the odd periodic extension of the function $a(x)$.

(b) Calculate the Fourier coefficients A_n from equation (4.6), and show that they are given by

$$A_n = \frac{1}{5\pi n}\left[\cos\left(\frac{\pi n}{3}\right) - \cos\left(\frac{2\pi n}{3}\right)\right].$$

(c) Show that A_n vanishes for even values of n.

(d) Any integer n can be written in a unique way as $n = 6p + q$, where p is an integer and $q \in \{0, 1, 2, 3, 4, 5\}$. The values of $\cos(\pi n/3)$ depend only on q and are given by

For instance, $n = 65$ gives $p = 10$ and $q = 5$.

$$\cos\left[(6p+q)\frac{\pi}{3}\right] = \cos\left(\frac{\pi q}{3}\right) = \begin{cases} 1 & \text{for } q = 0, \\ 1/2 & \text{for } q = 1 \text{ and } q = 5, \\ -1/2 & \text{for } q = 2 \text{ and } q = 4, \\ -1 & \text{for } q = 3. \end{cases}$$

Use this to calculate the values of A_{6p+q} for odd q.

(e) Insert the result into equation (4.2), and determine the solution of the initial-value problem.

Exercise 4.20

Consider a string of length l with free boundary conditions at both ends. The initial conditions are given by $u(x, 0) = a(x) = 0$ and $u_t(x, 0) = b(x)$, which is defined on the interval $[0, l]$ by

$$b(x) = \begin{cases} x & \text{for } 0 \leq x \leq l/2, \\ l - x & \text{for } l/2 < x \leq l. \end{cases}$$

For $0 \leq x \leq l$, this is the function considered previously for the string with fixed boundaries; compare equation (4.8).

(a) What is the even periodic extension of $b(x)$? What is its period?

(b) Sketch the function $b(x)$, and compare it with the odd periodic extension shown in Figure 4.3. How are the two functions related?

(c) Calculate the Fourier cosine coefficients B_n by applying equation (4.52) to $b(x)$.

(d) In this example, $b(x)$ is also periodic with period l. We can compute its Fourier coefficients with respect to this period, so

$$b(x) = \frac{\widetilde{B}_0}{2} + \sum_{n=1}^{\infty} \widetilde{B}_n \cos\left(\frac{2\pi n x}{l}\right),$$

where

$$\widetilde{B}_n = \frac{4}{l}\int_0^{l/2} dx\, b(x) \cos\left(\frac{2\pi n x}{l}\right), \quad \text{for } n \geq 0.$$

Show that $\widetilde{B}_m = B_{2m}$, and that this implies that the two Fourier series agree.

(e) Compare your result with that of Exercise 4.8. Verify the relation $b(x) = l/4 - a(2x)/2$, where $a(x)$ is the even periodic extension of equation (4.55). Check that this relation is obeyed by the respective Fourier expansions.

(f) Calculate the corresponding solution of the initial-value problem for a string of length l with two free ends.

Exercise 4.21

Consider the string with one free and one fixed end of Subsection 4.3.3. The general solution of the corresponding boundary-value problem is given in equation (4.40) on page 154.

(a) What is the period of the function $u(x,t)$ of equation (4.40) in the variable x? In other words, what is the smallest $L > 0$ such that $u(x+L, t) = u(x,t)$?

(b) Is the function $u(x,t)$ of equation (4.40) even or odd with respect to reflections at $x=0$ and $x=l$? In other words, how are $u(-x)$ and $u(2l-x)$ related to $u(x,t)$?

(c) Express the initial conditions $u(x,0) = a(x)$ and $u_t(x,0) = b(x)$ in terms of the normal mode expansion (4.40).

(d) Show that the orthogonality relation (4.51) implies

$$\frac{2}{l} \int_0^l dx \cos\left(\frac{\pi(2m-1)x}{2l}\right) \cos\left(\frac{\pi(2n-1)x}{2l}\right) = \delta_{mn}, \quad m,n \geq 1.$$

[Hint: Starting from equation (4.51) with l replaced by $2l$, split the integration into two parts, from 0 to l and from l to $2l$, and eventually use the identity $\cos[\pi(2n-1) - x] = -\cos x$ to combine the two parts into a single integral.]

(e) Show that the corresponding Fourier coefficients A_n and B_n, for $n = 1, 2, 3, \ldots$, of the functions $a(x)$ and $b(x)$ are given by

$$A_n = \frac{2}{l} \int_0^l dx\, a(x) \cos\left(\frac{\pi(2n-1)x}{2l}\right),$$

$$B_n = \frac{4}{\pi(2n-1)c} \int_0^l dx\, b(x) \cos\left(\frac{\pi(2n-1)x}{2l}\right),$$

respectively.

# Solutions to Exercises in Chapter 4

Solution 4.1

Integration by parts gives

$$\int_0^z dx\, x \sin(kx) = \left[-\frac{x\cos(kx)}{k}\right]_0^z + \frac{1}{k}\int_0^z dx\, \cos(kx)$$
$$= -\frac{z\cos(kz)}{k} + \frac{\sin(kz)}{k^2},$$

which gives equation (4.12).

Solution 4.2

(a) The initial displacement $a(x)$ corresponds to a tenth of the triangular displacement considered in Example 4.1. Therefore the corresponding Fourier coefficients are $A_n/10$ with A_n as given in equations (4.13), and the corresponding solution is $u(x,t)/10$ with $u(x,t)$ as in equation (4.14).

(b) Even though the two solutions differ only by a constant factor, there is a difference regarding their applicability to a real vibrating string. In the derivation of the wave equation for transverse vibrating strings in Chapter 1, we assumed that the amplitude of the vibrations is small, so we could neglect terms of second order in the displacement. The motion corresponding to the initial condition of this exercise satisfies that assumption better than the motion with the initial condition of Example 4.1, so we should expect it to give a more accurate description of the motion of a real string.

Solution 4.3

For $x = 0$ and $x = l$, we have $\sin(\pi(2m-1)0/l) = \sin(\pi(2m-1)l/l) = 0$, so the boundary conditions are satisfied. At $t = 0$, the other sine factors vanish, i.e. $\sin(\pi(2m-1)c0/l) = 0$, thus $u(x,0) = 0$. All that is left to check is the partial derivative at $t = 0$, which is given by

$$u_t(x,0) = \sum_{m=1}^\infty \frac{\pi(2m-1)c}{l}\,\frac{4l^2(-1)^{m-1}}{\pi^3(2m-1)^3 c}\sin\left(\frac{\pi(2m-1)x}{l}\right)$$
$$= \sum_{m=1}^\infty \frac{4l(-1)^{m-1}}{\pi^2(2m-1)^2}\sin\left(\frac{\pi(2m-1)x}{l}\right),$$

which is indeed the Fourier series of the function $b(x)$, which has coefficients given by equations (4.13).

Solution 4.4

First observe that $b(x) = -a(x)/2$, so the functions have the same Fourier coefficients apart from an overall factor $-\frac{1}{2}$. Moreover, we solved the initial-value problem with this function $a(x)$ and $b(x) = 0$ in Exercise 4.2, and calculated the Fourier coefficients of the function $10a(x)$ in equations (4.13).

For the coefficients A_n and B_n in equation (4.2), we thus obtain

$$A_{2m-1} = \frac{2l(-1)^{m-1}}{5\pi^2(2m-1)^2}, \quad B_{2m-1} = -\frac{l^2(-1)^{m-1}}{5\pi^3(2m-1)^3 c}, \quad A_{2m} = B_{2m} = 0,$$

for $m = 1, 2, 3, \ldots$. The solution of the initial-value problem is then

$$u(x,t) = \sum_{m=1}^\infty \frac{2l(-1)^{m-1}}{5\pi^2(2m-1)^2}\sin\left(\frac{\pi(2m-1)x}{l}\right)$$
$$\times \left[\cos\left(\frac{\pi(2m-1)ct}{l}\right) - \frac{l}{2\pi(2m-1)c}\sin\left(\frac{\pi(2m-1)ct}{l}\right)\right].$$

Solution 4.5

The net force acting on the ring is

$$\boldsymbol{F} = \boldsymbol{F}_1 + \boldsymbol{F}_2 + \boldsymbol{F}_4.$$

Its component in the $\mathbf{i}$ direction vanishes, so

$$\boldsymbol{F}(t) = [T\, u_x(0,t) - k\, u(0,t)]\, \mathbf{j}.$$

Newton's second law for the motion of the ring in the $\mathbf{j}$ direction gives

$$M\, u_{tt}(0,t) = T\, u_x(0,t) - k\, u(0,t),$$

which, in the limit as $M \to 0$, becomes

$$T\, u_x(0,t) - k\, u(0,t) = 0.$$

Dividing by $T > 0$ and using $\kappa = k/T$ yields equation (4.29).

Solution 4.6

(a) We consider the forces acting on the particle of mass M. These are the tension force $\boldsymbol{F}_1 = -T(\mathbf{i} + u_x(l,t)\mathbf{j})$, the force $\boldsymbol{F}_2 = T\mathbf{i}$ that keeps the end at $x = l$, and the spring force $\boldsymbol{F}_4 = -\kappa T u(l,t)\mathbf{j}$.

The total force is

$$\boldsymbol{F} = \boldsymbol{F}_1 + \boldsymbol{F}_2 + \boldsymbol{F}_4 = -T[u_x(l,t) + \kappa\, u(l,t)]\, \mathbf{j},$$

so Newton's second law for the motion in the $\mathbf{j}$ direction gives

$$M\, u_{tt}(l,t) = -T[u_x(l,t) + \kappa\, u(l,t)].$$

(b) For vanishing mass M, the condition becomes

$$u_x(l,t) + \kappa\, u(l,t) = 0,$$

which is equation (4.31).

Solution 4.7

(a) For $m \neq n$, using the relation given in the Hint, we obtain

$$\frac{2}{l}\int_0^l dx\, \cos\left(\frac{\pi m x}{l}\right) \cos\left(\frac{\pi n x}{l}\right)$$

$$= \frac{1}{l}\int_0^l dx\, \left[\cos\left(\frac{\pi(m-n)x}{l}\right) + \cos\left(\frac{\pi(m+n)x}{l}\right)\right]$$

$$= \frac{1}{l}\left[\frac{l}{\pi(m-n)}\sin\left(\frac{\pi(m-n)x}{l}\right) + \frac{l}{\pi(m+n)}\sin\left(\frac{\pi(m+n)x}{l}\right)\right]_0^l$$

$$= \frac{1}{\pi}\left(\frac{\sin[\pi(m-n)]}{m-n} + \frac{\sin[\pi(m+n)]}{m+n}\right)$$

$$= 0,$$

which vanishes because $\sin(k\pi) = 0$ for any integer k. For $m = n \geq 1$, we use the same trigonometric relation for $\alpha = \beta$, which gives $2\cos^2\alpha = 1 + \cos 2\alpha$. This yields

$$\frac{2}{l}\int_0^l dx\, \cos^2\left(\frac{\pi n x}{l}\right) = 1 + \frac{1}{l}\int_0^l dx\, \cos\left(\frac{2\pi n x}{l}\right)$$

$$= 1 + \frac{1}{l}\left[\frac{l}{2\pi n}\sin\left(\frac{2\pi n x}{l}\right)\right]_0^l$$

$$= 1,$$

because again the sine function vanishes at $x = 0$ and at $x = l$.

(b) To make use of the orthogonality relation, we take the infinite sum for $a(x)$ given in equation (4.49), multiply it by $(2/l)\cos(\pi m x/l)$, and integrate over x from $x = 0$ to $x = l$. This gives, for $m > 0$,

$$\frac{2}{l}\int_0^l dx\, a(x) \cos\left(\frac{\pi m x}{l}\right)$$

$$= \frac{2}{l}\int_0^l dx \cos\left(\frac{\pi m x}{l}\right)\left[\frac{A_0}{2} + \sum_{n=1}^{\infty} A_n \cos\left(\frac{\pi n x}{l}\right)\right]$$

$$= \frac{A_0}{l}\int_0^l dx \cos\left(\frac{\pi m x}{l}\right) + \sum_{n=1}^{\infty} A_n \frac{2}{l}\int_0^l dx \cos\left(\frac{\pi m x}{l}\right)\cos\left(\frac{\pi n x}{l}\right)$$

$$= \frac{A_0}{\pi m}\left[\sin\left(\frac{\pi m x}{l}\right)\right]_0^l + \sum_{n=1}^{\infty} A_n \delta_{mn} = A_m,$$

because the first term in the last line vanishes. This is precisely equation (4.52) for the coefficient A_m for $m > 0$. For $m = 0$, we obtain

$$\frac{2}{l}\int_0^l dx\, a(x) = \frac{2}{l}\int_0^l dx\left[\frac{A_0}{2} + \sum_{n=1}^{\infty} A_n \cos\left(\frac{\pi n x}{l}\right)\right]$$

$$= \frac{A_0}{l}\int_0^l dx + \sum_{n=1}^{\infty} A_n \frac{2}{l}\int_0^l dx \cos\left(\frac{\pi n x}{l}\right)$$

$$= A_0 + \sum_{n=1}^{\infty} \frac{2 A_n}{\pi n}\left[\sin\left(\frac{\pi n x}{l}\right)\right]_0^l = A_0,$$

because all terms in the infinite sum vanish. Note that we again interchanged the order of summation and integration, and it is tacitly assumed that the sum converges in an appropriate sense, so that this interchange is justifiable.

Solution 4.8

(a) Integration by parts gives

$$\int_0^z dx\, x \cos(kx) = \left[\frac{x \sin(kx)}{k}\right]_0^z - \frac{1}{k}\int_0^z dx \sin(kx)$$

$$= \frac{z \sin(kz)}{k} + \left[\frac{\cos(kx)}{k^2}\right]_0^z$$

$$= \frac{z \sin(kz)}{k} + \frac{\cos(kz) - 1}{k^2}.$$

(b) We obtain

$$A_0 = \frac{2}{l}\int_0^l dx \left(\frac{l}{2} - x\right) = \frac{2}{l}\left[\frac{lx - x^2}{2}\right]_0^l = 0,$$

and, for $n > 0$,

$$A_n = \frac{2}{l}\int_0^l dx \left(\frac{l}{2} - x\right) \cos\left(\frac{\pi n x}{l}\right)$$

$$= \int_0^l dx \cos\left(\frac{\pi n x}{l}\right) - \frac{2}{l}\int_0^l dx\, x \cos\left(\frac{\pi n x}{l}\right)$$

$$= \left[\frac{l}{\pi n}\sin\left(\frac{\pi n x}{l}\right)\right]_0^l - \frac{2}{l}\left(\frac{l^2 \sin(\pi n)}{\pi n} + l^2 \frac{\cos(\pi n) - 1}{\pi^2 n^2}\right)$$

$$= \frac{l}{\pi n}\sin(\pi n) - \frac{2l}{\pi n}\sin(\pi n) - 2l \frac{\cos(\pi n) - 1}{\pi^2 n^2}$$

$$= [1 - (-1)^n]\frac{2l}{\pi^2 n^2}.$$

Here, we used part (a) with $z = l$ and $k = \pi n/l$. Thus $A_n = 0$ for even n, and for odd $n = 2m - 1$ we find

$$A_{2m-1} = \frac{4l}{\pi^2 (2m - 1)^2}.$$

(c) The coefficients A_n are the absolute values of the corresponding coefficients A_n of equations (4.13) for the Fourier sine series.

(d) The solution of the initial-value problem is given by equation (4.48), with A_n as obtained above and $B_n = 0$. This yields
$$u(x,t) = \sum_{m=1}^{\infty} \frac{4l}{\pi^2(2m-1)^2} \cos\left(\frac{\pi(2m-1)x}{l}\right) \cos\left(\frac{\pi(2m-1)ct}{l}\right).$$
We recover precisely the result of equation (4.14) when we shift x by $l/2$ to the right, because
$$u\left(x - \frac{l}{2}, t\right) = \sum_{m=1}^{\infty} \frac{4l}{\pi^2(2m-1)^2} \cos\left(\frac{\pi(2m-1)x}{l} + \frac{\pi}{2} - m\pi\right) \cos\left(\frac{\pi(2m-1)ct}{l}\right)$$
$$= -\sum_{m=1}^{\infty} \frac{4l}{\pi^2(2m-1)^2} \sin\left(\frac{\pi(2m-1)x}{l} - m\pi\right) \cos\left(\frac{\pi(2m-1)ct}{l}\right)$$
$$= \sum_{m=1}^{\infty} \frac{4l(-1)^{m-1}}{\pi^2(2m-1)^2} \sin\left(\frac{\pi(2m-1)x}{l}\right) \cos\left(\frac{\pi(2m-1)ct}{l}\right),$$
where we used $\cos(y + \pi/2) = -\sin y$ and $\sin(y - m\pi) = (-1)^m \sin y$ for integer m.

Solution 4.9

(a) The boundary conditions are
$$u_x(0,t) = 0, \quad u_x(l,t) + \kappa\, u(l,t) = 0;$$
see equation (4.31) concerning the correct sign at the $x = l$ end. With $u(x,t) = f(x)g(t)$, the corresponding boundary conditions for $f(x)$ are
$$f'(0) = 0, \quad f'(l) + \kappa f(l) = 0.$$
The general solution for $f(x)$ was given in equation (4.35). The boundary condition at $x = 0$ implies that $f(x) = a\cos(kx)$, as in the case with two free ends. At $x = l$, the boundary condition yields
$$0 = -ak\sin(kl) + a\kappa\cos(kl) = a\cos(kl)\left(\kappa - k\tan(kl)\right)$$
(provided $\cos(kl) \neq 0$). This means that $\kappa - k\tan(kl) = 0$, which yields the desired condition.

(b) The functions $\tan z$ and $1/z$ are sketched in Figure 4.15. Solutions of the equation $\tan z = 1/z$ correspond to intersection points of the two graphs. Such points are marked by dots in Figure 4.15.

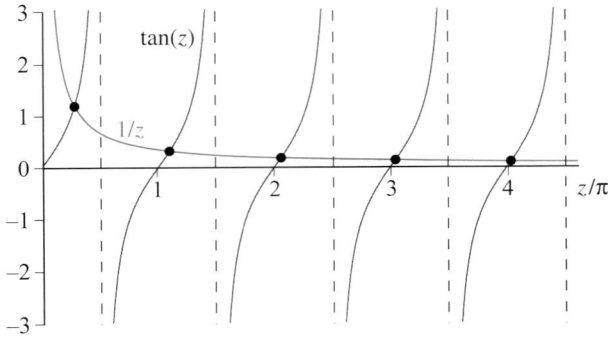

Figure 4.15 Plot of the functions $\tan z$ and $1/z$

Within any interval $(n - \frac{1}{2})\pi < z < (n + \frac{1}{2})\pi$, for $n = 0, 1, 2, 3, \ldots$, there is precisely one solution z_n of the equation $\tan z = 1/z$, because $\tan z$ in that interval is a strictly monotonically increasing function of z which maps the interval to the entire set of real numbers. So the graph of $1/z$ intersects that of $\tan z$ precisely once on each interval. As $1/z$ decreases with increasing z, the solutions move towards the zeros of the tangent function, so $z_n - n\pi \to 0$ as $n \to \infty$.

(c) For the limiting case where $\kappa = 0$, the condition becomes $\tan(kl) = 0$, hence the possible values of k are determined by the condition $\sin(kl) = 0$, which gives $k_n = n\pi/l$, $n = 1, 2, 3, \ldots$.

(d) For the other limiting case, $\kappa \to \infty$, the condition becomes $\cot(kl) = 0$, so $\cos(kl) = 0$, which gives $k_n = \pi(n + \frac{1}{2})/l$, $n = 0, 1, 2, \ldots$.

(e) With increasing n, the solutions z_n are of the form $\pi(n + \epsilon_n)/l$ with decreasing $\epsilon_n > 0$. So for large n, the values of k_n become closer and closer to $k_n = \pi n/l$, which was the limiting case for a weak spring at the end. Because the frequencies are $\nu_n = \omega_n/(2\pi) = k_n c/(2\pi)$, the lower-frequency modes are more affected by the presence of the spring. See equation (2.6).

Solution 4.10

With the parameters given, we have $\omega_{\text{ex}} = 25\pi$ and $k_{\text{ex}} = \omega_{\text{ex}}/c = \pi/8$. We again try a particular solution of the form
$$u^{(\text{par})}(x,t) = U(x)\sin(\omega_{\text{ex}}t) = U(x)\sin(25\pi t).$$
The function $U(x) = C\cos(k_{\text{ex}}x) = C\cos(\pi x/8)$, with some constant C, is a solution of the differential equation (4.62) that satisfies the free boundary condition at $x = 0$, because
$$u_x^{(\text{par})}(0,t) = U'(0)\sin(25\pi t) = 0.$$
At $x = l = 10$, we obtain
$$u^{(\text{par})}(10,t) = C\cos(\pi 5/4)\sin(25\pi t) = -\frac{C}{\sqrt{2}}\sin(25\pi t),$$
so we can satisfy the boundary condition at $x = 10$ by choosing $C = -2\sqrt{2}$. Thus a particular solution is
$$u^{(\text{par})}(x,t) = -2\sqrt{2}\cos(\pi x/8)\sin(25\pi t).$$

Solution 4.11

(a) For $\mu > 0$ and $\omega_0 = 0$, we have $\mu > \omega_0$, so the mode is strongly damped.

(b) The two eigenvalues of the auxiliary equation (4.75) are $\lambda_1 = 0$ and $\lambda_2 = -2\mu$, so equation (4.77) gives
$$g(t) = a + be^{-2\mu t}.$$

(c) For $C = 0$, the differential equation becomes
$$g''(t) + 2\mu g'(t) = 0.$$
Integrating once gives
$$g'(t) + 2\mu g(t) = D,$$
with an arbitrary integration constant D. The general solution of this inhomogeneous first-order differential equation is the sum of the general solution $b\exp(-2\mu t)$ of the homogeneous equation and a particular solution, which we can choose as a constant
$$g(t) = \frac{D}{2\mu} = a.$$
Thus the general solution is
$$g(t) = a + be^{-2\mu t},$$
as obtained above.

(Alternatively, notice that after one integration we obtain an ordinary linear first-order differential equation, and solve using the integrating factor method described in Block 0, Subsection 1.3.2.)

(d) The normal modes of an undamped string with two free ends are given in equation (4.47), and the solution for the case $n = 0$ in equation (4.43). The position-dependent part stays the same, given by $f(x) = \cos(\pi n x/l) = 1$ for $n = 0$. The corresponding solution for the damped string is
$$u_0(x,t) = f(x)g(t) = A_0 + B_0 e^{-2\mu t}.$$

Solution 4.12

We have $\omega_n = \pi nc/l = 10\pi n$.

(a) $\omega_1 = 10\pi < \mu$: strongly damped.
(b) $\omega_2 = 20\pi < \mu$: strongly damped.
(c) $\omega_3 = 30\pi = \mu$: critically damped.
(d) $\omega_4 = 40\pi > \mu$: weakly damped.
(e) $\omega_5 = 50\pi > \mu$: weakly damped.

Solution 4.13

We have $\omega_n = \pi nc/l$. The mode n is critically damped for $\mu = \omega_n = \pi nc/l$, and strongly damped for $\mu > \pi nc/l$.

(a) For a critically damped mode n, we have $\mu = \pi nc/l$. This gives a term

$$(A_n + B_n t)\exp\left(-\frac{\pi nct}{l}\right)\sin\left(\frac{\pi nx}{l}\right)$$

in the general solution.

(b) For a strongly damped mode n, the contribution is

$$\left(A_n e^{\lambda_1 t} + B_n e^{\lambda_2 t}\right)\sin\left(\frac{\pi nx}{l}\right),$$

where $\lambda_{1,2} = -\mu \pm \sqrt{\mu^2 - \omega_n^2} < 0$ are the two solutions of the auxiliary equation.

Solution 4.14

As seen in Exercise 4.12, the modes $n = 1$ and $n = 2$ are strongly damped, the mode $n = 3$ is critically damped, and all other modes $n \geq 4$ are weakly damped. We have $\omega_1 = 10\pi$, thus the solutions of the auxiliary equation are $\lambda_{1,2} = (-30 \pm 20\sqrt{2})\pi$. For $n = 2$, we obtain $\omega_1 = 20\pi$ and $\lambda_{1,2} = (-30 \pm 10\sqrt{5})\pi$. For $n \geq 4$, we have angular frequencies $\Omega_n = 10\pi\sqrt{n^2 - 9}$.

Thus the general solution is

$$u(x,t) = \left(A_1 e^{(-30+20\sqrt{2})\pi t} + B_1 e^{(-30-20\sqrt{2})\pi t}\right)\sin(\pi x)$$
$$+ \left(A_2 e^{(-30+10\sqrt{5})\pi t} + B_2 e^{(-30-10\sqrt{5})\pi t}\right)\sin(2\pi x)$$
$$+ (A_3 + B_3 t)e^{-30\pi t}\sin(3\pi x)$$
$$+ \sum_{n=4}^{\infty}\left[A_n \sin(10\pi\sqrt{n^2-9}\,t) + B_n \cos(10\pi\sqrt{n^2-9}\,t)\right]e^{-30\pi t}\sin(\pi nx).$$

Solution 4.15

Applying the product rule to equation (4.84), we obtain

$$u_t(x,t) = -\mu e^{-\mu t}\sum_{n=1}^{\infty}[A_n \cos(\Omega_n t) + B_n \sin(\Omega_n t)]\sin(k_n x)$$
$$+ e^{-\mu t}\sum_{n=1}^{\infty}[-\Omega_n A_n \sin(\Omega_n t) + \Omega_n B_n \cos(\Omega_n t)]\sin(k_n x),$$

hence

$$b(x) = u_t(x,0) = -\mu\sum_{n=1}^{\infty} A_n \sin(k_n x) + \sum_{n=1}^{\infty}\Omega_n B_n \sin(k_n x)$$
$$= \sum_{n=1}^{\infty}(\Omega_n B_n - \mu A_n)\sin(k_n x),$$

as required.

Solution 4.16

Since the ends are fixed, $\omega_n = \pi nc/l = 20\pi n$ for $n = 1, 2, 3, \ldots$, so $\omega_n > \mu$ and all modes are weakly damped. The general solution is given by

$$u(x,t) = \sum_{n=1}^{\infty} [A_n \cos(\Omega_n t) + B_n \sin(\Omega_n t)] e^{-\mu t} \sin(k_n x),$$

with $k_n = \omega_n/c = \pi n/10$ and $\Omega_n = 2\pi\sqrt{100n^2 - 1}$; see equations (4.84) and (4.83). The coefficients A_n and B_n are given in equations (4.88) and (4.89), respectively. Since $b(x) = 0$, $B_n = \mu A_n/\Omega_n$. The coefficients A_n are $1/10$ of those given in equations (4.13) with $l = 10$, so

See Exercise 4.2.

$$A_{2m-1} = \frac{4(-1)^{m-1}}{\pi^2 (2m-1)^2}, \quad A_{2m} = 0,$$

for $m = 1, 2, 3, \ldots$. The solution of the initial-value problem is thus

$$u(x,t) = \sum_{m=1}^{\infty} \frac{4(-1)^{m-1}}{\pi^2(2m-1)^2} \left[\cos(\Omega_{2m-1}t) + \frac{2\pi}{\Omega_{2m-1}} \sin(\Omega_{2m-1}t)\right] e^{-2\pi t} \sin\left(\frac{(2m-1)\pi x}{10}\right).$$

Solution 4.17

(a) For $\omega_{\text{ex}} = \omega_m$, equation (4.112) becomes

$$|A_m| = \frac{|p_m|}{\sqrt{4\mu^2\omega_{\text{ex}}^2}} = \frac{p_m}{2\mu\omega_m}.$$

(b) For the phase ϕ_m, equations (4.113) and (4.114) become

$$\cos\phi_m = 0,$$
$$\sin\phi_m = -\frac{2\mu\omega_m}{\sqrt{4\mu^2\omega_m^2}} = -1.$$

Hence $\phi_m = -\pi/2$ (up to multiples of 2π).

Solution 4.18

(a) First, the numerator $|p_m|$ of $|A_m|$ given by equation (4.112) is constant, which means that maxima of $|A_m|$ occur at minima of the square root in the denominator. Second, the square root function is monotonically increasing, so these minima coincide with those of the argument under the square root, which gives the quoted expression.

(b) Using the chain rule, the derivative is

$$\frac{d}{d\omega_{\text{ex}}}\left[(\omega_m^2 - \omega_{\text{ex}}^2)^2 + 4\mu^2\omega_{\text{ex}}^2\right] = -4\omega_{\text{ex}}(\omega_m^2 - \omega_{\text{ex}}^2) + 8\mu^2\omega_{\text{ex}}.$$

This vanishes if

$$4\omega_{\text{ex}}\left[-(\omega_m^2 - \omega_{\text{ex}}^2) + 2\mu^2\right] = 0,$$

which gives $\omega_{\text{ex}} = 0$ or $\omega_m^2 - \omega_{\text{ex}}^2 = 2\mu^2$; thus

$$\omega_{\text{ex}} = \pm\sqrt{\omega_m^2 - 2\mu^2}.$$

(c) For a real positive solution of the quadratic equation, we need $\omega_m^2 > 2\mu^2$, thus $\omega_m > \sqrt{2}\mu$. This is a stronger requirement than the weak damping condition $\omega_m > \mu$.

(d) The phase for ω_{ex} at the maximum satisfies

$$\tan\phi_m = \frac{2\mu\omega_{\text{ex}}}{\omega_{\text{ex}}^2 - \omega_m^2} = -\frac{2\mu\omega_{\text{ex}}}{2\mu^2} = -\frac{\omega_{\text{ex}}}{\mu}.$$

As μ becomes small, this ratio is a large negative number, thus ϕ_m approaches $-\pi/2$ (up to integer multiples of π).

Solution 4.19

(a) The extension of $a(x)$ is sketched in Figure 4.16.

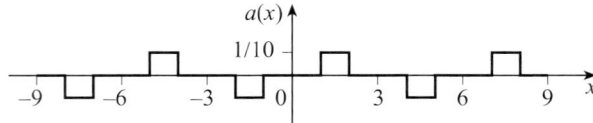

Figure 4.16 The odd periodic extension of $a(x)$

(b) We obtain

$$A_n = \frac{2}{l}\int_0^l dx\, a(x)\sin\left(\frac{\pi nx}{l}\right)$$
$$= \frac{1}{15}\int_1^2 dx\,\sin\left(\frac{\pi nx}{3}\right)$$
$$= -\frac{1}{5\pi n}\left[\cos\left(\frac{\pi nx}{3}\right)\right]_1^2$$
$$= \frac{1}{5\pi n}\left[\cos\left(\frac{\pi n}{3}\right) - \cos\left(\frac{2\pi n}{3}\right)\right],$$

as required.

(c) For even $n = 2m$, we obtain, using $\cos\alpha = \cos(2\pi m - \alpha)$, the result

$$\cos\left(\frac{2\pi m}{3}\right) - \cos\left(\frac{4\pi m}{3}\right) = \cos\left(\frac{2\pi m}{3}\right) - \cos\left(2\pi m - \frac{4\pi m}{3}\right)$$
$$= \cos\left(\frac{2\pi m}{3}\right) - \cos\left(\frac{2\pi m}{3}\right) = 0,$$

so $A_{2m} = 0$.

(d) The non-vanishing coefficients are

$$A_{6p+1} = \frac{1}{5\pi(6p+1)}\left[\cos\left(\frac{\pi}{3}\right) - \cos\left(\frac{2\pi}{3}\right)\right] = \frac{1}{5\pi(6p+1)},$$
$$A_{6p+3} = \frac{1}{5\pi(6p+3)}\left[\cos\pi - \cos 2\pi\right] = -\frac{2}{5\pi(6p+3)},$$
$$A_{6p+5} = \frac{1}{5\pi(6p+5)}\left[\cos\left(\frac{5\pi}{3}\right) - \cos\left(\frac{4\pi}{3}\right)\right] = \frac{1}{5\pi(6p+5)}.$$

(e) As $b(x) = 0$, we have $B_n = 0$ for all n. The solution is

$$u(x,t) = \sum_{p=0}^\infty \left[\frac{1}{5\pi(6p+1)}\sin\left(\frac{\pi(6p+1)x}{3}\right)\cos\left(\frac{\pi(6p+1)ct}{3}\right)\right.$$
$$-\frac{2}{5\pi(6p+3)}\sin\left(\frac{\pi(6p+3)x}{3}\right)\cos\left(\frac{\pi(6p+3)ct}{3}\right)$$
$$\left.+\frac{1}{5\pi(6p+5)}\sin\left(\frac{\pi(6p+5)x}{3}\right)\cos\left(\frac{\pi(6p+5)ct}{3}\right)\right].$$

Solution 4.20

(a) The even extension is obtained by setting $b(x) = b(-x)$ for $-l \leq x \leq 0$, and continuing periodically such that $b(x+2l) = b(x)$. However, in this case it turns out that the function $b(x)$ actually has period l.

(b) The function $b(x)$ is sketched in Figure 4.17.

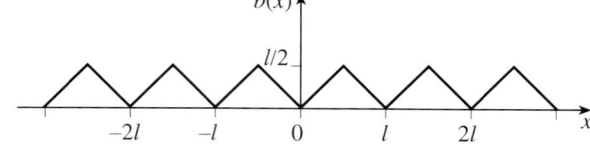

Figure 4.17 An even periodic extension of the function $b(x)$

This even extension is the absolute value $b(x) = |a(x)|$ of the odd extension $a(x)$ of Figure 4.3.

(c) The coefficients are
$$B_n = \frac{2}{l}\int_0^{l/2} dx\, x\cos\left(\frac{\pi n x}{l}\right) + \frac{2}{l}\int_{l/2}^{l} dx\,(l-x)\cos\left(\frac{\pi n x}{l}\right).$$

Substituting $y = l - x$ in the second integral gives
$$B_n = \frac{2}{l}\int_0^{l/2} dx\, x\cos\left(\frac{\pi n x}{l}\right) + \frac{2}{l}\int_0^{l/2} dy\, y\cos\left(\pi n - \frac{\pi n y}{l}\right).$$

Using $\cos(\pi n - y) = (-1)^n \cos y$, this becomes
$$B_n = [1 + (-1)^n]\frac{2}{l}\int_0^{l/2} dx\, x\cos\left(\frac{\pi n x}{l}\right).$$

This vanishes for odd values of n, so $B_{2m-1} = 0$ for $m = 1, 2, 3, \ldots$. For $n = 0$, we obtain
$$B_0 = \frac{4}{l}\int_0^{l/2} dx\, x = \frac{4}{l}\left[\frac{x^2}{2}\right]_0^{l/2} = \frac{l}{2}.$$

For even $n = 2m$ with $m = 1, 2, 3, \ldots$, we obtain
$$B_{2m} = \frac{4}{l}\int_0^{l/2} dx\, x\cos\left(\frac{2\pi m x}{l}\right) = \frac{\pi m l \sin(\pi m) + l[\cos(\pi m) - 1]}{\pi^2 m^2}$$
$$= \frac{l[(-1)^m - 1]}{\pi^2 m^2},$$

where we used the result of Exercise 4.8(a), and the relations $\sin(\pi m) = 0$ and $\cos(\pi m) = (-1)^m$. This vanishes for even m. Thus $B_{4m} = 0$ and
$$B_{4m-2} = \frac{-2l}{\pi^2(2m-1)^2}.$$

(d) The coefficient $\widetilde{B}_0$ is
$$\widetilde{B}_0 = \frac{4}{l}\int_0^{l/2} dx\, x = \frac{4}{l}\left[\frac{x^2}{2}\right]_0^{l/2} = \frac{l}{2},$$

thus $\widetilde{B}_0 = B_0$. For $n > 0$, we obtain
$$\widetilde{B}_n = \frac{4}{l}\int_0^{l/2} dx\, b(x)\cos\left(\frac{2\pi n x}{l}\right)$$
$$= \frac{4}{l}\int_0^{l/2} dx\, x\cos\left(\frac{2\pi n x}{l}\right)$$
$$= \frac{\pi n l \sin(\pi n) + l[\cos(\pi n) - 1]}{\pi^2 n^2}.$$

This is the same expression (with n in place of m) as we obtained for B_{2m}, thus $\widetilde{B}_m = B_{2m}$. As $B_{2m-1} = 0$ for $m = 1, 2, 3, \ldots$, all terms in the two expansions agree.

(e) It suffices to check the given relation for $0 \le x \le l$, due to the periodicity with period $2l$ of both functions, and their symmetry. For $0 \le x \le l/2$,
$$\frac{l}{4} - \frac{a(2x)}{2} = \frac{l}{4} - \frac{l - 4x}{4} = x = b(x),$$

and for $l/2 \le x \le l$, we have
$$a(2x) = a(2x - 2l) = a(2l - 2x) = l/2 - 2l + 2x,$$

thus
$$\frac{l}{4} - \frac{a(2x)}{2} = \frac{l}{4} - \frac{l - 4l + 4x}{4} = l - x = b(x),$$

so the relation holds. Using the series for $a(x)$ obtained in Exercise 4.8 gives
$$b(x) = \frac{l}{4} - \sum_{m=1}^{\infty} \frac{2l}{\pi^2(2m-1)^2}\cos\left(\frac{2\pi(2m-1)x}{l}\right),$$

and we recover the coefficients B_{4m-2} of part (c).

(f) Equation (4.48) gives the general solution for a string with two free ends. Since $u(x,0) = a(x) = 0$, we have $A_n = 0$ for $n = 0, 1, 2, \ldots$. Inserting the calculated values for B_n, taking into account the additional factor $\pi n c/l$ in equation (4.50), gives

$$u(x,t) = \frac{lt}{4} - \sum_{m=1}^{\infty} \frac{l^2}{\pi^3(2m-1)^3 c} \cos\left(\frac{2\pi(2m-1)x}{l}\right) \sin\left(\frac{2\pi(2m-1)ct}{l}\right).$$

Solution 4.21

(a) The solution $u(x,t)$ given in equation (4.40) is periodic in x with period $L = 4l$.

(b) Because cosine is an even function, and $u(x,t)$ of equation (4.40) involves only cosine functions of x, the function $u(-x,t) = u(x,t)$ is even with respect to $x = 0$.

With respect to $x = l$, for the position-dependent term in equation (4.40), replacing x by $2l - x$, we find

$$\cos\left(\frac{\pi(2n-1)(2l-x)}{2l}\right) = \cos\left(\pi(2n-1) - \frac{\pi(2n-1)x}{2l}\right)$$

$$= -\cos\left(\frac{\pi(2n-1)x}{2l}\right),$$

hence $u(2l-x,t) = -u(x,t)$. So the function is odd with respect to $x = l$.

(c) The normal mode expansions of the initial data are

$$a(x) = \sum_{n=1}^{\infty} A_n \cos\left(\frac{\pi(2n-1)x}{2l}\right),$$

$$b(x) = \frac{\pi c}{2l} \sum_{n=1}^{\infty} (2n-1) B_n \cos\left(\frac{\pi(2n-1)x}{2l}\right).$$

(d) We use the orthogonality relation (4.51) with l replaced by $2l$, and m and n renamed to p and q, respectively:

$$\frac{1}{l}\int_0^{2l} dx \cos\left(\frac{\pi p x}{2l}\right) \cos\left(\frac{\pi q x}{2l}\right) = \delta_{pq}.$$

We need only the case when p and q are odd, so

$$\frac{1}{l}\int_0^{2l} dx \cos\left(\frac{\pi(2m-1)x}{2l}\right) \cos\left(\frac{\pi(2n-1)x}{2l}\right) = \delta_{mn}.$$

We can split the integral into two parts, so

$$\frac{1}{l}\int_0^{l} dx \cos\left(\frac{\pi(2m-1)x}{2l}\right) \cos\left(\frac{\pi(2n-1)x}{2l}\right)$$

$$+ \frac{1}{l}\int_l^{2l} dx \cos\left(\frac{\pi(2m-1)x}{2l}\right) \cos\left(\frac{\pi(2n-1)x}{2l}\right) = \delta_{mn},$$

integrating from 0 to l and from l to $2l$, respectively. By substituting $y = 2l - x$, the second half of the integral from l to $2l$ can be rewritten as

$$\frac{1}{l}\int_l^{2l} dx \cos\left(\frac{\pi(2m-1)x}{2l}\right) \cos\left(\frac{\pi(2n-1)x}{2l}\right)$$

$$= \frac{1}{l}\int_l^0 (-dy) \cos\left(\frac{\pi(2m-1)(2l-y)}{2l}\right) \cos\left(\frac{\pi(2n-1)(2l-y)}{2l}\right)$$

$$= \frac{1}{l}\int_0^l dy \cos\left(\frac{\pi(2m-1)(2l-y)}{2l}\right) \cos\left(\frac{\pi(2n-1)(2l-y)}{2l}\right)$$

$$= \frac{1}{l}\int_0^l dy \cos\left(\pi(2m-1) - \frac{\pi(2m-1)y}{2l}\right) \cos\left(\pi(2n-1) - \frac{\pi(2n-1)y}{2l}\right).$$

Now, using $\cos[\pi(2p-1) - z] = -\cos z$, we have $z = \pi(2m-1)y/(2l)$ for $p = m$, and $z = \pi(2n-1)y/(2l)$ for $p = n$, so this becomes

$$\frac{1}{l}\int_0^l dy \cos\left(\frac{\pi(2m-1)y}{2l}\right)\cos\left(\frac{\pi(2n-1)y}{2l}\right),$$

which is the same as the integral from 0 to l. Therefore

$$\frac{1}{l}\int_0^{2l} dx \cos\left(\frac{\pi(2m-1)x}{2l}\right)\cos\left(\frac{\pi(2n-1)x}{2l}\right)$$

$$= \frac{2}{l}\int_0^l dx \cos\left(\frac{\pi(2m-1)x}{2l}\right)\cos\left(\frac{\pi(2n-1)x}{2l}\right) = \delta_{mn},$$

which is the orthogonality relation.

(e) Multiplying the Fourier series for $a(x)$ of part (c) by $(2/l)\cos[\pi(2m-1)x/(2l)]$, and integrating, yields

$$\frac{2}{l}\int_0^l dx\, a(x)\cos\left(\frac{\pi(2m-1)x}{2l}\right)$$

$$= \frac{2}{l}\int_0^l dx \sum_{n=1}^{\infty} A_n \cos\left(\frac{\pi(2n-1)x}{2l}\right)\cos\left(\frac{\pi(2m-1)x}{2l}\right)$$

$$= \sum_{n=1}^{\infty} A_n \frac{2}{l}\int_0^l dx \cos\left(\frac{\pi(2n-1)x}{2l}\right)\cos\left(\frac{\pi(2m-1)x}{2l}\right)$$

$$= \sum_{n=1}^{\infty} A_n \delta_{mn} = A_m,$$

where we have assumed that we may change the order of integration and summation, and used the orthogonality relation. We thus obtain the desired relation for the coefficients, with n replaced by m.

Analogously, from the Fourier series for $b(x)$ obtained in part (c), we obtain

$$\frac{2}{l}\int_0^l dx\, b(x)\cos\left(\frac{\pi(2m-1)x}{2l}\right)$$

$$= \frac{\pi c}{2l}\frac{2}{l}\int_0^l dx \sum_{n=1}^{\infty}(2n-1)B_n \cos\left(\frac{\pi(2n-1)x}{2l}\right)\cos\left(\frac{\pi(2m-1)x}{2l}\right)$$

$$= \frac{\pi c}{2l}\sum_{n=1}^{\infty}(2n-1)B_n \frac{2}{l}\int_0^l dx \cos\left(\frac{\pi(2n-1)x}{2l}\right)\cos\left(\frac{\pi(2m-1)x}{2l}\right)$$

$$= \frac{\pi c}{2l}\sum_{n=1}^{\infty}(2n-1)B_n \delta_{mn} = \frac{\pi(2m-1)c}{2l}B_m,$$

which yields the desired relation for the coefficients, again with n replaced by m.

CHAPTER 5
Waves in two and three dimensions

5.1 Introduction

So far, we have limited the discussion to one-dimensional wave motion. However, we live in a three-dimensional world, and wave phenomena can be two- or three-dimensional as well. Our main example so far has been a stretched string, which is one-dimensional. Water waves (albeit not well described by a linear wave equation) are an example of two-dimensional wave motion, where $u(\boldsymbol{r},t)$ corresponds to the height of the water at position $\boldsymbol{r} = x\mathbf{i} + y\mathbf{j}$ in the two-dimensional plane. Another example is the motion of a drum's membrane, where the function $u(\boldsymbol{r},t)$ describes the deformation of the membrane from its equilibrium position. A three-dimensional example is sound waves in air, where $u(\boldsymbol{r},t)$ may correspond to the pressure of the air at position $\boldsymbol{r} = x\mathbf{i} + y\mathbf{j} + z\mathbf{k}$ in three-dimensional space.

$\mathbf{i}$ and $\mathbf{j}$ denote the unit vectors in a Cartesian coordinate system, with coordinates x and y, respectively.

$\mathbf{k}$ denotes the Cartesian unit vector orthogonal to $\mathbf{i}$ and $\mathbf{j}$.

In this chapter, we consider waves in two and three dimensions. We limit the discussion to *scalar waves*, which means that the wave is described by a (scalar) function $u(\boldsymbol{r},t)$. Sound waves are of this type, where the pressure and density vary as functions of the position variable $\boldsymbol{r}$ and time t.

The main example considered in this chapter is that of transverse waves on a vibrating membrane. By separation of variables, the eigenmodes for two-dimensional waves on membranes of certain shapes are obtained. For rectangular membranes, these involve trigonometric functions as in the case of a string; however, for circular membranes, we find a different set of functions, called Bessel functions. Some properties of these functions are discussed.

We then consider isotropic waves in two and three dimensions. We discuss plane wave solutions, and demonstrate how Fourier transforms can be used to solve partial differential equations.

5.2 Vibrating membranes

You may think of a vibrating membrane as a two-dimensional analogue of a vibrating string. An example of a vibrating membrane is the head of a drum. If we consider small transverse vibrations of a uniform membrane, the wave equation can be derived in a similar way as for the stretched string.

The role of the string tension is now played by an isotropic *surface tension* $T(\boldsymbol{r})$. The physical meaning of this quantity can be understood as being similar to the string tension in the one-dimensional case. If you imagine cutting the membrane along a straight line through the point $\boldsymbol{r}$, the surface tension $T(\boldsymbol{r})$ determines the force (per unit length) required to hold the two parts together. The term *isotropic* means that this force does not depend on the direction of the imagined cut. Furthermore, if the membrane is uniform, the surface tension does not depend on position, so $T(\boldsymbol{r}) = T$ is constant.

We shall not go through the derivation of the wave equation again. The arguments are very similar to the discussion of the transverse vibrations of a stretched string, but technically more tedious due to the second dimension involved. The wave equation for small transverse vibrations of a uniform membrane turns out to be

Two-dimensional isotropic wave equation
$$u_{tt}(\boldsymbol{r},t) = c^2 \left[u_{xx}(\boldsymbol{r},t) + u_{yy}(\boldsymbol{r},t) \right], \tag{5.1}$$

where $u(\boldsymbol{r},t)$ denotes the transverse displacement of the membrane at position $\boldsymbol{r}$ and time t. The wave speed is given by $c^2 = T/\rho_A$, where ρ_A denotes the planar density (mass per unit area) of the membrane, and T (in units of N m^{-1}) is the uniform isotropic surface tension.

In general, the membrane may have any shape. We assume that it is finite and that its boundary forms a closed curve $\mathcal{C}$ in the two-dimensional plane. For a membrane that is fixed at its boundary, like the membrane of a drum, the boundary condition on the function $u(\boldsymbol{r},t)$ is thus $u(\boldsymbol{r},t) = 0$ for all points $\boldsymbol{r}$ on the curve $\mathcal{C}$ and for all times. This is another example of a *Dirichlet boundary condition*, which determines the value of the function on the boundary, in this case a certain curve $\mathcal{C}$. We shall consider mainly two particular shapes of membranes: rectangular and circular.

> We introduced Dirichlet boundary conditions in Subsection 2.1.1.

5.2.1 Free transverse vibrations of a rectangular membrane

We consider a rectangular membrane with dimensions L in the $\mathbf{i}$ direction and M in the $\mathbf{j}$ direction. So we want to solve the wave equation (5.1) in the domain $0 \le x \le L$, $0 \le y \le M$. The boundary conditions are

$$\begin{aligned} u(y\mathbf{j},t) = u(L\mathbf{i}+y\mathbf{j},t) &= 0 \quad \text{for } 0 \le y \le M \text{ and any } t, \\ u(x\mathbf{i},t) = u(x\mathbf{i}+M\mathbf{j},t) &= 0 \quad \text{for } 0 \le x \le L \text{ and any } t. \end{aligned} \tag{5.2}$$

We defer the introduction of initial conditions until later.

Separation of variables

The first step in the solution is the same as in the one-dimensional case. We separate the position and time dependence, and consider the product

$$u(\boldsymbol{r},t) = f(\boldsymbol{r})\, g(t), \tag{5.3}$$

as we did in Chapter 2 for the one-dimensional wave equation. Inserting equation (5.3) in the wave equation (5.1) and rearranging the terms yields

$$\frac{1}{c^2} \frac{g''(t)}{g(t)} = \frac{f_{xx}(\boldsymbol{r}) + f_{yy}(\boldsymbol{r})}{f(\boldsymbol{r})}. \tag{5.4}$$

Again the variables have been separated, as the left-hand side depends only on time t, and the right-hand side depends only on the two-dimensional position variable $\boldsymbol{r}$. Hence both sides must equal a constant C.

5.2 Vibrating membranes

The equation for the time-dependent part becomes

$$g''(t) - c^2 C g(t) = 0, \tag{5.5}$$

which is precisely the same as in the one-dimensional case of a vibrating string; compare equation (2.51). For the function $f(\mathbf{r})$, we are left with a partial differential equation

$$f_{xx}(\mathbf{r}) + f_{yy}(\mathbf{r}) - C f(\mathbf{r}) = 0; \tag{5.6}$$

compare the analogous equation (2.50) of the one-dimensional case. The boundary conditions (5.2) result in boundary conditions

$$f(y\mathbf{j}) = f(L\mathbf{i} + y\mathbf{j}) = f(x\mathbf{i}) = f(x\mathbf{i} + M\mathbf{j}) = 0, \tag{5.7}$$

for all values of x and y in the intervals $0 \leq x \leq L$ and $0 \leq y \leq M$.

Separation of variables yet again

Now we play the same trick again: we separate the x and y dependence by considering a product

$$f(\mathbf{r}) = X(x) Y(y) \tag{5.8}$$

and inserting this into equation (5.6). This gives, after dividing by $X(x)Y(y)$,

$$\frac{X''(x)}{X(x)} + \frac{Y''(y)}{Y(y)} = C. \tag{5.9}$$

This implies that X''/X and Y''/Y are both constant, hence

$$X''(x) - C_1 X(x) = 0, \quad Y''(y) - C_2 Y(y) = 0, \tag{5.10}$$

with $C_1 + C_2 = C$. From equations (5.7), the boundary conditions for X and Y are

$$X(0) = X(L) = 0, \quad Y(0) = Y(M) = 0. \tag{5.11}$$

Solution of the position-dependent part

Comparing equations (5.10) and (5.11) with their respective counterparts (2.50) and (2.54) in the one-dimensional case, it follows that both constants C_1 and C_2 must be negative for non-trivial solutions. Setting $C_1 = -p^2 < 0$ and $C_2 = -q^2 < 0$, equations (5.10) become

$$X''(x) + p^2 X(x) = 0, \quad Y''(y) + q^2 Y(y) = 0. \tag{5.12}$$

The solutions of these equations that satisfy the boundary conditions (5.11) are

$$X_n(x) = a_n \sin(p_n x), \quad p_n = \frac{n\pi}{L}, \quad n = 1, 2, 3, \ldots, \tag{5.13}$$

and

$$Y_m(y) = b_m \sin(q_m y), \quad q_m = \frac{m\pi}{M}, \quad m = 1, 2, 3, \ldots; \tag{5.14}$$

compare equations (2.55). For the rectangular membrane, the position-dependent part of the eigenmodes thus consists of the product of two solutions of the vibrating string problem, one for the x and one for the y direction, so

$$\begin{aligned} f_{n,m}(\mathbf{r}) &= X_n(x) Y_m(y) \\ &= a_{n,m} \sin\left(\frac{n\pi x}{L}\right) \sin\left(\frac{m\pi y}{M}\right), \quad n, m = 1, 2, 3, \ldots. \end{aligned} \tag{5.15}$$

Solution of the time-dependent part

The equation for the time-dependent part is (5.5), which is the same as obtained in the one-dimensional system discussed in Chapter 2.

The separation constant C now takes the values

$$\begin{aligned}C_{n,m} &= -(p_n^2 + q_m^2) \\ &= -\left[\left(\frac{n\pi}{L}\right)^2 + \left(\frac{m\pi}{M}\right)^2\right], \quad n,m = 1,2,3,\ldots.\end{aligned} \quad (5.16)$$

Denoting $C_{n,m} = -k_{n,m}^2 = -\omega_{n,m}^2/c^2$, the solution for $g(t)$ is

$$g_{n,m}(t) = A_{n,m}\cos(\omega_{n,m}t) + B_{n,m}\sin(\omega_{n,m}t), \quad (5.17)$$

Compare equation (2.57).

with

$$\omega_{n,m} = k_{n,m}c = c\sqrt{\left(\frac{n\pi}{L}\right)^2 + \left(\frac{m\pi}{M}\right)^2}, \quad n,m = 1,2,3,\ldots. \quad (5.18)$$

Normal modes of a rectangular membrane

The normal modes have the form

$$\begin{aligned}u_{n,m}(\boldsymbol{r},t) &= f_{n,m}(\boldsymbol{r})\,g_{n,m}(t) \\ &= \sin\left(\frac{\pi n x}{L}\right)\sin\left(\frac{\pi m y}{M}\right)[A_{n,m}\cos(\omega_{n,m}t) + B_{n,m}\sin(\omega_{n,m}t)] \\ &= D_{n,m}\sin\left(\frac{\pi n x}{L}\right)\sin\left(\frac{\pi m y}{M}\right)\cos(\omega_{n,m}t + \phi_{n,m}), \quad (5.19)\end{aligned}$$

with $\omega_{n,m}$ as given in equations (5.18). The last line of equation (5.19) gives an alternative form for $g_{n,m}(t)$ which will be used below. However, in what follows we mainly concentrate on the position-dependent parts $f_{n,m}(\boldsymbol{r})$ of the normal modes $u_{n,m}(\boldsymbol{r},t)$, and refer to these as the *eigenmodes* of the membrane. The time-dependent part is determined by the corresponding eigenvalues $k_{n,m}^2 = \omega_{n,m}^2/c^2$ given in equations (5.18).

We saw in equations (2.52) and (2.53), and Exercise 2.24, that the two expressions given for $g_{n,m}(t)$ are equivalent.

The term 'eigenvalues' is used in the same sense as in the one-dimensional case: $-k_{n,m}^2 = C$ in equation (5.6) are the eigenvalues of the two-dimensional differential operator $\frac{\partial^2}{\partial x^2} + \frac{\partial^2}{\partial y^2}$ with the appropriate boundary conditions. We usually forget about the negative sign and refer to $k_{n,m}^2$ as the eigenvalues, and to $\omega_{n,m}$ as the (angular) eigenfrequencies.

We note that instead of the separated nodes in the one-dimensional case, we now find entire curves on which the normal modes vanish. Such curves are called *nodal lines*. The normal modes $f_{n,m}(\boldsymbol{r})$ of equations (5.15), being products of two sine functions, vanish on the straight lines defined by $\boldsymbol{r} = x_P\mathbf{i} + y\mathbf{j}$ with $x_P = PL/n$ and $y \in [0, M]$, which are along the $\mathbf{j}$ direction, and on the lines defined by $\boldsymbol{r} = x\mathbf{i} + y_Q\mathbf{j}$ with $y_Q = QM/m$ and $x \in [0, L]$, along the $\mathbf{i}$ direction, where $P = 0, 1, 2, \ldots, n$ and $Q = 0, 1, 2, \ldots, m$. The nodal lines in this case form a rectangular pattern themselves.

The cases $P = 0, n$ and $Q = 0, m$ correspond to the boundaries of the rectangular membrane, and would not be considered as nodal lines.

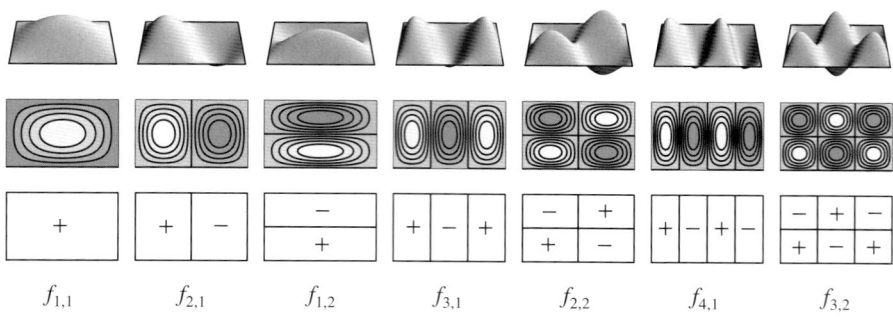

Figure 5.1 Snapshots of the lowest-frequency eigenmodes of a rectangular membrane of side lengths $L = (1+\sqrt{5})l/2$ and $M = l$: perspective views (top row), contour plots (middle row) and the corresponding nodal line patterns (bottom row)

5.2 Vibrating membranes

Figure 5.1 shows the lowest-frequency eigenmodes of a rectangular membrane for a 'golden' rectangle with side lengths satisfying $L/M = (1+\sqrt{5})/2$, the *golden mean* or *golden section*, which is a number that is of major importance in art, architecture and music as well as mathematics. Here, we choose it as an *irrational* number, in order to consider a 'general' rectangle without particular symmetry properties. The eigenmodes are displayed in three different ways: (i) as snapshots of the maximum deformation in perspective view; (ii) as contour plots, with contour lines along constant deformation u, and shading indicating parts with positive u (bright) and negative u (dark); (iii) as patterns of nodal lines, with signs indicating positive and negative values of u. The sign of u changes if a nodal line is crossed, and the nodal line pattern gives a good indication of what the normal mode looks like. As the membrane moves with time, according to a sinusoidal function of the argument $\omega_{n,m} t$, all signs will change simultaneously, which happens when the time-dependent part $g_{n,m}(t)$ of the normal mode motion vanishes. So the given signs apply to half the period of the motion; during the other half, all signs are reversed. The motion in time for a single normal mode is shown in Figure 5.2.

> An irrational number, such as $\sqrt{2}$, is a number that is not rational, i.e. not a ratio of two integers.

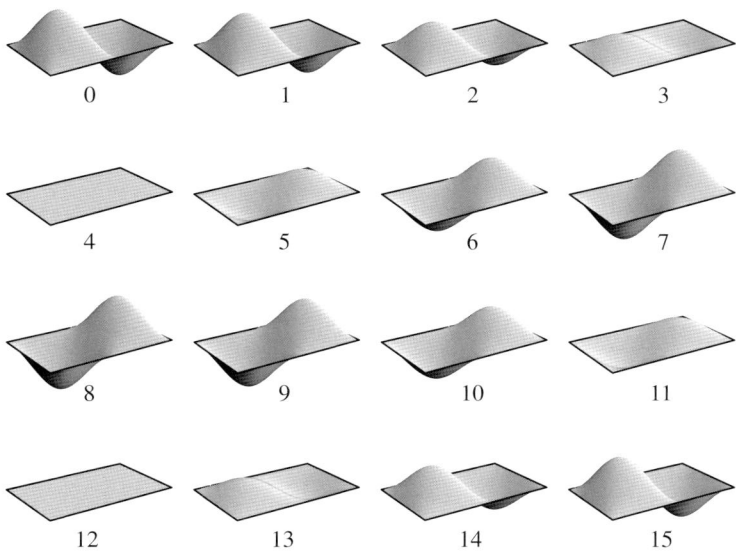

Figure 5.2 One period of the motion for the normal mode $u_{2,1}(\mathbf{r}, t) = f_{2,1}(\mathbf{r}) g_{2,1}(t)$ of the rectangular membrane of side lengths $L = (1+\sqrt{5})l/2$ and $M = l$. The shape of the membrane is depicted at $t = j\tau/16$, where τ is the period of the eigenmode and $j = 0, 1, \ldots, 15$.

What is the significance of the eigenvalue spectrum? The eigenvalues determine the eigenfrequencies $\omega_{n,m}$ according to equations (5.18), so each eigenmode has its own eigenfrequency. These govern the time-dependence of the vibration of the membrane, and also correspond to the frequency of the sound waves generated by a drum, say. As discussed in Chapter 4, a driving force, such as hitting a drum with a stick, will generally excite not just one, but various eigenmodes. So the sound of a drum consists of not a single frequency, but a mixture of frequencies – they all have to be among the eigenfrequencies of the drum, which depend on its shape and material properties.

For a general rectangle, the eigenvalues and hence frequencies for different normal modes are, in general, different from each other. This is the case for the current example, which is the reason why an irrational length ratio was chosen. If the ratio of the squared side lengths L^2/M^2 happens to be a rational number, this statement is no longer true, and some eigenvalues occur (at least) twice, for different normal modes – they are said to be *degenerate.* Whereas there may always be a few 'accidental' degeneracies, a systematic occurrence of degeneracies among eigenvalues invariably points at an underlying (though not always apparent) symmetry of the system.

The square membrane

Let us consider a square membrane as an example that shows ample degeneracies. For a square membrane, we have $L = M = l$, and equations (5.18) simplify to

$$k_{n,m}^2 = \frac{\pi^2}{l^2}(n^2 + m^2) = \frac{\pi^2 N}{l^2}, \quad n, m = 1, 2, 3, \ldots, \tag{5.20}$$

where $N = n^2 + m^2$. So the eigenvalue depends only on the sum N of the squares of the integers n and m. Note that the smallest value of N is $N = 1 + 1 = 2$, which is non-degenerate. The second-smallest eigenvalue corresponds to $N = 1 + 4 = 5$, which occurs twice, because $k_{1,2}^2 = k_{2,1}^2$. Any eigenvalue $k_{n,m}^2$ with different n and m is degenerate, because $k_{n,m}^2 = k_{m,n}^2$. But there are still more degeneracies in this case. As an example, consider the case $N = 65$. This can be written as a sum of squares in four ways, namely as $65 = 1^2 + 8^2 = 8^2 + 1^2 = 4^2 + 7^2 = 7^2 + 4^2$, so there are four different normal mode patterns for the corresponding eigenvalue. The nodal line patterns of the lowest-frequency eigenmodes of the square membrane are shown in Figure 5.3.

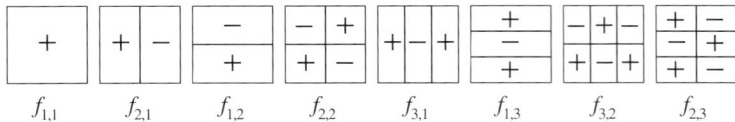

Figure 5.3 Nodal line patterns of the eight lowest-frequency eigenmodes of a square membrane of side lengths $L = M = l$

Exercise 5.1

Consider the values $N = 3, 8, 13, 36, 50, 85$ in equations (5.20). Decide whether or not they correspond to eigenvalues. If a value does correspond to an eigenvalue, determine the number of degenerate normal modes. [Hint: You have to find all ways to write N as a sum of two squares $N = n^2 + m^2$ with integers $n, m = 1, 2, 3, \ldots$.]

What does it mean if an eigenvalue is degenerate? It means that we have several eigenmodes $f_{n,m}(\mathbf{r})$ that satisfy equation (5.6) for the same eigenvalue $k_{n,m}^2$. Therefore they all have the same period and frequency. But then, an arbitrary linear combination of these functions also satisfies the same equation with the same eigenvalue, so there is a certain kind of arbitrariness about the choice of eigenmodes.

> This is analogous to the eigenvalues and eigenvectors of matrices. If an eigenvalue of a matrix is degenerate, there are several linearly independent eigenvectors corresponding to this eigenvalue, and any linear combination of these is again an eigenvector for the same eigenvalue.

We implicitly made the choice of the eigenmodes for the degenerate eigenvalues when we imposed the separation (5.8). An example is considered in Exercise 5.2.

Exercise 5.2

For a square membrane with $L = M = l$, consider the eigenvalue $k_{1,2}^2 = k_{2,1}^2 = 5\pi^2/l^2$ and the corresponding eigenmodes $f_{1,2}(\mathbf{r})$ and $f_{2,1}(\mathbf{r})$ of equations (5.15). (Assume $a_{1,2} = a_{2,1} = 1$ for this exercise.)

(a) What are the nodal lines of $f_{1,2}(\mathbf{r})$ and $f_{2,1}(\mathbf{r})$?

5.2 Vibrating membranes

(b) Verify that the linear combination $f(\mathbf{r}) = A f_{1,2}(\mathbf{r}) + B f_{2,1}(\mathbf{r})$ satisfies equation (5.6) on page 187 with $-C = \omega^2/c^2 = 5\pi^2/l^2$.

(c) Show that the nodal lines of $f(\mathbf{r})$ are determined by the equation $A\cos(\pi y/l) + B\cos(\pi x/l) = 0$. [Hint: Use the trigonometric identity $\sin 2\alpha = 2\sin\alpha\cos\alpha$.]

(d) Calculate the nodal lines for the case $A = -B$. [Hint: Use $\cos 2\beta - \cos 2\alpha = 2\sin(\alpha-\beta)\sin(\alpha+\beta)$.]

(e) Calculate the nodal lines for the case $A = B$. [Hint: Use $\cos 2\beta + \cos 2\alpha = 2\cos(\alpha-\beta)\cos(\alpha+\beta)$.]

A general motion of the membrane is a superposition of normal modes $u(\mathbf{r},t) = \sum_{n,m} a_{n,m}\, u_{n,m}(\mathbf{r},t)$. Whereas the motion in a single normal mode $u_{n,m}(\mathbf{r},t) = f_{n,m}(\mathbf{r})\, g_{n,m}(t)$ is periodic in time, with angular frequency $\omega_{n,m}$, this is, in general, no longer the case for linear combinations of different normal modes. More precisely, a combination of two eigenmodes gives rise to a periodic motion in time if and only if the ratio of the two eigenfrequencies is a rational number. In that case, the two periods of the eigenmodes are *commensurate*, and the period of the combined motion is given by the smallest period that is an integer multiple of both 'eigenperiods'. So not all motions that are periodic in time correspond to a single normal mode; such motions can also be superpositions of normal modes with commensurate frequencies.

As an example, consider the eigenmodes $f_{1,1}(\mathbf{r})$ and $f_{2,2}(\mathbf{r})$ of a square of side $l = \pi$, which have eigenvalues $k_{1,1}^2 = 1^2 + 1^2 = 2$ and $k_{2,2}^2 = 2^2 + 2^2 = 8$, respectively. This means $k_{2,2} = 2k_{1,1}$ and hence $\omega_{2,2} = 2\omega_{1,1}$. So the frequency of the eigenmode $f_{2,2}(\mathbf{r})$ is precisely twice that of the eigenmode $f_{1,1}(\mathbf{r})$. A general motion in these two eigenmodes is of the form

$$u(\mathbf{r},t) = A f_{1,1}(\mathbf{r})\cos(\omega_{1,1}t + \phi_{1,1}) + B f_{2,2}(\mathbf{r})\cos(\omega_{2,2}t + \phi_{2,2}); \quad (5.21)$$

see equation (5.19). Because $\omega_{2,2} = 2\omega_{1,1}$, we have $u(\mathbf{r},t+\tau) = u(\mathbf{r},t)$ for $\tau = \tau_{1,1} = 2\pi/\omega_{1,1} = 2\tau_{2,2}$. Hence $u(\mathbf{r},t)$ has period τ. The motion of this superposition of two eigenmodes is shown in Figure 5.4.

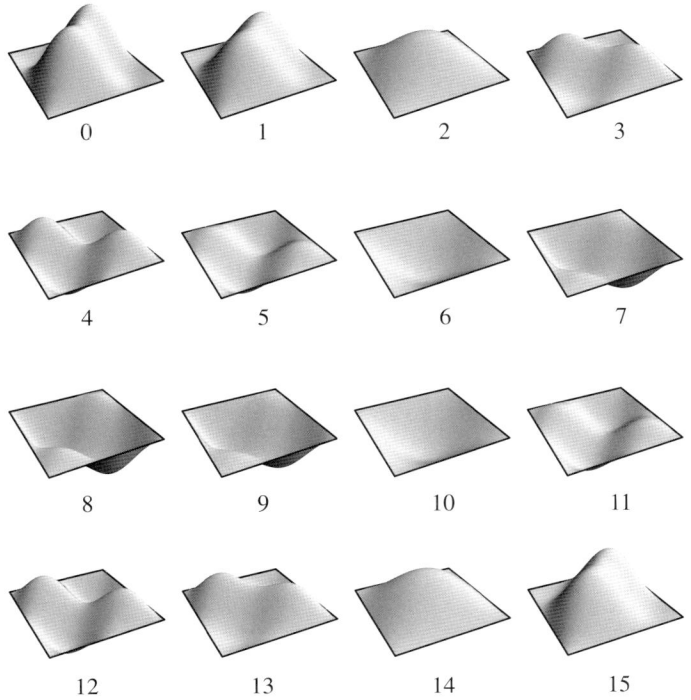

Figure 5.4 One period of the motion of the square membrane in a superposition of the eigenmodes $f_{1,1}(\mathbf{r})$ and $f_{2,2}(\mathbf{r})$. The shape of the membrane is shown at times $t = j\tau/16$ with $j = 1, 2, \ldots, 15$.

The 2 × 1 rectangle

As a second example of a particular rectangle, consider the case where $L = 2M = 2l$. In this 'double-square' case, the eigenvalues $k_{n,m}^2$ of equations (5.18) are given by

$$k_{n,m}^2 = \frac{\pi^2}{4l^2}(n^2 + 4m^2) = \frac{\pi^2 N}{4l^2}, \tag{5.22}$$

where now $N = n^2 + 4m^2$. In this case, the lowest eigenvalue is again non-degenerate, and given by $N = 1^2 + 4 \times 1^2 = 5$. The smallest degenerate eigenvalue corresponds to $N = 20 = 2^2 + 4 \times 2^2 = 4^2 + 4 \times 1^2$. Eigenvalues $k_{n,m}^2$ with even n satisfy $k_{n,m}^2 = k_{2m,n/2}^2$, so they are at least double degenerate for $n \neq 2m$. So again there are systematic degeneracies in the spectrum. The nodal line patterns of the lowest-frequency eigenmodes are shown in Figure 5.5.

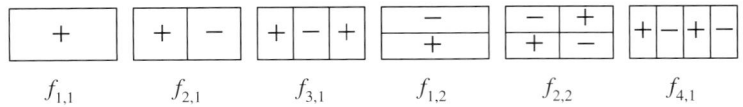

Figure 5.5 Nodal line patterns of the lowest-frequency eigenmodes of a rectangular membrane of side lengths $L = 2M = 2l$

Exercise 5.3

Consider the eigenvalues $k_{n,m}^2$ of equation (5.22) for the $2l \times l$ rectangular membrane.

(a) Show that the eigenfrequencies of all modes with $m = 0$ or $n = 0$ are commensurate.

(b) Show that the frequencies of the modes $n = 2$, $m = 1$ and $n = 2$, $m = 2$ are incommensurate.

(c) What is the period of a motion in a superposition of the modes with $n = 3$, $m = 0$ and $n = 0$, $m = 2$?

Exercise 5.4

(a) Write down the eigenmodes $f_{n,m}$ for the $2l \times l$ rectangular membrane.

(b) Find all eigenmodes which have a nodal line at $x = l$, dividing the $2l \times l$ rectangle into two squares.

(c) Why can these modes also be interpreted as modes of an $l \times l$ square membrane?

(d) Verify that this yields all modes of the $l \times l$ square membrane.

Initial-value problem for the rectangular membrane

We briefly discuss the initial-value problem for rectangular membranes. The initial data are

$$u(\boldsymbol{r}, 0) = a(\boldsymbol{r}), \quad u_t(\boldsymbol{r}, 0) = b(\boldsymbol{r}), \tag{5.23}$$

and we are interested in the solution $u(\boldsymbol{r}, t)$ that satisfies the boundary conditions (5.2) of an $L \times M$ rectangular membrane.

5.2 Vibrating membranes

We proceed as in the one-dimensional case. The first step is to write the general solution of the boundary-value problem as a linear combination of the normal modes (5.19). This gives

General solution for the rectangular membrane

$$u(\mathbf{r},t) = \sum_{n=1}^{\infty}\sum_{m=1}^{\infty} \sin\left(\frac{\pi n x}{L}\right) \sin\left(\frac{\pi m y}{M}\right)$$
$$\times [A_{n,m}\cos(\omega_{n,m}t) + B_{n,m}\sin(\omega_{n,m}t)], \quad (5.24)$$

with real constants $A_{n,m}$ and $B_{n,m}$, and $\omega_{n,m}^2 = \pi^2 c^2(n^2/L^2 + m^2/M^2)$; compare equations (5.18). This may look complicated on first view. However, if you compare it to the one-dimensional case of a vibrating string, you will realise that it is very similar, except that now it involves a double series, one for the x-coordinate, and the other for the y-coordinate. So we can again solve the initial-value problem by Fourier series, as in the one-dimensional case.

See Section 4.2 of Chapter 4.

Inserting $t=0$ into the general solution gives

$$a(\mathbf{r}) = u(\mathbf{r},0) = \sum_{n=1}^{\infty}\sum_{m=1}^{\infty} A_{n,m} \sin\left(\frac{\pi n x}{L}\right)\sin\left(\frac{\pi m y}{M}\right), \quad (5.25)$$

which is a double Fourier series with coefficients $A_{n,m}$. The partial derivative at $t=0$ gives

$$b(\mathbf{r}) = u_t(\mathbf{r},0)$$
$$= \pi c \sum_{n=1}^{\infty}\sum_{m=1}^{\infty} B_{n,m}\sqrt{\frac{n^2}{L^2}+\frac{m^2}{M^2}} \sin\left(\frac{\pi n x}{L}\right)\sin\left(\frac{\pi m y}{M}\right). \quad (5.26)$$

We can calculate the coefficients in a way that is completely analogous to that for the vibrating string, but we need to integrate twice, over x and y.

For instance, consider equation (5.25). Take x fixed for the moment, and consider this as a Fourier series in $\sin(\pi m y/M)$, so

$$a(\mathbf{r}) = \sum_{m=1}^{\infty} A_m(x) \sin\left(\frac{\pi m y}{M}\right), \quad (5.27)$$

where $\quad A_m(x) = \sum_{n=1}^{\infty} A_{n,m} \sin\left(\frac{\pi n x}{L}\right).$

We know how to calculate the coefficients $A_m(x)$; they are given by

$$A_m(x) = \frac{2}{M}\int_0^M dy\, a(\mathbf{r}) \sin\left(\frac{\pi m y}{M}\right). \quad (5.28)$$

Compare equation (4.6).

Now, in equation (5.27) the coefficients $A_m(x)$ are in turn expressed as a Fourier series in $\sin(\pi n x/L)$, so we integrate once more to obtain

$$A_{n,m} = \frac{2}{L}\int_0^L dx\, A_m(x) \sin\left(\frac{\pi n x}{L}\right)$$
$$= \frac{4}{LM}\int_0^L dx \int_0^M dy\, a(\mathbf{r}) \sin\left(\frac{\pi n x}{L}\right)\sin\left(\frac{\pi m y}{M}\right). \quad (5.29)$$

In the same way, we obtain

$$B_{n,m} = \frac{4}{\pi c \sqrt{n^2 M^2 + m^2 L^2}}$$
$$\times \int_0^L dx \int_0^M dy\, b(\mathbf{r}) \sin\left(\frac{\pi n x}{L}\right)\sin\left(\frac{\pi m y}{M}\right). \quad (5.30)$$

With this, we have the complete solution of the initial-value problem for rectangular membranes. Once more, the solution is expressed as a sum over normal modes in equation (5.24), with coefficients $A_{n,m}$ and $B_{n,m}$ which are given by the Fourier series of the initial data $a(\mathbf{r})$ and $b(\mathbf{r})$, respectively.

Exercise 5.5

Consider a rectangular membrane with $L = 3$ and $M = 1$. The initial conditions are $a(\boldsymbol{r}) = 4\sin(2\pi x)\sin(2\pi y)$ and $b(\boldsymbol{r}) = 0$.

(a) Verify that the initial data satisfy the boundary conditions.

(b) Calculate the solution $u(\boldsymbol{r}, t)$ of the initial-value problem.

5.2.2 A triangular membrane

Returning to the square membrane, as $k_{n,m}^2 = k_{m,n}^2$, any eigenvalue where n and m differ occurs at least twice. As we shall see below, the differences between the corresponding eigenmodes (5.15), with equal coefficients, are the eigenmodes of certain triangular membranes. They have the form

Eigenmodes for right-angled isosceles triangular membrane

$$f_{n,m}^{(-)}(\boldsymbol{r}) = f_{n,m}(\boldsymbol{r}) - f_{m,n}(\boldsymbol{r})$$
$$= \sin\left(\frac{n\pi x}{l}\right)\sin\left(\frac{m\pi y}{l}\right) - \sin\left(\frac{m\pi x}{l}\right)\sin\left(\frac{n\pi y}{l}\right). \quad (5.31)$$

These functions vanish on the diagonal $y = x$ of the square,

$$f_{n,m}^{(-)}(x, x) = 0, \quad (5.32)$$

so this diagonal is a nodal line.

This means that the function $f_{n,m}^{(-)}(\boldsymbol{r})$, for any pair of different positive integers n and m, is also a solution for the boundary-value problem of a *triangular membrane*, where the triangle is an isosceles right-angled triangle with edge lengths l and $\sqrt{2}l$ – just half the square we started with. Which of the two halves of the square we choose is irrelevant; for definiteness, say $\boldsymbol{r} \in \{x\mathbf{i} + y\mathbf{j} : x, y \in [0, l], x \geq y\}$.

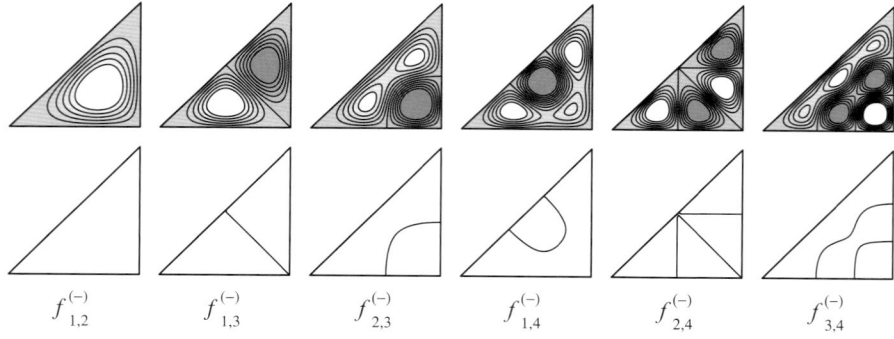

Figure 5.6 The lowest-frequency eigenmodes of the triangular membrane, shown as a contour plot (top row) and as a pattern of nodal lines (bottom row)

The lowest-frequency eigenmodes are shown in Figure 5.6. Due to the linear combinations, the nodal lines need not be straight. For the square, we could also use these linear combinations of degenerate eigenmodes, so nodal lines need not be straight lines in that case either, but for these modes of the triangle we have no choice – there is no degeneracy left, so there is no 'simpler' choice of eigenmodes available.

The question arises whether *all* eigenmodes of this particular triangular membrane are obtained in this way? The answer is affirmative, which can be seen as follows. Given an arbitrary eigenmode $f(\boldsymbol{r})$ of the triangular membrane, we can extend it to an eigenmode of the square membrane by setting $f(\boldsymbol{r}') = -f(\boldsymbol{r})$, where $\boldsymbol{r}' = y\mathbf{i} + x\mathbf{j}$ is the vector that corresponds to the mirror position of $\boldsymbol{r} = x\mathbf{i} + y\mathbf{j}$ in the other half of the square $x, y \in [0, l]$. This extended function satisfies the wave equation and the boundary conditions for the square membrane, is continuous and differentiable everywhere, and in addition has a nodal line for $x = y$. The corresponding eigenvalue must therefore also be an eigenvalue for the square membrane, and the corresponding normal mode must be a normal mode or a linear combination of normal modes for the square membrane that corresponds to the particular eigenvalue. Therefore there can be no normal modes of the isosceles right-angled triangle besides those which we can obtain from *all* linear combinations of normal modes of the square membrane with the same frequency which have a nodal line at $x = y$. They also need to be anti-symmetric under the exchange of x and y, which then limits us to the modes of equation (5.31).

There is one non-trivial step hidden in this argument. We must ensure that the solution for the square made up by 'gluing' together the two parts for the triangles indeed satisfies the wave equation along the 'gluing' line. The function itself vanishes by assumption, but the derivatives need not behave appropriately. However, it works for the case at hand, as can be seen from symmetry arguments. The definition with the 'mirror' solution guarantees that derivatives from both sides of the 'gluing' line agree. This is why we chose $f(\boldsymbol{r}') = -f(\boldsymbol{r})$ rather than $f(\boldsymbol{r}') = f(\boldsymbol{r})$.

5.2.3 Non-separable boundaries

For a general triangle (or a membrane of general shape), the method of separation of variables fails. This is because the boundary condition on the function $f(\boldsymbol{r})$, namely that $f(\boldsymbol{r}) = 0$ on the triangle, does not translate into boundary conditions on the functions $X(x)$ and $Y(y)$ of equation (5.8), which each depend on only a single variable, x or y. In general, such boundaries lead to *non-separable* boundary-value problems, which cannot be solved by the method of separation of variables. Only for a number of rather special shapes can the complete set of eigenmodes be written down as a closed expression. Besides the rectangles and the isosceles right-angled triangle discussed above, these include equilateral triangles, and circular and elliptic membranes.

If the methods introduced here allow us to solve only these special cases, and cannot be applied in a more general setting, why do we bother to discuss them in so much detail? With today's computing power, it is, in principle, possible to calculate normal mode patterns for rather complicated shapes, at least for low frequencies. Nevertheless, these computations are by no means easy to perform, and require sophisticated numerical methods. But as far as the low-frequency modes – often those of interest – are concerned, these are not so different if the shape of the membrane is deformed slightly. For such cases, the results obtained for our 'nice' separable membranes can at least give us a qualitative understanding of the motion. In essence, the general picture of how a membrane moves can be understood by investigating simple examples.

5.2.4 Circular membranes

We now consider a circular membrane of radius l. In this case, we use polar coordinates r and θ instead of Cartesian coordinates $x = r\cos\theta$ and $y = r\sin\theta$. For simplicity, in what follows we write the solution explicitly as a function of the three variables r, θ and t, using the same letter u; so $u(r, \theta, t) = u(\boldsymbol{r}, t)$ denotes the function expressed in polar coordinates. The homogeneous wave equation for the displacement $u(r, \theta, t)$ then becomes

$$\frac{\partial^2}{\partial t^2} u(r,\theta,t) - c^2 \left[\frac{1}{r} \frac{\partial}{\partial r} \left(r \frac{\partial}{\partial r} u(r,\theta,t) \right) + \frac{1}{r^2} \frac{\partial^2}{\partial \theta^2} u(r,\theta,t) \right] = 0, \quad (5.33)$$

where we have used the Laplacian operator in polar coordinates which was introduced in Block 0, Section 2.4. For the circular membrane, the boundary condition is simple in polar coordinates:

$$u(l, \theta, t) = 0, \quad (5.34)$$

for all angles $0 \le \theta < 2\pi$ and any time t. This is the reason why polar coordinates are preferable here; in Cartesian coordinates the boundary condition is $u(\boldsymbol{r}, t) = 0$ for $r^2 = x^2 + y^2 = l^2$, which mixes the position coordinates x and y in an awkward way.

Even though there is no actual boundary condition on the angle variable, we need to impose a condition in order to ensure that our solution makes sense. As θ is an angle, the function $u(r, \theta, t)$ must return to its original value after a full rotation. In other words, we require

$$u(r, \theta + 2\pi, t) = u(r, \theta, t), \quad (5.35)$$

so the function must be *periodic* in the angle variable θ. Note that this condition originates from our choice of coordinates rather than being a physical boundary condition of the membrane.

In equation (5.35), we extend the range of θ outside the interval $0 \le \theta < 2\pi$. Alternatively, we could impose the conditions $u(r, 0, t) = \lim_{\theta \to 2\pi} u(r, \theta, t)$ and $u_\theta(r, 0, t) = \lim_{\theta \to 2\pi} u_\theta(r, \theta, t)$.

As a first step, we once more separate the time dependence by considering

$$u(r, \theta, t) = v(r, \theta)\, g(t); \quad (5.36)$$

compare equation (5.3). The function $g(t)$ is once more a solution of equation (5.5). For normal mode solutions, the separation constant C must be negative, so we once more choose $C = -k^2 = -\omega^2/c^2$, which gives

$$g(t) = A \cos(\omega t) + B \sin(\omega t) \quad (5.37)$$

for the time-dependent part.

For the position-dependent part $v(r, \theta)$, the equation becomes

$$\frac{1}{r} \frac{\partial}{\partial r} \left(r \frac{\partial}{\partial r} v(r,\theta) \right) + \frac{1}{r^2} \frac{\partial^2}{\partial \theta^2} v(r,\theta) = C\, v(r,\theta) = -k^2\, v(r,\theta). \quad (5.38)$$

The periodicity and boundary conditions on $v(r, \theta)$ are

$$v(r, \theta + 2\pi) = v(r, \theta), \quad v(l, \theta) = 0, \quad (5.39)$$

for $0 \le r \le l$ and $0 \le \theta < 2\pi$.

As a second step, we now separate the two variables r and θ, by considering the product

$$v(r, \theta) = R(r)\, \Theta(\theta), \quad (5.40)$$

and substituting this into equation (5.38). This gives

$$\frac{r}{R(r)} \frac{d}{dr} \left(r \frac{d}{dr} R(r) \right) + k^2 r^2 = -\frac{1}{\Theta(\theta)} \frac{d^2}{d\theta^2} \Theta(\theta). \quad (5.41)$$

Now we use the same argument as before – the left-hand side of equation (5.41) depends only on r, and the right-hand side depends only on θ.

5.2 Vibrating membranes

The two sides can be equal for arbitrary values of r and θ only if both are individually constant. If we call this constant $-C_1$, then we arrive at the two ordinary differential equations

$$\Theta''(\theta) - C_1 \Theta(\theta) = 0 \tag{5.42}$$

and

$$r\frac{d}{dr}\left(r\frac{d}{dr}R(r)\right) + \left(k^2 r^2 + C_1\right) R(r) = 0 \tag{5.43}$$

for $\Theta(\theta)$ and $R(r)$, respectively. The periodicity and boundary conditions are

$$\Theta(\theta + 2\pi) = \Theta(\theta), \quad R(l) = 0. \tag{5.44}$$

The angular equation

We start with the angular equation (5.42). This is again our favourite second-order differential equation. In order to have non-trivial solutions, we once more require $C_1 \leq 0$, because the hyperbolic functions sinh and cosh that appear for positive C_1 do not satisfy the periodicity condition $\Theta(\theta + 2\pi) = \Theta(\theta)$. Setting $-C_1 = n^2$ with $n \geq 0$, the general solution is

$$\Theta_n(\theta) = A_n \cos(n\theta) + B_n \sin(n\theta) = D_n \sin(n\theta + \phi_n) \tag{5.45}$$

for $n > 0$, and $\Theta_0(\theta) = A_0 + B_0 \theta$ for $n = 0$. As before (compare equation (5.19)), we use two alternative expressions for the solution. Periodicity requires

$$\begin{aligned} 0 &= \Theta_n(\theta + 2\pi) - \Theta_n(\theta) \\ &= D_n[\sin(n\theta + 2n\pi + \phi_n) - \sin(n\theta + \phi_n)], \end{aligned} \tag{5.46}$$

which is satisfied when n is an integer, because $\sin(\alpha + 2\pi n) = \sin \alpha$ for integer n. The case $n = 0$ yields a constant function $\Theta_0(\theta) = A_0$. So the angular eigenmodes are

$$\begin{aligned} \Theta_0(\theta) &= A_0, \\ \Theta_n(\theta) &= A_n \cos(n\theta) + B_n \sin(n\theta) \\ &= D_n \sin(n\theta + \phi_n), \quad n = 1, 2, 3, \ldots. \end{aligned} \tag{5.47}$$

The solution (5.47) determines the angular variation of the eigenmodes for the circular membrane. For $n = 0$, the angular part is a constant. For higher values of n, the function $\Theta_n(\theta)$ vanishes at $2n$ values θ_p, $\theta_p + \pi$, where $\theta_p = -\phi_n + p\pi/n$ with $p = 0, 1, 2, \ldots, n-1$. The corresponding eigenmodes therefore have n nodal lines which divide the circular membrane into $2n$ identical sections; see Figure 5.7. The actual positions of the nodal lines, i.e. their angles with respect to the coordinate axes, depend on the choice of the coefficients A_n, B_n, or the angle ϕ_n, respectively, in equations (5.47). By a proper choice, any position can be achieved. This makes sense, because no point on the circle is distinguished, and our choice of axes is arbitrary. So we should expect that with any eigenmode we also have its rotated versions.

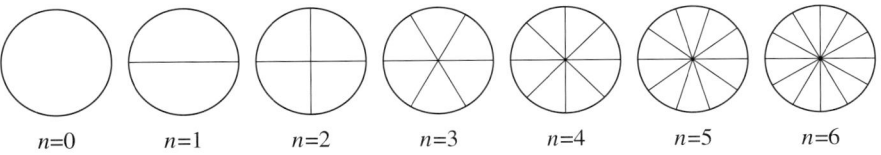

Figure 5.7 Angular nodal line patterns of eigenmodes of a circular membrane. For $n > 0$, the nodal lines shown are those of $\Theta_n(\theta)$ of equation (5.47) with $A_n = 0$ (or $\phi_n = 0$ for $n > 0$).

In other words, the angular part already shows us that all eigenmodes, apart from the fundamental mode with $n = 0$, are degenerate, because solutions with the same value of n, and hence the same value of C_1, will correspond to the same eigenvalue. For a degenerate eigenvalue, there are infinitely many such eigenmodes, because we are free to choose the two coefficients in equations (5.47). How does one count the degree of degeneracy of an eigenvalue? This is done in exactly the same way as in the case of matrix eigenvalues. The degree of degeneracy is the maximum number of *linearly independent* solutions for a given eigenvalue. In other words, it is the *minimum* number of functions that we have to take into account such that *any* eigenmode for the given eigenvalue can be written as a linear combination of these 'basis functions'. In this example, the degree of degeneracy can be read off from equations (5.47), which for each value of n involves two arbitrary constants. Any solution is a linear combination of *two* functions, so the eigenvalues with $n > 0$ are twice degenerate.

The radial equation

Let us now move on to the radial equation (5.43). We can rewrite it in the form

$$r^2 R''(r) + r R'(r) + (k^2 r^2 - n^2) R(r) = 0. \tag{5.48}$$

Introducing $z = kr$ as the new variable, and defining $Z(z) = Z(kr) = R(r)$, we have

$$Z'(z) = \frac{1}{k} R'(r), \quad Z''(z) = \frac{1}{k^2} R''(r), \tag{5.49}$$

so equation (5.48) becomes

Bessel differential equation

$$z^2 Z''(z) + z Z'(z) + (z^2 - n^2) Z(z) = 0. \tag{5.50}$$

In principle, additional degeneracies may occur via the solution of the radial equation; this does not happen for the case of the circular membrane.

This is known as the *Bessel differential equation*. As it is a second-order linear ordinary differential equation, it has two linearly independent solutions. A common choice is to use linearly independent functions $J_n(z)$ and $Y_n(z)$, known as *Bessel functions* of the first and second kind, respectively. Expressing the general solution of equation (5.50) as a linear combination of these, we obtain

$$Z_n(z) = a_n J_n(z) + b_n Y_n(z). \tag{5.51}$$

Therefore the general solution of the radial equation (5.48) can be written as

$$R_n(r) = a_n J_n(kr) + b_n Y_n(kr). \tag{5.52}$$

Before we discuss the normal modes of the circular membrane, we briefly introduce some properties of Bessel functions.

5.2.5 Bessel functions

Bessel functions are named after Friedrich Wilhelm Bessel (1784–1846), who introduced them when studying planetary motion. Bessel functions are examples of *special functions* that arise in mathematical physics as solutions of differential equations. While you will be reasonably familiar with trigonometric and hyperbolic functions such as sine, cosine, hyperbolic sine and hyperbolic cosine, it is quite probable that you have not encountered Bessel functions previously.

Bessel functions are sometimes called cylinder functions or cylindrical harmonics.

5.2 Vibrating membranes

In mathematical physics, Bessel functions occur naturally when one considers two-dimensional systems with rotational symmetry, such as our example of a vibrating circular membrane. For this reason, we embark on a slight detour at this stage, and spend a few pages on Bessel functions, quoting some properties that will be used later on, in most cases without proof. Traditionally, values of Bessel functions, and other special functions of mathematical physics which cannot be calculated by basic arithmetic operations, were collated in mathematical tables. Nevertheless, you should not think of these functions as being in any way more mysterious or unusual than the trigonometric and hyperbolic functions mentioned above; just as for the latter, Bessel functions can be defined as solutions of second-order ordinary differential equations. The reason that a standard calculator has buttons for trigonometric functions but not for Bessel functions (although more sophisticated computer packages do) is just that the former are more commonly used.

Bessel functions of the first kind, $J_n(z)$, are solutions of the differential equation (5.50) which are finite at $z = 0$. For the case that $n \geq 0$ is an integer, they can also be expressed as

Bessel functions of the first kind
$$J_n(z) = \frac{1}{\pi i^n} \int_0^\pi d\theta\, e^{iz\cos\theta} \cos(n\theta), \quad n = 0, 1, 2, \ldots. \tag{5.53}$$

Despite the complex numbers appearing in this formula, the Bessel functions are real, as can be seen from the expression

$$J_n(z) = \frac{1}{\pi} \int_0^\pi d\theta\, \cos[z\sin\theta - n\theta], \tag{5.54}$$

which can be shown to be equivalent to equation (5.53). The Bessel functions are normalised such that

$$\int_0^\infty dz\, J_n(z) = 1. \tag{5.55}$$

For $n = 0, 1, 2$, graphs of the Bessel functions J_n are shown in Figure 5.8.

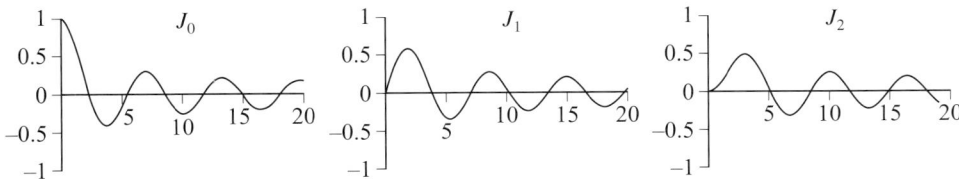

Figure 5.8 The Bessel functions J_0, J_1 and J_2

In fact, for $n = 0$ we have $J_0(0) = 1$, and for $n = 1, 2, \ldots$, equation (5.53) yields

$$J_n(0) = \frac{1}{\pi i^n} \int_0^\pi d\theta\, \cos(n\theta)$$
$$= \frac{1}{n\pi i^n} \big[\sin(n\theta)\big]_0^\pi = 0. \tag{5.56}$$

A set of linearly independent solutions which diverge for $z = 0$ are the *Bessel functions of the second kind*, $Y_n(z)$, so the general solution of equation (5.50) for integer values of n has the form $Z(z) = aJ_n(z) + bY_n(z)$, with $b = 0$ if $Z(0)$ is finite.

These Bessel functions are sometimes called *Neumann functions*.

Exercise 5.6

Consider equation (5.53) for $n = 0$.

(a) Write J_0 as a sum of its real and imaginary parts.

(b) Show that the imaginary part vanishes. [Hint: Split the integration at $\theta = \pi/2$, and substitute $\phi = \pi - \theta$ in the integral from $\pi/2$ to π.]

(c) Show that the real part agrees with equation (5.54). [Hint: Split the integration at $\theta = \pi/2$, and substitute $\phi = \pi/2 - \theta$ and $\phi = 3\pi/2 - \theta$, respectively.]

Behaviour for small arguments

To investigate the behaviour of solutions of the differential equation (5.50) for small z, we consider a solution that satisfies

$$Z(z) = z^m + O(z^{m+1}) \tag{5.57}$$

for a suitable integer $m \geq 0$. We can think of this as the first term in a Taylor series expansion of $Z(z)$ about $z = 0$. Inserting this into the differential equation (5.50) gives

$$z^2 m(m-1) z^{m-2} + zm z^{m-1} + (z^2 - n^2) z^m = O(z^{m+1}). \tag{5.58}$$

Keeping only terms of order z^m gives

$$[m(m-1) + m - n^2] z^m = [m^2 - n^2] z^m = 0, \tag{5.59}$$

so $m^2 = n^2$, and it follows that $J_n(z) = C_n z^n + O(z^{n+1})$ for a suitable constant C_n. The other solution, $m = -n$, is dismissed because $J_n(0)$ is finite by definition. Without proof, we quote the result for the coefficient C_n, which is determined by the normalisation condition (5.55), giving

$$J_n(z) = \frac{z^n}{2^n \, n!} + O(z^{n+1}) \tag{5.60}$$

for the small-argument behaviour of the Bessel functions $J_n(z)$.

Exercise 5.7

In this exercise, we demonstrate that the leading term in the Taylor expansion of (5.53) agrees with our result (5.60).

(a) Use $\cos\theta = (e^{i\theta} + e^{-i\theta})/2$ to derive the relation

$$2\cos\theta \cos(n\theta) = \cos[(n+1)\theta] + \cos[(n-1)\theta].$$

(b) Use this relation to express $\cos^2\theta \cos(n\theta)$ as a sum of cosines.

(c) Hence find which values of m enter terms $\cos(m\theta)$ in the corresponding expression for $\cos^k \theta \cos(n\theta)$. Show that $\cos^n \theta \cos(n\theta) = 2^{-n} + \cdots$, where the other terms involve functions $\cos(m\theta)$ with $m = 2, 4, \ldots, 2n$.

(d) From equation (5.53), calculate the kth derivative $J_n^{(k)}(0)$ at $z = 0$. Use the previous result to argue that $J_n^{(k)}(0) = 0$ for $k < n$.

(e) Calculate the first non-vanishing term in the Taylor expansion of $J_n(z)$ about $z = 0$, and compare with equation (5.60).

This argument can be extended to obtain a power series for $J_n(z)$, which is done in an appendix at the end of this chapter.

See page 220.

Behaviour for large arguments

In order to investigate the asymptotic behaviour of the Bessel functions for large z, consider the function $B(z)$ defined by $Z(z) = z^{-\frac{1}{2}} B(z)$. Using the product rule, the derivatives of $Z(z)$ are

$$Z'(z) = z^{-\frac{1}{2}} B'(z) - \tfrac{1}{2} z^{-\frac{3}{2}} B(z) \qquad (5.61)$$

and

$$Z''(z) = z^{-\frac{1}{2}} B''(z) - z^{-\frac{3}{2}} B'(z) + \tfrac{3}{4} z^{-\frac{5}{2}} B(z). \qquad (5.62)$$

By inserting these into equation (5.50), we obtain a differential equation for $B(z)$:

$$\begin{aligned}
0 &= z^2 Z''(z) + z Z'(z) + (z^2 - n^2) Z(z) \\
&= z^{\frac{3}{2}} B''(z) - z^{\frac{1}{2}} B'(z) + \tfrac{3}{4} z^{-\frac{1}{2}} B(z) + z^{\frac{1}{2}} B'(z) - \tfrac{1}{2} z^{-\frac{1}{2}} B(z) \\
&\quad + z^{\frac{3}{2}} B(z) - n^2 z^{-\frac{1}{2}} B(z) \\
&= z^{\frac{3}{2}} \left[B''(z) + \left(\tfrac{1}{4} z^{-2} + 1 - n^2 z^{-2} \right) B(z) \right] \\
&= z^{\frac{3}{2}} \left[B''(z) + \left(1 + \frac{1 - 4n^2}{4 z^2} \right) B(z) \right]. \qquad (5.63)
\end{aligned}$$

Hence $B(z)$ satisfies the differential equation

$$B''(z) + \left(1 + \frac{1 - 4n^2}{4 z^2} \right) B(z) = 0. \qquad (5.64)$$

For sufficiently large values of z, in particular $z \gg |n|$, the second term in the coefficient of $B(z)$ is small compared to the first, and the solutions resemble those of the differential equation

> The symbol $\gg$ means 'much larger than'.

$$B''(z) + B(z) = 0, \qquad (5.65)$$

which are of the form $a \cos z + b \sin z$. So, for large arguments, the Bessel functions $J_n(z)$ and $Y_n(z)$ behave like $z^{-\frac{1}{2}}[a_n \cos z + b_n \sin z]$. Without proof, we quote the result that the coefficients are such that, for large arguments z, the normalised Bessel functions are asymptotically given by

$$\begin{aligned}
J_n(z) &\simeq \sqrt{\frac{2}{\pi z}} \cos\left(z - \frac{n\pi}{2} - \frac{\pi}{4} \right) \\
&= \begin{cases} (-1)^{\frac{n}{2}} \sqrt{\frac{2}{\pi z}} \cos\left(z - \frac{\pi}{4} \right) & \text{for even } n, \\ (-1)^{\frac{n-1}{2}} \sqrt{\frac{2}{\pi z}} \sin\left(z - \frac{\pi}{4} \right) & \text{for odd } n. \end{cases}
\end{aligned} \qquad (5.66)$$

A comparison between a Bessel function and its asymptotic form is shown in Figure 5.9 for the case $n = 7$; the figure also includes the corresponding small-z approximation of equation (5.60).

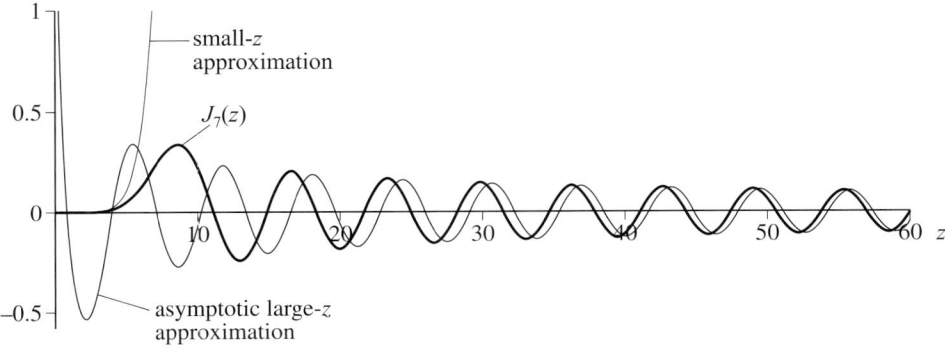

Figure 5.9 The Bessel function $J_7(z)$, its small-z approximation (5.60) and its asymptotic approximation (5.66)

Zeros of the Bessel functions

Each Bessel function $J_n(z)$ has infinitely many zeros. In Figure 5.10, the locations of the first nine zeros $z_{7,m}$ (where $m = 1, 2, 3, \ldots$ enumerates the zeros) of the Bessel function $J_7(z)$ for $z > 0$ are indicated. Approximate numerical values for a few zeros $z_{n,m}$ of the Bessel functions J_n, with $n = 0, 1, \ldots, 7$, are given in Table 5.1.

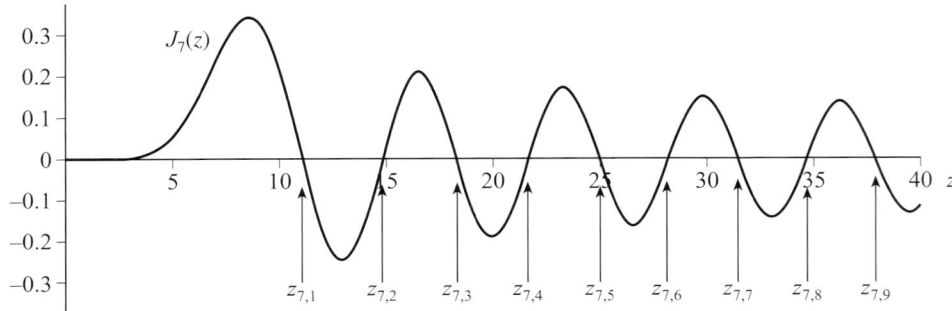

Figure 5.10 Locations of the zeros $z_{7,m}$ of the Bessel function J_7

Table 5.1 Approximate values of the smallest positive zeros $z_{n,m} > 0$ of the Bessel functions $J_n(z)$

m	$n = 0$	$n = 1$	$n = 2$	$n = 3$	$n = 4$	$n = 5$	$n = 6$	$n = 7$
1	2.405	3.832	5.136	6.380	7.588	8.771	9.936	11.086
2	5.520	7.016	8.417	9.761	11.065	12.339	13.589	14.821
3	8.654	10.173	11.620	13.015	14.373	15.700	17.004	18.288
4	11.792	13.324	14.796	16.223	17.616	18.980	20.321	21.642
5	14.931	16.471	17.960	19.409	20.827	22.218	23.586	24.935
6	18.071	19.616	21.117	22.583	24.019	25.430	26.820	28.191
7	21.212	22.760	24.270	25.748	27.199	28.627	30.034	31.423
8	24.352	25.904	27.421	28.908	30.371	31.812	33.233	34.637
9	27.493	29.047	30.569	32.065	33.537	34.989	36.422	37.839
10	30.635	32.190	33.717	35.219	36.699	38.160	39.603	41.031

For large arguments z, the Bessel functions $J_n(z)$ are asymptotically given by equations (5.66). Thus the locations $z_{n,m}/\pi$ of the zeros divided by π become asymptotically integer-spaced, behaving like

$$z_{n,m} \simeq \begin{cases} \pi(\widetilde{m} - \frac{1}{4}) & \text{for even } n, \\ \pi(\widetilde{m} + \frac{1}{4}) & \text{for odd } n, \end{cases} \tag{5.67}$$

for large values of m. Here $\widetilde{m}$ differs from m by an integer that depends on n. This is due to the fact that the relation (5.66) is approximately correct only for large arguments z. So for small integers $\widetilde{m}$ in equations (5.67), we may not actually find a zero of the Bessel function J_n. This affects the counting of the zeros, resulting in an integer shift between m, which counts the zeros successively, and the integer $\widetilde{m}$ that enters in equations (5.67). An example of this shift is shown in Figure 5.11.

5.2 Vibrating membranes

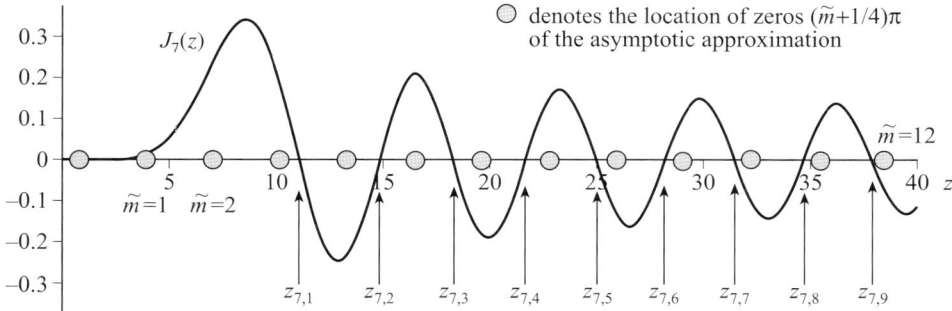

Figure 5.11 The zeros of the Bessel function $J_7(z)$, and the asymptotic locations of zeros $(\widetilde{m}+1/4)\pi$. In this example, the zeros $z_{n,m}$ are asymptotically close to $(m+3+1/4)\pi$, so $\widetilde{m} = m+3$ for $n=7$.

Figure 5.12 shows the Bessel functions J_0, J_6 and J_{12}. For large arguments, the functions resemble the asymptotic form (5.66), and the zeros start to move closer together near the zeros of the trigonometric expressions of equations (5.66). For small arguments, the Bessel functions with larger n have fewer zeros, which explains the different enumeration of zeros.

Note that the prefactor $(-1)^{n/2}$ in equations (5.66) is positive for $n=0$ and $n=12$, but negative for $n=6$.

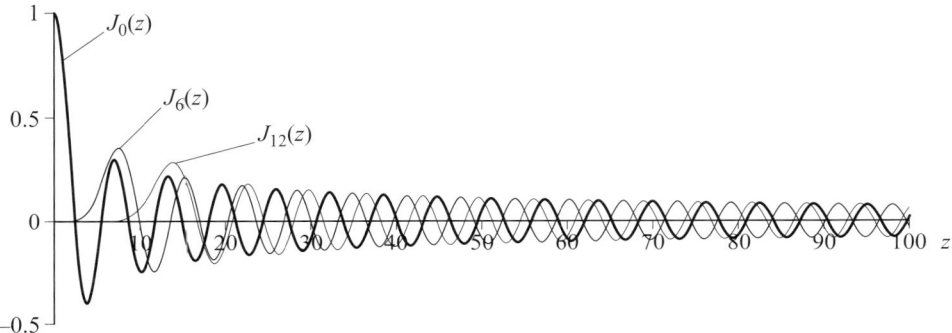

Figure 5.12 The Bessel functions $J_0(z)$, $J_6(z)$ and $J_{12}(z)$

Exercise 5.8

From Table 5.1, compute how much the actual values of the zeros differ from the corresponding asymptotic values of equations (5.67) for $n=0$ and $n=7$, and $m=5$ and $m=10$.

5.2.6 Normal mode solutions for circular membranes

We now return to the radial equation (5.48). As n is an integer, and $Y_n(z)$ diverges as $z \to 0$, only the Bessel functions $J_n(z)$ yield proper solutions for the vibrating membrane. In other words, we require that the functions $R_n(r)$ stay finite on our membrane, which implies $b_n = 0$ in equation (5.52).

The solution thus reads

$$R_n(r) = a_n J_n(kr). \tag{5.68}$$

The boundary condition for R of equations (5.44) becomes

$$R_n(l) = 0 = J_n(kl), \tag{5.69}$$

which determines the possible values of k in terms of the zeros of the Bessel function J_n. This may be compared to the condition $\sin(kl) = 0$ which determined the possible values of k for the string with fixed ends, and also for the rectangular membrane. Here, the condition $J_n(kl) = 0$ restricts k to a discrete set of values $k_{n,m}$, where $k_{n,m} = z_{n,m}/l$ in terms of the zeros of the Bessel function J_n, enumerated by $m = 1, 2, 3, \ldots$. In contrast to the previous case, where the boundary condition was of the form $\sin(kl) = 0$, we cannot write down a simple closed expression for the zeros in this case.

We now return to the solutions of the position-dependent part of the wave equation (5.38). According to equation (5.40), the eigenmodes are products of the angular parts $\Theta_n(\theta)$ discussed previously, and the radial parts $R_{n,m}(r)$, so

Eigenmodes of circular membranes

$$\begin{aligned} v_{n,m}(r,\theta) &= R_{n,m}(r)\,\Theta_n(\theta) \\ &= J_n(k_{n,m}r)[A_n \cos(n\theta) + B_n \sin(n\theta)], \end{aligned} \tag{5.70}$$

where $n = 0, 1, 2, \ldots$ and $m = 1, 2, 3, \ldots$, and $k_{n,m} = z_{n,m}/l$. As the Bessel functions J_n have infinitely many zeros, for each angular mode $\Theta_n(\theta)$, $n = 0, 1, 2, \ldots$, we have infinitely many radial modes at our disposal.

The eigenmodes have two types of nodal lines. The first arises when the angular part $\Theta_n(\theta)$ vanishes, as discussed above. The second type stems from the radial part. These form concentric circles of radii $0 < r < l$, where r is a solution of $J_n(k_{n,m}r) = 0$. In other words, the mode $v_{n,m}$ has precisely $m-1$ radial nodal lines, corresponding to the $m-1$ zeros of the Bessel function $J_n(k_{n,m}r)$ for $0 < r < l$ (the mth zero of that function being at $r = l$ by definition of $k_{n,m}$). The nodal line patterns of the eigenmodes of lowest frequencies are shown in Figure 5.13.

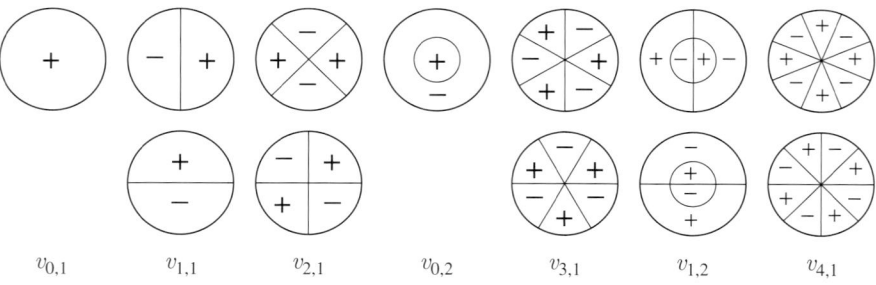

Figure 5.13 Nodal line patterns of the lowest-frequency eigenmodes of a circular membrane. The patterns shown correspond to choosing either $A_n = 0$ (bottom row) or $B_n = 0$ (top row) in equation (5.70).

The eigenvalues $k_{n,m}^2$ determine the frequencies of the corresponding eigenmodes via $\omega_{n,m} = k_{n,m}c$, where c is the wave speed that enters the wave equation. If we want to sort the eigenmodes by their frequency, which eventually is what makes up the sound from a drum, say, then we have to sort them according to their eigenvalues $k_{n,m}^2$. These depend on both n and m, as $k_{n,m}$ is the mth solution of the equation $J_n(kl) = 0$. All but those involving the fundamental angular mode are twice degenerate, as we pointed out above when considering the angular part, and no additional degeneracies arise because there are no coincidences between zeros of Bessel functions. In Figure 5.13, the nodal line patterns shown for the degenerate cases correspond to the choices $A_n = 0$ and $B_n = 0$ in equation (5.70).

5.2 Vibrating membranes

For large arguments z, the Bessel functions $J_n(z)$ are asymptotically given by equations (5.66), and the zeros are located near the positions given in equations (5.67). Thus the values of $k_{n,m} = z_{n,m}/l$ become asymptotically integer-spaced in π/l, behaving like

$$k_{n,m} \simeq \begin{cases} (\tilde{m} - \frac{1}{4})\pi/l & \text{for even } n, \\ (\tilde{m} + \frac{1}{4})\pi/l & \text{for odd } n, \end{cases} \quad (5.71)$$

for large values of m. As in equations (5.67), $\tilde{m}$ differs from m by an integer that depends on n.

> In deriving the eigenmodes, we must be careful because we use polar coordinates. For $n = 0$, when the eigenmode has no angular dependency, it is sufficient that the radial part stays finite at the origin $r = 0$. However, for $n \geq 1$, when there is a non-trivial dependency on the angle variable θ, we actually need more than this to ensure that the solution represents a continuous membrane in the vicinity of the origin $r = 0$. In this case, we need to demand that the radial function $R_n(r)$ vanishes, so $R_n(0) = 0$ for $n \geq 1$. In fact, for the Bessel functions we have $J_0(0) = 1$ and $J_n(0) = 0$ for $n \geq 1$ (compare equation (5.56)), so this is automatically satisfied by our solutions (5.68). But it is worth keeping in mind that coordinate systems that contain singular points, in this case the origin $r = 0$ where the angle θ is not defined, may impose additional constraints on the solutions of the separated ordinary differential equations.

Exercise 5.9

Using the data of Table 5.1, determine the approximate values of the nine lowest eigenvalues $k_{n,m}$ for a circular membrane of radius $l = 10$.

Exercise 5.10

(a) Write down the general form for the eigenmodes of a circular membrane of radius l.

(b) Now choose the angular function Θ such that one angular nodal line (provided there is one) lies in the **i** direction, and give the corresponding eigenmode.

The eigenmodes for the wave equation for a circular membrane are given by equation (5.70), where J_n is the nth Bessel function. $v_{n,m}(r,\theta)$ represents an eigenmode with n angular nodes (yielding straight nodal lines at constant angle θ) and $m-1$ radial nodes (corresponding to circular nodal lines at constant radius r), not counting the circular boundary as a node. The positions of the lowest radial nodes can be extracted from Table 5.1.

5.2.7 Can one 'hear' the shape of a drum?

A celebrated problem related to vibrating membranes was posed by the mathematician Mark Kac in 1966: he asked whether one can 'hear' the shape of a drum. The motion of a drum is a combination of its normal modes, so the sound contains a spectrum of frequencies which are the eigenfrequencies of the drum. So, more precisely, the mathematical question is whether the list of eigenvalues, the eigenfrequencies of a membrane, determines its shape. Conversely, if that is not the case, then there should be differently shaped membranes which have the same spectrum of eigenvalues. Can one find a counter-example, that is, two different shapes that have precisely the same spectrum?

The answer to this question is that there are indeed different shapes that give rise to the same set of eigenfrequencies, so you cannot determine the shape of a membrane from the list of its eigenfrequencies. This was shown only in the early 1990s by C. Gordon, D. Webb and S. Wolpert, who explicitly constructed *isospectral* drums, meaning drums which have the same spectrum. A simple example of two isospectral drums is shown in Figure 5.14.

See, for instance, T. A. Driscoll, 'Eigenmodes of isospectral drums', *SIAM Rev.* **39** (1997) pp. 1–17.

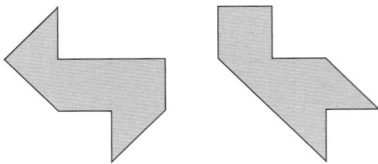

Figure 5.14   Two isospectral polygonal drums

Even though it is possible to prove mathematically that these two membranes have precisely the same spectrum, there are no techniques available to actually calculate the eigenvalues or corresponding eigenmodes analytically. This may sound astounding to you if you are not familiar with mathematical proofs. However, it may be much easier to decide whether two quantities are equal or different than it is to actually calculate them – and this is the case for the problem at hand, at least for the particular examples which were constructed precisely for that purpose.

It is beyond the scope of this course to go into more detail. This discussion of isospectral drums was included to provide you with a vague idea about the extent and limitations of current knowledge on this subject. Whereas most of the methods and results discussed so far date back several centuries, there still exist many problems and open questions in this classical field of mathematical physics.

5.3 Isotropic waves in two and three dimensions

In the previous section, we considered vibrating membranes, deriving the normal mode solutions for rectangular, circular and some triangular membranes. As for the vibrating string with fixed ends, boundary conditions play an important role, in that they restrict the parameter k, the wave number, to a set of discrete values, which label the eigenmodes. Now we want to take a step back and look at solutions of the wave equation in two and three dimensions in the infinite domain, in a spirit similar to that in the discussion of infinite strings at the end of Chapter 2. In particular, this means that no boundary conditions need to be considered.

We consider wave motion in two and three dimensions, where the position coordinate x in the wave equation is replaced by a vector $\boldsymbol{r}$ in the Euclidean space $\mathbb{R}^2$ or $\mathbb{R}^3$. For simplicity, we restrict ourselves to the case where the wave is described by a (scalar) real function $u(\boldsymbol{r}, t)$. This is now a function of position in space, given by the two- or three-dimensional vector $\boldsymbol{r}$, and time t. So at each point in space and at any given time, the wave is described by the value of the function u at that position and time.

5.3 Isotropic waves in two and three dimensions

The natural generalisation of the one- and two-dimensional isotropic wave equations (1.1) and (5.1) to three dimensions, expressed in Cartesian coordinates, has the form

$$u_{tt}(\boldsymbol{r},t) = c^2\left[u_{xx}(\boldsymbol{r},t) + u_{yy}(\boldsymbol{r},t) + u_{zz}(\boldsymbol{r},t)\right], \tag{5.72}$$

with $\boldsymbol{r} = x\mathbf{i} + y\mathbf{j} + z\mathbf{k}$. This wave equation is appropriate for *isotropic* systems, which have the property that they do not behave differently in different directions of space. The isotropic wave equation is often written as

Three-dimensional isotropic wave equation

$$u_{tt}(\boldsymbol{r},t) = c^2\,\boldsymbol{\nabla}^2 u(\boldsymbol{r},t), \tag{5.73}$$

where $\boldsymbol{\nabla}$ is the *gradient* operator, and $\boldsymbol{\nabla}^2 = \boldsymbol{\nabla}\cdot\boldsymbol{\nabla}$ is called the *Laplacian operator*.

See Block 0, Section 2.3.

The symbol Δ is also commonly used for the Laplacian operator.

In three-dimensional Cartesian coordinates $\boldsymbol{r} = x\mathbf{i} + y\mathbf{j} + z\mathbf{k}$, the gradient operator is $\boldsymbol{\nabla} = \mathbf{i}\frac{\partial}{\partial x} + \mathbf{j}\frac{\partial}{\partial y} + \mathbf{k}\frac{\partial}{\partial z}$, and the Laplacian operator is given by

$$\boldsymbol{\nabla}^2 = \frac{\partial^2}{\partial x^2} + \frac{\partial^2}{\partial y^2} + \frac{\partial^2}{\partial z^2}. \tag{5.74}$$

Sometimes, it may be preferable to use different coordinate systems, such as cylindrical polar or spherical coordinates. In this case, equation (5.73) can still be used, but the Laplacian operator has to be rewritten in terms of these coordinates and, in general, will look different from equation (5.74).

See Block 0, Section 2.4.

5.3.1 Generalisation of d'Alembert's solution

You might wonder whether we can generalise d'Alembert's solution of the one-dimensional wave equation to obtain a general solution for the two- and three-dimensional cases. Indeed, this can be done for isotropic wave equations such as (5.1) and (5.72) in two and three dimensions, respectively, as we shall see below.

In the one-dimensional case, we started from two arbitrary real functions, f and g, and the general solution had the form $u(x,t) = F(x - ct) + G(x + ct)$, which we could interpret as a sum of two pulses, one moving to the right and one moving to the left, both with velocity c. Now, the role of the position variable x is played by the vector $\boldsymbol{r}$ in Euclidean space $\mathbb{R}^2$ or $\mathbb{R}^3$, but the time variable is just a single real variable as before. In order to see how we can generalise d'Alembert's solution, it might be instructive to replace the arguments $x \pm ct$ by a multiple $k(x \pm ct) = kx \pm \omega t$, which is a combination of variables that has appeared many times before. Now, in two or three dimensions, a natural generalisation is $\boldsymbol{k}\cdot\boldsymbol{r} \pm \omega t = k(\boldsymbol{k}\cdot\boldsymbol{r}/k \pm ct)$, which is also a scalar variable. Note that $\boldsymbol{k} = k_x\mathbf{i} + k_y\mathbf{j} + k_z\mathbf{k}$ here is the *wave vector*, not the Cartesian unit vector $\mathbf{k}$, and that $k = |\boldsymbol{k}|$. In the following exercise, we ask you to verify that indeed $f(\boldsymbol{k}\cdot\boldsymbol{r} \pm \omega t)$, with $\boldsymbol{k}\cdot\boldsymbol{k} = k^2 = \omega^2/c^2$, is a solution of the wave equation, for an arbitrary, at least twice differentiable function f.

It is customary to denote the wave vector by the letter $\boldsymbol{k}$, and we follow this convention here.

Exercise 5.11

Verify that $f(\boldsymbol{k}\cdot\boldsymbol{r} \pm \omega t)$, for an arbitrary, at least twice differentiable function f, satisfies the three-dimensional wave equation (5.73) for $\boldsymbol{k}\cdot\boldsymbol{k} = k^2 = \omega^2/c^2$.

How can we interpret this solution? In the one-dimensional case, we had two arbitrary functions in the general solution, and the arguments $x \pm ct$ involved only the constant c, which appears in the wave equation. Now we have a vector $\boldsymbol{k}$ and a number ω, and the latter can be expressed as $\omega = kc$. So for *any* vector $\boldsymbol{k}$, and an arbitrary function f, we now have a solution of the wave equation of the form

$$u(\boldsymbol{r},t) = f(\boldsymbol{k} \cdot \boldsymbol{r} - \omega t), \tag{5.75}$$

and any linear combinations of these are solutions as well. It is not necessary to consider $\boldsymbol{k} \cdot \boldsymbol{r} + \omega t$ separately here, because the vector $\boldsymbol{k}$ can be chosen in any direction. For example, in the one-dimensional case, choosing $\boldsymbol{k}/k = \mathbf{i}$ gives $\boldsymbol{k} \cdot \boldsymbol{r} - \omega t = k(x - ct)$, and $\boldsymbol{k}/k = -\mathbf{i}$ gives $\boldsymbol{k} \cdot \boldsymbol{r} - \omega t = k(-x - ct) = -k(x + ct)$, so both terms $F(x - ct)$ and $G(x + ct)$ in d'Alembert's solution (2.14) can be expressed in this form.

How can a solution of the type (5.75), with a fixed vector $\boldsymbol{k}$, be interpreted? We first observe that it depends on the position $\boldsymbol{r}$ only via the scalar product $\boldsymbol{k} \cdot \boldsymbol{r}$. Thus $u(\boldsymbol{r},t)$, at a given instant of time t, is constant for all positions $\boldsymbol{r}$ with $\boldsymbol{k} \cdot \boldsymbol{r} = kR$, where R is an arbitrary constant and $k = |\boldsymbol{k}|$. The equation $\boldsymbol{k} \cdot \boldsymbol{r} = kR$ defines a line in two dimensions and a plane in three dimensions, perpendicular to $\boldsymbol{k}$ and passing through the point $\boldsymbol{r}_R = \boldsymbol{k}R/k$. To see this, note that if $\boldsymbol{s}$ is orthogonal to $\boldsymbol{k}$, so that $\boldsymbol{k} \cdot \boldsymbol{s} = 0$, then $\boldsymbol{k} \cdot (\boldsymbol{r}_R + \boldsymbol{s}) = \boldsymbol{k} \cdot \boldsymbol{r}_R = kR$; see Figure 5.15.

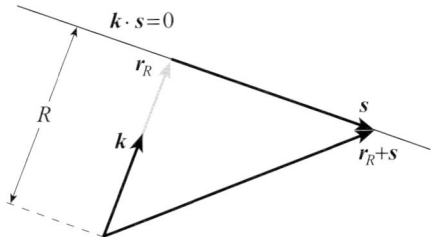

Figure 5.15 The vectors $\boldsymbol{r}$ with constant scalar product $\boldsymbol{k} \cdot \boldsymbol{r} = kR$ define a line in two-dimensional space

So $u(\boldsymbol{r},t)$ is constant along lines or planes perpendicular to $\boldsymbol{k}$. For this reason, such solutions are called *plane waves*; see Figure 5.16.

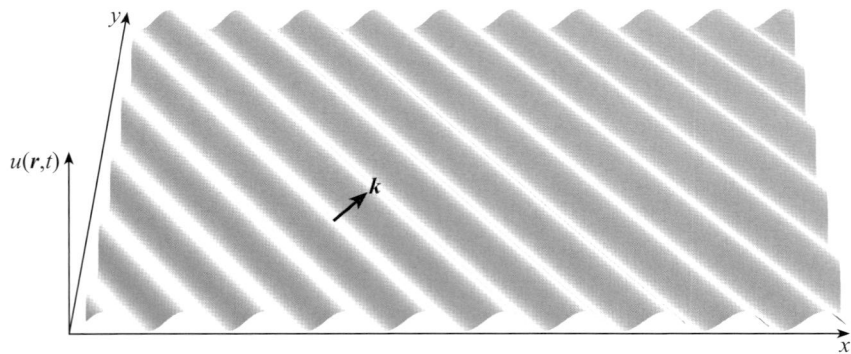

Figure 5.16 Sketch of a plane wave solution $u(\boldsymbol{r},t)$ with a wave vector $\boldsymbol{k}$

Exercise 5.12

Consider the solution $f(\boldsymbol{k}\cdot\boldsymbol{r} - \omega t)$ of the wave equation for the wave vector $\boldsymbol{k} = k\mathbf{i}$, with a given value of $k > 0$, and $\omega = kc$.

(a) Show that this function depends only on x and t, where $\boldsymbol{r} = x\mathbf{i} + y\mathbf{j} + z\mathbf{k}$, and, moreover, depends only on the combination $x - ct$.

(b) Argue that this solution represents a wave moving along the $\mathbf{i}$ direction.

The plane wave is constant on the lines or planes with $\boldsymbol{k}\cdot\boldsymbol{r} = kR$. We now consider how these planes move with time t. For $\boldsymbol{k}\cdot\boldsymbol{r} = kR$, the term $f(\boldsymbol{k}\cdot\boldsymbol{r} - \omega t)$ becomes $f[k(R - ct)]$. Essentially, this is the same as in the one-dimensional case, except that the argument now involves $R - ct$ in place of $x - ct$. It means that the wave pattern on a plane (or a line in the two-dimensional case) moves in the direction of the vector $\boldsymbol{k}$ with a constant velocity c. So the vector $\boldsymbol{k}$, the wave vector, determines the direction of motion, and the plane wave solutions are constant on lines or planes orthogonal to the direction of motion.

Particular plane wave solutions of the form $u(\boldsymbol{r}, t) = e^{i(\boldsymbol{k}\cdot\boldsymbol{r} - \omega t)}$ are called *harmonic plane waves*. As mentioned at the end of Chapter 2, we can expand a general solution in terms of harmonic plane waves, and this expansion leads to Fourier transforms. Looking at this from a different point of view, we can use Fourier transforms to solve differential equations. We now discuss how this applies to the wave equation, starting with the one-dimensional case.

5.3.2 The Fourier transform of the wave equation

So far, we have exploited Fourier series to derive solutions of the wave equation for finite strings. However, we can also make use of the Fourier transform to solve the wave equation, in particular for the case where no boundary conditions need be taken into account. Before we use this method to solve the two- and three-dimensional wave equation, we illustrate it for the one-dimensional case of an infinite string.

Essentially, this is analogous to the use of Fourier series for finite strings or membranes. As was pointed out at the end of Chapter 2, Fourier transforms arise naturally as expansions in terms of harmonic solutions $e^{\pm ikx}$ of the wave equation, with the Fourier transformed functions playing the role of the coefficients. In the approach that we discuss now, we first derive an equation for these coefficients, then solve it for the coefficients, and transform back to obtain the corresponding solution of the wave equation.

Starting again from the homogeneous wave equation

$$u_{tt}(x, t) = c^2\, u_{xx}(x, t), \tag{5.76}$$

and considering the Fourier transform with respect to the position coordinate x,

$$\widetilde{u}(k, t) = \frac{1}{\sqrt{2\pi}} \int_{-\infty}^{\infty} dx\, u(x, t)\, e^{-ikx}, \tag{5.77}$$

See equation (3.25).

we derive the equation

Fourier-transformed wave equation

$$\widetilde{u}_{tt}(k, t) = -c^2 k^2\, \widetilde{u}(k, t). \tag{5.78}$$

Here, we have used the Fourier transform of derivatives discussed in Section 3.6, which gives $ik\,\tilde{u}(k,t)$ for the Fourier transform of $u_x(x,t)$, and thus $(ik)^2\tilde{u}(k,t) = -k^2\tilde{u}(k,t)$ for the Fourier transform of $u_{xx}(x,t)$.

Effectively, we have now reduced the problem of solving a partial differential equation (5.76) to that of solving an ordinary differential equation (5.78). Equation (5.78) contains only derivatives with respect to time t; the variable k enters only algebraically, as a factor k^2. Thus k can be treated as a parameter, and we can solve equation (5.78) for any given value of the parameter k. The general solution is

$$\tilde{u}(k,t) = A(k)e^{i\omega t} + A^*(k)e^{-i\omega t}, \quad \omega = kc, \tag{5.79}$$

with (in general) complex coefficients $A(k)$ determined by the initial conditions. Because the inverse transform is

$$u(x,t) = \frac{1}{\sqrt{2\pi}} \int_{-\infty}^{\infty} dk\,\tilde{u}(k,t)\,e^{ikx}, \tag{5.80}$$

See equation (3.24).

we actually need to know the appropriate solution $\tilde{u}(k,t)$, and hence the coefficients $A(k)$, for all real values of k, in order to be able to transform back to the position variable x.

Exercise 5.13

Consider the wave equation (5.76). Instead of taking the Fourier transform with respect to the position variable x, consider now the Fourier transform with respect to time

$$\tilde{u}(x,\omega) = \frac{1}{\sqrt{2\pi}} \int_{-\infty}^{\infty} dt\,u(x,t)\,e^{-i\omega t},$$

and calculate the corresponding differential equation for $\tilde{u}(x,\omega)$.

Although this function differs from that defined in equation (5.77), it is customary to use the same symbol $\tilde{u}$ here. The arguments indicate in which variable(s) we transform.

Exercise 5.14

Consider now the two-dimensional Fourier transform with respect to both position and time, as discussed in Section 3.10. Defining

$$\tilde{u}(k,\omega) = \frac{1}{2\pi} \int_{-\infty}^{\infty} dx \int_{-\infty}^{\infty} dt\,u(x,t)\,e^{-i(kx+\omega t)},$$

what is the condition on the variables k and ω obtained from the wave equation (5.76)?

Exercise 5.15

Consider the one-dimensional wave equation with initial conditions $u(x,0) = a(x)$ and $u_t(x,0) = 0$.

(a) What are the corresponding initial conditions for the Fourier transform $\tilde{u}(k,t)$?

(b) Use the initial conditions to determine the coefficient $A(k)$ in equation (5.79).

(c) Perform the integral in the inverse Fourier transformation (5.80), and deduce the solution of the initial-value problem. Do you recognise this solution?

Note that by assuming that the Fourier transforms of the function $u(x,t)$ with respect to the position or time coordinates exist, we implicitly impose conditions on the solution $u(x,t)$; essentially, the function has to decay to zero sufficiently rapidly in the relevant variable to make the Fourier integrals converge. There are solutions of the wave equation (5.76) for which the Fourier integrals do not exist. The simplest example is a constant solution $u(x,t) = C$

5.3 Isotropic waves in two and three dimensions

where C is non-zero, which clearly satisfies the wave equation (5.76) because all derivatives vanish. Nevertheless, it is possible to make sense of the Fourier integral for this case, in terms of so-called generalised functions. This is not of concern here or within the scope of this course, but it is worth keeping in mind that by using integral transforms to solve differential equations, one may have imposed additional conditions on the solutions.

The Fourier transform method can be applied to multi-dimensional wave motion, using the multi-dimensional Fourier transform discussed in Section 3.10. We apply the d-dimensional Fourier transform from equations (3.59) and (3.58),

$$\widetilde{u}(\boldsymbol{k},t) = \frac{1}{(2\pi)^{d/2}} \int_{\mathbb{R}^d} d\boldsymbol{r}\, u(\boldsymbol{r},t)\, e^{-i\boldsymbol{k}\cdot\boldsymbol{r}}, \tag{5.81}$$

$$u(\boldsymbol{r},t) = \frac{1}{(2\pi)^{d/2}} \int_{\mathbb{R}^d} d\boldsymbol{k}\, \widetilde{u}(\boldsymbol{k},t)\, e^{i\boldsymbol{k}\cdot\boldsymbol{r}}, \tag{5.82}$$

for $d=2$ or $d=3$, where $d\boldsymbol{r}$ and $d\boldsymbol{k}$ denote the d-dimensional volume elements, and the subscript $\mathbb{R}^d$ on the integration sign indicates that we integrate over the entire d-dimensional space. The isotropic wave equation (5.73) becomes

$$\widetilde{u}_{tt}(\boldsymbol{k},t) = -c^2 k^2\, \widetilde{u}(\boldsymbol{k},t), \tag{5.83}$$

Here $k^2 = \boldsymbol{k}\cdot\boldsymbol{k}$.

because the Fourier transform of $\boldsymbol{\nabla}^2 u(\boldsymbol{r},t)$ is $-(\boldsymbol{k}\cdot\boldsymbol{k})\,\widetilde{u}(\boldsymbol{k},t)$. This is shown in Example 5.1 below.

Apart from the fact that the function $\widetilde{u}(\boldsymbol{k},t)$ now depends on the vector $\boldsymbol{k}$, we obtain the same equation as for the Fourier transform of the one-dimensional wave equation (compare equation (5.78)), where k^2 is now the squared length of the vector $\boldsymbol{k}$. The general solution of (5.83) is

$$\widetilde{u}(\boldsymbol{k},t) = A(\boldsymbol{k})e^{i\omega t} + B(\boldsymbol{k})e^{-i\omega t}, \tag{5.84}$$

where $A(\boldsymbol{k})$ and $B(\boldsymbol{k})$ are complex. The general real solution, expressed in terms of complex exponential functions, has the form

$$\widetilde{u}(\boldsymbol{k},t) = A(\boldsymbol{k})e^{i\omega t} + A^*(\boldsymbol{k})e^{-i\omega t}, \tag{5.85}$$

where $A^*(\boldsymbol{k})$ is the complex conjugate of $A(\boldsymbol{k})$. As usual, we use $\omega = kc$, where $k = |\boldsymbol{k}|$.

The coefficients $A(\boldsymbol{k})$ in the solution (5.85) are determined by the initial conditions. In general, they depend on the magnitude and direction of $\boldsymbol{k}$. However, if the initial conditions are independent of direction, then $A(\boldsymbol{k})$ depends only on $k = |\boldsymbol{k}|$, and we obtain the same expression as in the one-dimensional case; compare equation (5.79). The only difference is that the real number k of the one-dimensional solutions is now interpreted as the length of the two- or three-dimensional vector $\boldsymbol{k}$.

As an aside, note that this simplification is actually a consequence of the isotropy of the wave equation (5.73). The isotropy means that the problem possesses rotational symmetry. If the initial conditions do not introduce a direction dependence, then there is no preferred direction of space, and nothing changes if the system is rotated. The presence of such symmetries often allows us to reduce the problem to a lower-dimensional one. It is important to understand the symmetries of a system, because they may indicate how to solve the corresponding equations. In practice, this is usually done by changing to a coordinate system that is adapted to the symmetry of the problem, which in this case would be spherical coordinates.

Example 5.1

Fourier transforms of derivatives were considered in Section 3.6. The one-dimensional Fourier transform of a derivative $f'(x)$ is $ik\widetilde{f}(k)$.

(a) Use equation (5.81) with $d = 3$ to show that the Fourier transform of $u_x(\boldsymbol{r},t)$ is $ik_x\widetilde{u}(\boldsymbol{k},t)$.

(b) Further, show that the Fourier transform of $\boldsymbol{\nabla}u(\boldsymbol{r},t)$ is $i\boldsymbol{k}\,\widetilde{u}(\boldsymbol{k},t)$.

(c) Deduce that the Fourier transform of $\boldsymbol{\nabla}^2 u(\boldsymbol{r},t)$ is $-(\boldsymbol{k}\cdot\boldsymbol{k})\,\widetilde{u}(\boldsymbol{k},t)$.

Solution

(a) We use Cartesian coordinates

$$\boldsymbol{r} = x\mathbf{i} + y\mathbf{j} + z\mathbf{k} \quad \text{and} \quad \boldsymbol{k} = k_x\mathbf{i} + k_y\mathbf{j} + k_z\mathbf{k}, \tag{5.86}$$

and write $u(x,y,z,t) = u(\boldsymbol{r},t)$ for the function u expressed in terms of these coordinates.

The three-dimensional Fourier transform of $u_x(x,y,z,t)$ is

$$\begin{aligned}
\widetilde{u}_x(\boldsymbol{k},t) &= \frac{1}{(2\pi)^{3/2}}\int_{\mathbb{R}^3} d\boldsymbol{r}\, u_x(\boldsymbol{r},t)\, e^{-i\boldsymbol{k}\cdot\boldsymbol{r}} \\
&= \frac{1}{(2\pi)^{3/2}}\int_{-\infty}^{\infty} dx \int_{-\infty}^{\infty} dy \int_{-\infty}^{\infty} dz\, u_x(x,y,z,t)\, e^{-ik_x x}\, e^{-ik_y y}\, e^{-ik_z z} \\
&= \frac{1}{2\pi}\int_{-\infty}^{\infty} dx \int_{-\infty}^{\infty} dy\, \widetilde{u}_x(x,y,k_z,t)\, e^{-ik_x x}\, e^{-ik_y y} \\
&= \frac{1}{\sqrt{2\pi}}\int_{-\infty}^{\infty} dx\, \widetilde{u}_x(x,k_y,k_z,t)\, e^{-ik_x x} \\
&= ik_x\widetilde{u}(k_x,k_y,k_z,t) \\
&= ik_x\widetilde{u}(\boldsymbol{k},t), \tag{5.87}
\end{aligned}$$

where the notation is such that $\widetilde{u}(x,y,k_z,t)$ denotes the one-dimensional Fourier transform of the function $u(x,y,z,t)$ with respect to z, and similarly $\widetilde{u}(x,k_y,k_z,t)$ denotes the two-dimensional Fourier transform with respect to y and z. In the last step, we used the one-dimensional Fourier transform of the derivative with respect to x.

It is unfortunate but common to use $\widetilde{u}$ to denote the Fourier transform of u, irrespective of the argument that is transformed. Here we use it only in these intermediate steps where it is clear which variables have been transformed.

(b) We have

$$\boldsymbol{\nabla}u(\boldsymbol{r},t) = u_x(\boldsymbol{r},t)\,\mathbf{i} + u_y(\boldsymbol{r},t)\,\mathbf{j} + u_z(\boldsymbol{r},t)\,\mathbf{k}. \tag{5.88}$$

Using the previous result for the three-dimensional Fourier transform of $u_x(\boldsymbol{r},t)$ and the analogous relations for the partial derivatives with respect to y and z, we obtain

$$\begin{aligned}
\frac{1}{(2\pi)^{3/2}}\int_{\mathbb{R}^3} d\boldsymbol{r}\, [\boldsymbol{\nabla}u(\boldsymbol{r},t)]\, e^{-i\boldsymbol{k}\cdot\boldsymbol{r}} &= (ik_x\mathbf{i} + ik_y\mathbf{j} + ik_z\mathbf{k})\,\widetilde{u}(\boldsymbol{k},t) \\
&= i\boldsymbol{k}\,\widetilde{u}(\boldsymbol{k},t). \tag{5.89}
\end{aligned}$$

(c) By applying the result of part (a) twice, the three-dimensional Fourier transform of the second partial derivative $u_{xx}(\boldsymbol{r},t)$ is

$$(ik_x)^2\widetilde{u}(\boldsymbol{k},t) = -k_x^2\widetilde{u}(\boldsymbol{k},t). \tag{5.90}$$

Hence the Fourier transform of $\boldsymbol{\nabla}^2 u(\boldsymbol{r},t)$ is

$$\begin{aligned}
-k_x^2\widetilde{u}(\boldsymbol{k},t) - k_y^2\widetilde{u}(\boldsymbol{k},t) - k_z^2\widetilde{u}(\boldsymbol{k},t) &= -(k_x^2 + k_y^2 + k_z^2)\,\widetilde{u}(\boldsymbol{k},t) \\
&= -(\boldsymbol{k}\cdot\boldsymbol{k})\,\widetilde{u}(\boldsymbol{k},t), \tag{5.91}
\end{aligned}$$

as required. ∎

5.3.3 Isotropic waves in two dimensions

We first consider the two-dimensional case, which we encountered when we discussed circular membranes. We start with an exercise where you are asked to express the solution of the wave equation as a Fourier integral.

Exercise 5.16

Consider the two-dimensional wave equation (5.1), where $r = x\mathbf{i} + y\mathbf{j}$.

(a) Consider the initial conditions $u(r,0) = a(r)$ and $u_t(r,0) = b(r)$. What are the corresponding initial conditions for $\tilde{u}(k,t)$?

(b) Compute the coefficients of the complex exponential functions in the solution (5.85) from the initial conditions.

(c) Express the corresponding solution $u(r,t)$ as the inverse Fourier transform of $\tilde{u}(k,t)$, without attempting to evaluate the integral.

The Fourier transform of a rotationally symmetric function is itself rotationally symmetric. In the following example, we express the Fourier integral in terms of the Bessel function J_0 by performing the angular integration in polar coordinates.

Example 5.2

Consider a function $f(r)$, with $r \in \mathbb{R}^2$, that depends only on the distance $r = |r|$ from the origin. We write $f(r)$ to emphasise that f can be considered as a function of the single variable r.

(a) Use $r = x\mathbf{i} + y\mathbf{j}$ and $k = k_x\mathbf{i} + k_y\mathbf{j}$ to write $\tilde{f}(k)$ as a double integral over the position variables.

(b) Transform the double integral to polar coordinates r, ϕ and k, ψ, where

$$x = r\cos\phi, \quad y = r\sin\phi, \quad k_x = k\cos\psi, \quad k_y = k\sin\psi, \quad (5.92)$$

and perform the angular integration. [Hint: Use the trigonometric identity $\cos(\alpha - \beta) = \cos\alpha\cos\beta + \sin\alpha\sin\beta$, and express $\tilde{f}$ as an integral over the Bessel function J_0.]

Solution

(a) In Cartesian coordinates, the Fourier transform is

$$\tilde{f}(k) = \frac{1}{2\pi} \int_{-\infty}^{\infty} dx \int_{-\infty}^{\infty} dy\, f(r)\, e^{-i(k_x x + k_y y)} \quad (5.93)$$

with $r = \sqrt{x^2 + y^2}$.

(b) In polar coordinates, we have

See Block 0, Subsection 2.2.9.

$$\tilde{f}(k) = \frac{1}{2\pi} \int_0^{\infty} dr \int_0^{2\pi} d\phi\, r\, f(r)\, e^{-ikr[\cos\psi\cos\phi + \sin\psi\sin\phi]}$$

$$= \frac{1}{2\pi} \int_0^{\infty} dr \int_0^{2\pi} d\phi\, r\, f(r)\, e^{-ikr\cos(\psi-\phi)}$$

$$= \frac{1}{2\pi} \int_0^{\infty} dr \int_0^{2\pi} d\phi\, r\, f(r)\, e^{ikr\cos(\pi+\phi-\psi)}, \quad (5.94)$$

where we have used the given trigonometric identity and $\cos(\pi - \alpha) = -\cos\alpha$.

Now, substituting $\theta = \pi + \phi - \psi$ gives

$$\widetilde{f}(\boldsymbol{k}) = \frac{1}{2\pi} \int_0^\infty dr \int_{\pi-\psi}^{3\pi-\psi} d\theta \, r \, f(r) \, e^{ikr\cos\theta}$$

$$= \frac{1}{2\pi} \int_0^\infty dr \int_0^{2\pi} d\theta \, r \, f(r) \, e^{ikr\cos\theta}$$

$$= \frac{1}{\pi} \int_0^\infty dr \int_0^{\pi} d\theta \, r \, f(r) \, e^{ikr\cos\theta}, \tag{5.95}$$

where we have used the periodicity, and in the last step the symmetry about $\theta = \pi$, of the cosine function. This now gives, using equation (5.53),

Fourier transform of a two-dimensional rotationally symmetric function

$$\widetilde{f}(k) = \int_0^\infty dr \, r \, f(r) \, J_0(kr), \tag{5.96}$$

which now depends only on $k = |\boldsymbol{k}|$ as the angle ψ has disappeared. ∎

Equation (5.96), which is sometimes referred to as the *Fourier–Bessel transform*, expresses the two-dimensional Fourier transform $\widetilde{f}$ of a rotationally symmetric function f (which means that f depends only on r, not on the angle ϕ) as an integral involving the Bessel function $J_0(kr)$. You are asked to apply this in Exercise 5.17 to express the solution of the two-dimensional wave equation as a Fourier–Bessel integral. An example is discussed in Exercise 5.18.

Exercise 5.17

Consider the same situation as in Exercise 5.16, but with initial conditions $u(\boldsymbol{r}, 0) = a(\boldsymbol{r}) = a(r)$ and $u_t(\boldsymbol{r}, 0) = b(\boldsymbol{r}) = b(r)$ which depend only on the modulus $r = |\boldsymbol{r}|$.

(a) Why does this imply $\widetilde{u}(\boldsymbol{k}, t) = \widetilde{u}(k, t)$, so $\widetilde{u}$ depends only on the modulus k as well?

(b) Use the same argument as in Example 5.2 to express the inverse Fourier transform of $\widetilde{u}(k, t)$, which is our solution $u(\boldsymbol{r}, t)$, as an integral over J_0. [Hint: As in Example 5.2, introduce polar coordinates to perform the angular integration.]

Exercise 5.18

(a) Calculate $\widetilde{f}(k)$, the Fourier–Bessel transform (5.96) of the function $f(r) = 1/r$.

(b) Consider the two-dimensional wave equation with initial conditions $u(\boldsymbol{r}, 0) = a(r) = 1/r$ and $u_t(\boldsymbol{r}, 0) = b(r) = 0$. Calculate $\widetilde{u}(k, t)$ from Exercises 5.16 and 5.17, and express the corresponding solution $u(r, t)$ as an integral over the Bessel function J_0.

This example has been chosen because the transform can easily be evaluated. Due to the divergence at $r = 0$, this solution cannot be applied to physical situations such as vibrating membranes.

In Exercise 5.18, the function $u(r, t)$ is given as an integral which can be evaluated to give

$$u(r, t) = \frac{1}{\sqrt{r^2 - (ct)^2}}, \quad \text{for } r > ct. \tag{5.97}$$

At time t, the solution diverges at a radius $r = ct$.

Exercise 5.17 shows that for rotationally symmetric initial conditions

$$u(\boldsymbol{r}, 0) = a(r), \quad u_t(\boldsymbol{r}, 0) = b(r), \tag{5.98}$$

the solution of the two-dimensional isotropic wave equation (5.1) is

Solution of two-dimensional isotropic wave equation with rotationally symmetric initial conditions

$$u(\boldsymbol{r},t) = \int_0^\infty dk\, k\, [A(k)e^{ikct} + A^*(k)e^{-ikct}]\, J_0(kr), \qquad (5.99)$$

with

$$A(k) = \frac{1}{2}\widetilde{a}(k) + \frac{1}{2ikc}\widetilde{b}(k). \qquad (5.100)$$

In general, as for the Fourier transform, the integral in equation (5.99) cannot be evaluated directly. However, such forms can be useful for numerical or further analytical analysis of the solutions.

5.3.4 Isotropic waves in three dimensions

In general, the situation in three dimensions resembles that in two dimensions; it is difficult to perform the inverse Fourier transformation of $\widetilde{u}(\boldsymbol{k},t)$ due to the presence of $\omega = kc = c\sqrt{\boldsymbol{k}\cdot\boldsymbol{k}}$ in the integrand (see Exercise 5.16(c)).

We consider the case where the initial data, encapsulated in two functions, a and b, depend only on the modulus $r = |\boldsymbol{r}|$, as in equation (5.98). Because the wave equation is isotropic, the solution has to stay spherically symmetric for all t. In the two-dimensional situation, the spherically symmetric case could be reduced to integrals involving the Bessel function $J_0(kr)$; see Example 5.2 and Exercise 5.17. What about the three-dimensional case? Maybe surprisingly, this turns out to give a simpler result, which we now derive.

In the situation at hand, the same argument as in the two-dimensional case shows that the Fourier transform $\widetilde{u}$ of the solution depends only on $k = |\boldsymbol{k}|$ and is given by

$$\widetilde{u}(k,t) = A(k)\, e^{i\omega t} + A^*(k)\, e^{-i\omega t}, \qquad (5.101)$$

Here, $A^*(k)$ denotes the complex conjugate of $A(k)$.

with $\omega = kc$ and

$$A(k) = \frac{1}{2}\widetilde{a}(k) + \frac{1}{2i\omega}\widetilde{b}(k); \qquad (5.102)$$

compare Exercises 5.16 and 5.17, and equations (5.85) and (5.100). Note that $\widetilde{a}(k)$ and $\widetilde{b}(k)$ are now the *three-dimensional* Fourier transforms of the functions $a(r)$ and $b(r)$; we write k and r because A, $\widetilde{a}$, $\widetilde{b}$ and a, b depend only on $k = |\boldsymbol{k}|$ and $r = |\boldsymbol{r}|$, respectively. For the inverse transformation, we need to evaluate the integral

$$u(\boldsymbol{r},t) = \frac{1}{(2\pi)^{3/2}} \int_{\mathbb{R}^3} d\boldsymbol{k}\, \widetilde{u}(k,t)\, e^{i\boldsymbol{k}\cdot\boldsymbol{r}}. \qquad (5.103)$$

The only dependence on the vector $\boldsymbol{k}$ arises in the term $\exp(i\boldsymbol{k}\cdot\boldsymbol{r})$. As in the two-dimensional case, it is possible to use suitable coordinates, in this case spherical polar coordinates, to perform two integrations, and be left with a single integral over the modulus k. However, this is technically more complicated in the three-dimensional case, and rather than going through the calculation, we take it for granted that the result depends only on $r = |\boldsymbol{r}|$. In that case, we can simplify the task by choosing $\boldsymbol{r}$ along a particular direction, say the $\mathbf{k}$ direction, so $\boldsymbol{r} = r\mathbf{k}$, and evaluate the integral for this case. With $\boldsymbol{k} = k_x\mathbf{i} + k_y\mathbf{j} + k_z\mathbf{k}$, we obtain

Note the difference between $\boldsymbol{k}$ denoting the wave vector and the Cartesian unit vector $\mathbf{k}$.

$$u(r,t) = \frac{1}{(2\pi)^{3/2}} \int_{\mathbb{R}^3} d\boldsymbol{k}\, \widetilde{u}(k,t)\, e^{i\boldsymbol{k}\cdot\boldsymbol{r}} = \frac{1}{(2\pi)^{3/2}} \int_{\mathbb{R}^3} d\boldsymbol{k}\, \widetilde{u}(k,t)\, e^{ik_z r}. \quad (5.104)$$

Spherical coordinates for $\boldsymbol{k}$ are k and two angles, ϕ and θ. Due to our choice of $\boldsymbol{r}$, we need to consider only $k_z = k \cos\theta$. The volume element in the three-dimensional $\boldsymbol{k}$ space is then $\delta V = k^2 \sin\theta\, \delta k\, \delta\phi\, \delta\theta$ (see Figure 5.17), and integration is over $k \geq 0$, $0 \leq \theta \leq \pi$ and $0 \leq \phi \leq 2\pi$.

Spherical coordinates for $\boldsymbol{k}$ are the radius k, the azimuthal angle ϕ (the angle between the x-axis and $\boldsymbol{k}$ in the (x,y)-plane), and the polar angle θ (the angle between the x-axis and $\boldsymbol{k}$).

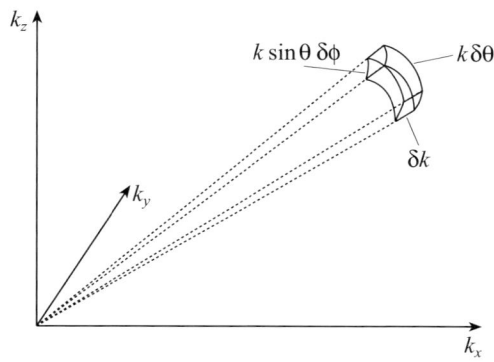

Figure 5.17 A volume element $\delta V = k^2 \sin\theta\, \delta k\, \delta\phi\, \delta\theta$ in three-dimensional spherical coordinates for $\boldsymbol{k}$

This yields

$$u(r,t) = \frac{1}{(2\pi)^{3/2}} \int_0^\infty dk \int_0^{2\pi} d\phi \int_0^\pi d\theta\, k^2 \sin\theta\, \widetilde{u}(k,t)\, e^{ikr\cos\theta}. \qquad (5.105)$$

The integration over ϕ just gives a factor of 2π, because the integrand does not depend on ϕ. Substituting $w = \cos\theta$ gives

$$u(r,t) = \frac{1}{\sqrt{2\pi}} \int_0^\infty dk \int_1^{-1} (-dw)\, k^2\, \widetilde{u}(k,t)\, e^{ikrw}$$

$$= \frac{1}{\sqrt{2\pi}} \int_0^\infty dk \int_{-1}^1 dw\, k^2\, \widetilde{u}(k,t)\, e^{ikrw}, \qquad (5.106)$$

because $dw/d\theta = -\sin\theta$, $w = 1$ for $\theta = 0$, and $w = -1$ for $\theta = \pi$. The integration over w gives

$$\int_{-1}^1 dw\, e^{ikrw} = \left[\frac{e^{ikrw}}{ikr}\right]_{-1}^1 = \frac{e^{ikr} - e^{-ikr}}{ikr} = \frac{2\sin(kr)}{kr}. \qquad (5.107)$$

Recalling the definition of the sinc function, $\mathrm{sinc}(x) = (\sin x)/x$ for $x \neq 0$, the final result is thus

See Chapter 3, page 111.

$$u(r,t) = \sqrt{\frac{2}{\pi}} \int_0^\infty dk\, \widetilde{u}(k,t)\, k^2\, \mathrm{sinc}(kr), \qquad (5.108)$$

which gives the solution of the initial-value problem as an integral over the Fourier transform $\widetilde{u}(k,t)$ given by equation (5.101), with coefficients given in equation (5.102). For this three-dimensional case, it is not the Bessel function $J_0(kr)$, but rather the sinc function $\mathrm{sinc}(kr) = \sin(kr)/(kr)$ that enters the integral.

Thus the solution of the three-dimensional isotropic wave equation (5.73) with spherically symmetric initial conditions $u(\boldsymbol{r},0) = a(r)$ and $u_t(\boldsymbol{r},0) = b(r)$ is

Solution of three-dimensional isotropic wave equation with spherically symmetric initial conditions

$$u(\boldsymbol{r},t) = \int_0^\infty dk\, k^2 \left[A(k)e^{ikct} + A^*(k)e^{-ikct}\right] \mathrm{sinc}(kr), \qquad (5.109)$$

with

$$A(k) = \frac{1}{2}\widetilde{a}(k) + \frac{1}{2ikc}\widetilde{b}(k). \qquad (5.110)$$

5.3 Isotropic waves in two and three dimensions

Exercise 5.19 shows that the three-dimensional Fourier transform of a spherically symmetric function $f(r)$ is given by

Fourier transform of a three-dimensional spherically symmetric function

$$\widetilde{f}(k) = \sqrt{\frac{2}{\pi}} \int_0^\infty dr\, r^2\, f(r) \operatorname{sinc}(kr). \tag{5.111}$$

In Exercise 5.20, we calculate the Fourier transform of a Gaussian function, then consider the solution of the three-dimensional wave equation with a Gaussian initial form.

Exercise 5.19

Consider a function $f(\mathbf{r})$, where $\mathbf{r} \in \mathbb{R}^3$, which depends only on the modulus $r = |\mathbf{r}|$, so we may write it as $f(r)$. By symmetry arguments, we may assume that the Fourier transform $\widetilde{f}(\mathbf{k}) = \widetilde{f}(k)$ depends only on the modulus $k = |\mathbf{k}|$.

Express $\widetilde{f}$ as a one-dimensional integral over r, by rewriting the Fourier transform in spherical coordinates r, θ, ϕ, and performing the angular integrations. [Hint: This is what we did in equations (5.103)–(5.108) for the inverse transformation.]

Exercise 5.20

Calculate the three-dimensional Fourier transform $\widetilde{a}(\mathbf{k})$ of the Gaussian $a(\mathbf{r}) = \exp(-r^2/2)$. [Hint: Use $\sin(kr) = -\operatorname{Im}[\exp(-ikr)]$ in the result of Exercise 5.19, express $r \exp(-r^2/2)$ in terms of the derivative of $\exp(-r^2/2)$, and use the properties of the Fourier transform.]

Exercise 5.21

Consider the three-dimensional wave equation with initial data given by $a(\mathbf{r}) = \exp(-r^2/2)$ and $b(\mathbf{r}) = 0$.

(a) Calculate $\widetilde{u}(k,t)$ from equations (5.101) and (5.102).

(b) Use equation (5.108) to write the solution $u(r,t)$ as an integral over k. Use the addition formula $2\sin\alpha\cos\beta = \sin(\alpha - \beta) + \sin(\alpha + \beta)$ to write the integral as the sum of two sine Fourier integrals, then evaluate these as in Exercise 5.20. [Hint: Remember that $\omega = kc$ depends on k.]

As Exercise 5.20 shows, the three-dimensional Fourier transform of the Gaussian $\exp(-r^2/2)$ is, amazingly, just the same function as its one-dimensional Fourier transform. Note that this is not true in general, because we used the identity $-rf(r) = f'(r)$ which is true only for the Gaussian.

5.4 Summary and Outcomes

In this chapter, we considered the motion of waves in two and three dimensions, where the independent position variable that enters the wave equation becomes a vector in two or three dimensions. This situation was, once more, addressed by means of Fourier methods.

Normal modes and Fourier series also apply in two and three dimensions, for special choices of boundary conditions, as shown in the example of rectangular and circular membranes. The latter example shows that not only sinusoidal functions play a role; in this case we encountered Bessel functions. For a generally shaped membrane, the boundary conditions are not separable. Higher-dimensional problems where separation of variables fails often cannot be solved in closed form. In such situations, one is limited to approximate solutions, for instance by the variational methods discussed later in this course, or to numerical solutions by computer.

In the separable case, with each eigenmode of the system, we obtain a characteristic number, the eigenvalue or eigenfrequency. This determines the time-dependence of the solution. If the system possesses a symmetry, such as a circular membrane which does not change if we rotate around its centre, this symmetry leads to systematic degeneracies in the spectrum of eigenvalues. In the degenerate case, there exists more than a single eigenmode corresponding to a fixed eigenfrequency, thus any linear combination of these modes is also an eigenmode.

At the end of this chapter, it was demonstrated how Fourier transforms can be used to solve partial differential equations. Fourier transforms turn differential equations into algebraic expressions for the Fourier-transformed functions. Solving these and transforming back yields integral expressions for the solutions of the differential equation.

After working through this chapter, you should:
- know how to generalise the wave equation to more than one dimension;
- be able to write down wave equations in two and three dimensions;
- know what isotropic and spherical symmetry mean;
- be able to write down plane wave solutions of the wave equation;
- understand the role of the wave vector of a plane wave;
- know how to use Fourier transforms of derivatives to solve some linear partial differential equations;
- understand how to address the initial-value problem by means of Fourier transforms;
- be able to calculate normal modes for rectangular and circular membranes;
- understand the notion of degeneracy of an eigenvalue;
- understand the meaning of nodal line patterns, and be able to draw nodal line patterns of normal modes in simple cases;
- understand how separation of variables is used, and appreciate that for most boundaries this method does not work;
- know some basic facts about Bessel functions;
- be able to apply Bessel functions in calculations of two-dimensional Fourier transforms of rotationally symmetric functions;
- be able to apply three-dimensional Fourier transforms of spherically symmetric functions;
- know how to solve the isotropic wave equation in two and three dimensions by Fourier transform for rotationally and spherically symmetric initial conditions.

5.5 Further Exercises

Exercise 5.22

Consider a rectangular membrane with edge lengths $L = 4$ and $M = 1$.

(a) What are the eigenvalues $k_{n,m}^2$ and eigenfrequencies $\omega_{n,m}$ of this membrane?

(b) Write down the eigenmodes $f_{n,m}(\boldsymbol{r})$ for this membrane.

(c) Sketch the nodal lines for the 5 eigenmodes of lowest frequency.

(d) Consider the 12 lowest eigenfrequencies. Are there any degeneracies?

(e) Find the systematic degeneracies in this case. In other words, determine values $n'(n,m)$ and $m'(n,m)$ such that $k_{n',m'}^2 = k_{n,m}^2$ for all n and m.

(f) Write down an equation determining the nodal lines for the differences of degenerate eigenmodes. Do they have a nodal line of the form $x = 4y$, which is the diagonal in the rectangle? In other words, can you derive eigenmodes of the triangular membrane obtained by cutting the rectangle along the diagonal, in the same way as we did for the isosceles right-angled triangle discussed in Subsection 5.2.2?

Exercise 5.23

In this exercise, we ask you to derive the normal modes of a semi-circular membrane, using the same approach as for the triangular membrane; compare Subsection 5.2.2.

(a) Give a complete set of eigenmodes for the circular membrane of radius l. In particular, you need two functions, which are not multiples of each other, for degenerate eigenvalues.

(b) Determine all eigenmodes of a circular membrane of radius l which have a nodal line in the $\mathbf{i}$ direction, thus for $\theta = 0$ (and thus also for $\theta = \pi$).

(c) Argue that these give *all* eigenmodes of the semi-circular membrane of radius l.

(d) Are there any degeneracies in the spectrum for the semi-circular membrane?

5.6 Appendix: Power series for Bessel functions

In Subsection 5.2.5, it was shown that the leading term of the Bessel function $J_n(x)$ for small argument x is of order x^n. Here, we use the same argument to calculate a power series for $J_n(x)$ of the form

$$J_n(x) = x^n \sum_{s=0}^{\infty} c_s x^s = \sum_{s=0}^{\infty} c_s x^{n+s} \qquad (5.112)$$

See page 200.

with $c_0 \neq 0$. Inserting this series into the differential equation (5.50) yields an infinite sum of terms, which must vanish. This means that each coefficient of a power x^{n+s} in the sum must vanish individually.

Substituting (5.112) into (5.50), the coefficient of x^n is

$$n(n-1)c_0 + nc_0 - n^2 c_0 = 0, \qquad (5.113)$$

because this result was used to fix the leading power in Subsection 5.2.5. The coefficient $c_0 \neq 0$ is arbitrary; it can be fixed by the additional normalisation condition (5.55). The coefficient of x^{n+1} is

$$(n+1)nc_1 + (n+1)c_1 - n^2 c_1 = (2n+1)c_1 = 0, \qquad (5.114)$$

which (assuming $n \neq -\tfrac{1}{2}$) implies $c_1 = 0$.

Let us consider the coefficient of x^{n+s} for $s > 1$. It is

$$(n+s)(n+s-1)c_s + (n+s)c_s + c_{s-2} - n^2 c_s$$
$$= [(n+s)^2 - n^2]c_s + c_{s-2} = 0, \qquad (5.115)$$

which yields a recurrence relation for the coefficients c_s:

$$c_s = -\frac{c_{s-2}}{(n+s)^2 - n^2} = -\frac{c_{s-2}}{s(2n+s)}. \qquad (5.116)$$

With $c_1 = 0$, it follows that all coefficients with odd index s vanish, so $c_{2t+1} = 0$ for all integers t. For the even coefficients, we obtain

$$c_{2t} = -\frac{1}{4t(n+t)} c_{2(t-1)} = c_0 \prod_{j=1}^{t} \left(-\frac{1}{4j(n+j)}\right)$$
$$= c_0 \frac{(-1)^t n!}{2^{2t} t! (n+t)!} \qquad (5.117)$$

because $\prod_{j=1}^{t} j = t!$ and

$$\prod_{j=1}^{t} (n+j) = (n+1)(n+2)\cdots(n+t) = \frac{(n+t)!}{n!}. \qquad (5.118)$$

Hence we obtain

$$J_n(x) = c_0 \sum_{t=0}^{\infty} \frac{(-1)^t n!}{2^{2t} t! (n+t)!} x^{n+2t} \qquad (5.119)$$

as the power series for Bessel functions of the first kind.

Using the normalisation constant $c_0 = 1/(2^n n!)$ from equation (5.60), the power series for the Bessel functions becomes

Power series for Bessel functions of the first kind

$$J_n(x) = \sum_{t=0}^{\infty} \frac{(-1)^t}{2^{n+2t} t! (n+t)!} x^{n+2t} = \sum_{t=0}^{\infty} \frac{(-1)^t}{t! (n+t)!} \left(\frac{x}{2}\right)^{n+2t}. \qquad (5.120)$$

Although we are interested only in the case where n is a non-negative integer, the argument works for any $n \neq -\tfrac{1}{2}$, provided that we give a proper interpretation to the 'factorials' $n!$ and $(n+t)!$ in the expression.

Solutions to Exercises in Chapter 5

Solution 5.1

From the list, $N = 3$ and $N = 36$ do not correspond to eigenvalues, as they cannot be written as the sum of two squares. The degeneracies for the other cases are as follows.

$N = 8 = 2^2 + 2^2$ is not degenerate,
$N = 13 = 2^2 + 3^2 = 3^2 + 2^2$ occurs twice,
$N = 50 = 1^2 + 7^2 = 5^2 + 5^2 = 7^2 + 1^2$ occurs three times,
$N = 85 = 2^2 + 9^2 = 6^2 + 7^2 = 7^2 + 6^2 = 9^2 + 2^2$ occurs four times.

Solution 5.2

(a) The function $f_{1,2}(\mathbf{r})$ vanishes within the square for $y = l/2$. So this solution has one nodal line that crosses the square in the middle, parallel to the $\mathbf{i}$ direction.

The function $f_{2,1}(\mathbf{r})$ vanishes within the square for $x = l/2$. In this case, there is one nodal line that crosses the square in the middle, but now parallel to the $\mathbf{j}$ direction.

(b) The partial derivatives of $f_{1,2}(\mathbf{r})$ are

$$\frac{\partial^2}{\partial x^2} f_{1,2}(\mathbf{r}) = -\left(\frac{\pi}{l}\right)^2 \sin\left(\frac{\pi x}{l}\right) \sin\left(\frac{2\pi y}{l}\right) = -\left(\frac{\pi}{l}\right)^2 f_{1,2}(\mathbf{r}),$$

$$\frac{\partial^2}{\partial y^2} f_{1,2}(\mathbf{r}) = -\left(\frac{2\pi}{l}\right)^2 \sin\left(\frac{\pi x}{l}\right) \sin\left(\frac{2\pi y}{l}\right) = -4\left(\frac{\pi}{l}\right)^2 f_{1,2}(\mathbf{r}),$$

so $f_{1,2}(\mathbf{r})$ satisfies equation (5.6) with $-C = \omega^2/c^2 = 5\pi^2/l^2$.

For $f_{2,1}(\mathbf{r})$, we obtain the same equations but with the factors π^2/l^2 and $4\pi^2/l^2$ interchanged.

Thus both $f_{1,2}(\mathbf{r})$ and $f_{2,1}(\mathbf{r})$ satisfy equation (5.6) with the same value of ω^2, and so does any linear combination.

(c) We have (using the suggested identity)

$$f(\mathbf{r}) = A \sin\left(\frac{\pi x}{l}\right) \sin\left(\frac{2\pi y}{l}\right) + B \sin\left(\frac{2\pi x}{l}\right) \sin\left(\frac{\pi y}{l}\right)$$

$$= 2A \sin\left(\frac{\pi x}{l}\right) \sin\left(\frac{\pi y}{l}\right) \cos\left(\frac{\pi y}{l}\right) + 2B \sin\left(\frac{\pi x}{l}\right) \cos\left(\frac{\pi x}{l}\right) \sin\left(\frac{\pi y}{l}\right)$$

$$= 2 \sin\left(\frac{\pi x}{l}\right) \sin\left(\frac{\pi y}{l}\right) \left[A \cos\left(\frac{\pi y}{l}\right) + B \cos\left(\frac{\pi x}{l}\right)\right].$$

Within the square, for $0 < x < l$ and $0 < y < l$, this vanishes only if the last factor vanishes, which yields the desired equation.

(d) For $A = -B$, the equation for the nodal lines becomes

$$0 = \cos\left(\frac{\pi x}{l}\right) - \cos\left(\frac{\pi y}{l}\right) = -2 \sin\left(\frac{\pi}{2l}(x+y)\right) \sin\left(\frac{\pi}{2l}(x-y)\right).$$

Within the square, $|x+y| < 2l$ and $|x-y| < l$, so this is satisfied only for $x+y = 0$ and $x-y = 0$. As both x and y are positive, there is only one nodal line, given by $x = y$. This is one of the diagonals of the square.

(e) For $A = B$, the equation for the nodal lines becomes

$$0 = \cos\left(\frac{\pi x}{l}\right) + \cos\left(\frac{\pi y}{l}\right) = 2 \cos\left(\frac{\pi}{2l}(x+y)\right) \cos\left(\frac{\pi}{2l}(x-y)\right).$$

The cosine function vanishes for odd multiples of $\pi/2$. Here, the only possibility is $x + y = l$, because $|x - y| < l$ inside the square. So again there is a single nodal line, given by $y = l - x$. This is the other diagonal of the square.

Solution 5.3

(a) For $m = 0$ or $n = 0$, the eigenfrequencies $\omega_{n,m}$ are
$$\omega_{n,0} = k_{n,0}c = \frac{\pi c}{2l}n, \quad \omega_{0,m} = k_{0,m}c = \frac{\pi c}{l}m;$$
the ratio of any two of these is rational.

(b) We have $N = n^2 + 4m^2 = 8$ for $n = 2$, $m = 1$, and $N = n^2 + 4m^2 = 20$ for $n = 2$, $m = 2$. So the ratio of the eigenfrequencies is
$$\frac{\omega_{2,2}}{\omega_{2,1}} = \sqrt{\frac{20}{8}} = \sqrt{\frac{10}{4}} = \frac{1}{2}\sqrt{10},$$
which is irrational.

(c) We have
$$\tau_{3,0} = \frac{2\pi}{\omega_{3,0}} = \frac{4l}{3c}, \quad \tau_{0,2} = \frac{2\pi}{\omega_{0,2}} = \frac{l}{c},$$
so we obtain $\tau = 3\tau_{3,0} = 4\tau_{0,2} = 4l/c$ for the period of a superposition of these two modes.

Solution 5.4

(a) The eigenmodes are
$$f_{n,m}(\mathbf{r}) = \sin\left(\frac{\pi n}{L}x\right)\sin\left(\frac{\pi m}{M}y\right) = \sin\left(\frac{\pi n}{2l}x\right)\sin\left(\frac{\pi m}{l}y\right).$$

(b) For $x = l$, we have
$$f_{n,m}(l\mathbf{i} + y\mathbf{j}) = \sin\left(\frac{\pi n}{2}\right)\sin\left(\frac{\pi m}{l}y\right).$$
This vanishes if $\pi n/2$ is an integer multiple of π, i.e. if n is even. All modes $f_{2n,m}(\mathbf{r})$, with $n, m = 1, 2, 3, \ldots$, thus have the required nodal line.

(c) On the nodal line, the mode is zero. Therefore these modes satisfy the wave equation and the boundary conditions for a square membrane with $0 \leq x, y \leq l$.

(d) The modes are
$$f_{2n,m}(\mathbf{r}) = \sin\left(\frac{\pi 2n}{2l}x\right)\sin\left(\frac{\pi m}{l}y\right) = \sin\left(\frac{\pi n}{l}x\right)\sin\left(\frac{\pi m}{l}y\right),$$
for $n, m = 1, 2, 3, \ldots$. These are all normal modes of the $l \times l$ square membrane, corresponding to $L = M = l$ in equations (5.15).

Solution 5.5

(a) The function $a(\mathbf{r}) = u(\mathbf{r}, 0)$ must satisfy the boundary condition that u vanishes on the boundary of the rectangle. Clearly, $a(\mathbf{r}) = 0$ for $x = 0$ or $y = 0$. For $x = L = 3$ we have $a(\mathbf{r}) = 4\sin(6\pi)\sin(2\pi y) = 0$, and for $y = M = 1$ we have $a(\mathbf{r}) = 4\sin(2\pi x)\sin(2\pi) = 0$. As $b(\mathbf{r}) = 0$, the initial data satisfy the boundary conditions.

(b) The solution $u(\mathbf{r}, t)$ is given in terms of normal modes by
$$u(\mathbf{r}, t) = \sum_{n=1}^{\infty}\sum_{m=1}^{\infty}\sin\left(\frac{\pi nx}{3}\right)\sin(\pi my)\left[A_{n,m}\cos(\omega_{n,m}t) + B_{n,m}\sin(\omega_{n,m}t)\right],$$
with $\omega_{n,m} = \pi c\sqrt{n^2 + 9m^2}/3$. The coefficients are given in equations (5.29) and (5.30). We have $B_{n,m} = 0$ for all values of n and m, because $b(\mathbf{r}) = 0$. We can calculate $A_{n,m}$ from equation (5.29) (using the orthogonality relation for the sine functions). However, we can also read the coefficients directly from the expansion (5.25), because the function $a(\mathbf{r})$ happens to be just a single term of this expansion. This gives $A_{6,2} = 4$, and $A_{n,m} = 0$ for all other values of n and m. The motion is thus in a single normal mode with frequency $\omega_{6,2} = \pi c\sqrt{36 + 36}/3 = 2\pi c\sqrt{2}$, and the solution is
$$u(\mathbf{r}, t) = 4\sin(2\pi x)\sin(2\pi y)\cos(2\pi c\sqrt{2}t).$$

See equation (2.62) in Chapter 2.

Solution 5.6

(a) Using $e^{ix} = \cos x + i\sin x$, we obtain
$$J_0(z) = \frac{1}{\pi}\int_0^\pi d\theta \cos(z\cos\theta) + \frac{i}{\pi}\int_0^\pi d\theta \sin(z\cos\theta).$$

(b) The imaginary part is
$$\frac{1}{\pi}\int_0^\pi d\theta \sin(z\cos\theta) = \frac{1}{\pi}\int_0^{\pi/2} d\theta \sin(z\cos\theta) + \frac{1}{\pi}\int_{\pi/2}^\pi d\theta \sin(z\cos\theta).$$

Substituting $\phi = \pi - \theta$ in the second integral and using $\cos(\pi - \phi) = -\cos\phi$ gives
$$\frac{1}{\pi}\int_{\pi/2}^\pi d\theta \sin(z\cos\theta) = -\frac{1}{\pi}\int_{\pi/2}^0 d\phi \sin[z\cos(\pi - \phi)]$$
$$= \frac{1}{\pi}\int_0^{\pi/2} d\phi \sin(-z\cos\phi)$$
$$= -\frac{1}{\pi}\int_0^{\pi/2} d\phi \sin(z\cos\phi),$$

which apart from the negative sign equals the integral from 0 to $\pi/2$, so the imaginary part vanishes.

(c) The real part is
$$J_0(z) = \frac{1}{\pi}\int_0^\pi d\theta \cos(z\cos\theta)$$
$$= \frac{1}{\pi}\int_0^{\pi/2} d\theta \cos(z\cos\theta) + \frac{1}{\pi}\int_{\pi/2}^\pi d\theta \cos(z\cos\theta).$$

Substituting $\phi = \pi/2 - \theta$ and $\phi = 3\pi/2 - \theta$, respectively, yields
$$J_0(z) = -\frac{1}{\pi}\int_{\pi/2}^0 d\phi \cos\left[z\cos\left(\frac{\pi}{2} - \phi\right)\right] - \frac{1}{\pi}\int_\pi^{\pi/2} d\phi \cos\left[z\cos\left(\frac{3\pi}{2} - \phi\right)\right]$$
$$= \frac{1}{\pi}\int_0^{\pi/2} d\phi \cos(z\sin\phi) + \frac{1}{\pi}\int_{\pi/2}^\pi d\phi \cos(-z\sin\phi)$$
$$= \frac{1}{\pi}\int_0^\pi d\phi \cos(z\sin\phi),$$

because cosine is an even function. This agrees with equation (5.54) for $n = 0$.

Solution 5.7

(a) We obtain
$$2\cos\theta\cos(n\theta) = \frac{(e^{i\theta} + e^{-i\theta})(e^{in\theta} + e^{-in\theta})}{2}$$
$$= \frac{e^{i(n+1)\theta} + e^{i(n-1)\theta} + e^{-i(n-1)\theta} + e^{-i(n+1)\theta}}{2}$$
$$= \frac{e^{i(n+1)\theta} + e^{-i(n+1)\theta}}{2} + \frac{e^{i(n-1)\theta} + e^{-i(n-1)\theta}}{2}$$
$$= \cos[(n+1)\theta] + \cos[(n-1)\theta].$$

(b) Using the above relation repeatedly gives
$$\cos^2\theta \cos(n\theta) = \cos\theta[\cos\theta\cos(n\theta)]$$
$$= \cos\theta \frac{\cos[(n+1)\theta] + \cos[(n-1)\theta]}{2}$$
$$= \frac{\cos[(n+2)\theta] + 2\cos[n\theta] + \cos[(n-2)\theta]}{4}.$$

(c) Writing $\cos^k\theta\cos(n\theta)$ as a sum of cosine terms involves terms $\cos(m\theta)$ with $m = n-k, n-k+2, \ldots, n+k-2, n+k$. For $k = n$, we have $m = 0, 2, 4, \ldots, 2n-2, 2n$, and the terms corresponding to $m = 0$ and $m = 2n$ occur with prefactor 2^{-n}.

(d) The derivative is
$$J_n^{(k)}(z) = \frac{1}{\pi i^n} \int_0^\pi d\theta \, (i\cos\theta)^k \, e^{iz\cos\theta} \cos(n\theta),$$
so at $z = 0$ we have
$$J_n^{(k)}(0) = \frac{1}{\pi i^{n-k}} \int_0^\pi d\theta \, \cos^k\theta \cos(n\theta).$$
We can write the product $\cos^k\theta \cos(n\theta)$ as a sum of cosines $\cos(m\theta)$ with $m = n-k, n-k+2, \ldots, n+k-2, n+k$, which are non-zero for $k < n$. These integrate to $\sin(m\theta)$, and since $\sin 0 = \sin(m\pi) = 0$, the definite integral vanishes; compare equation (5.56), which corresponds to the case $k = 0$.

(e) The Taylor expansion of $J_n(z)$ about $z = 0$ is
$$J_n(z) = \sum_{k=0}^\infty J_n^{(k)}(0) \frac{z^k}{k!}.$$
Since $J_n^{(k)}(0) = 0$ for $k < n$, the first non-vanishing term is at $k = n$, with $J_n^{(n)}(0) = 2^{-n}$. Hence
$$J_n(z) = \frac{z^n}{2^n \, n!} + O(z^{n+1}),$$
in agreement with equation (5.60).

Solution 5.8

For $n = 0$, we have $\tilde{m} = m$; compare Figure 5.8. The asymptotic values are $19\pi/4 \simeq 14.923$ for $\tilde{m} = m = 5$, and $39\pi/4 \simeq 30.631$ for $\tilde{m} = m = 10$. The differences between these and the values quoted in Table 5.1 are 0.008 and 0.004, which correspond to deviations of less than 0.1%.

For $n = 7$, we have $\tilde{m} = m + 3$. The asymptotic values are $33\pi/4 \simeq 25.918$ for $\tilde{m} = m + 3 = 8$, and $53\pi/4 \simeq 41.626$ for $\tilde{m} = m + 3 = 13$. Here, the differences are 0.983 and 0.595, corresponding to deviations of about 4% and 1.5%, respectively.

Solution 5.9

For $l = 10$, the eigenvalues are given by $k_{n,m} = z_{n,m}/10$. From Table 5.1, the lowest eigenvalues are $k_{0,1} \simeq 0.2405$, $k_{1,1} \simeq 0.3832$, $k_{2,1} \simeq 0.5136$, $k_{0,2} \simeq 0.5520$, $k_{3,1} \simeq 0.6380$, $k_{1,2} \simeq 0.7016$, $k_{4,1} \simeq 0.7588$, $k_{2,2} \simeq 0.8417$ and $k_{0,3} \simeq 0.8654$.

Solution 5.10

(a) The general form is
$$v_{n,m}(r,\theta) = J_n(k_{n,m}r)[A_n \cos(n\theta) + B_n \sin(n\theta)],$$
with arbitrary coefficients A_n and B_n. Here, m labels the zeros of the equation $J_n(k_{n,m}l) = 0$.

(b) For $n = 0$, there is no nodal line. The eigenmodes have the form
$$v_{0,m}(r,\theta) = J_0(k_{0,m}r),$$
where we set the coefficient $A_0 = 1$. For $n > 0$, there is at least one nodal line. By choosing $A_n = 0$, we can make sure that there is a nodal line for $\theta = 0$, which is in the $\mathbf{i}$ direction. The corresponding eigenmodes are
$$v_{n,m}(r,\theta) = J_n(k_{n,m}r) \sin(n\theta),$$
where we have set $B_n = 1$.

Solution 5.11

With $\mathbf{k} = k_x \mathbf{i} + k_y \mathbf{j} + k_z \mathbf{k}$ and $\mathbf{r} = x\mathbf{i} + y\mathbf{j} + z\mathbf{k}$, we have
$$f(\mathbf{k} \cdot \mathbf{r} \pm \omega t) = f(k_x x + k_y y + k_z z \pm \omega t).$$

The derivatives are

$$\nabla^2 f(\mathbf{k}\cdot\mathbf{r}\pm\omega t) = \left(\frac{\partial^2}{\partial x^2}+\frac{\partial^2}{\partial y^2}+\frac{\partial^2}{\partial z^2}\right) f(\mathbf{k}\cdot\mathbf{r}\pm\omega t)$$
$$= k_x^2\, f''(\mathbf{k}\cdot\mathbf{r}\pm\omega t)+k_y^2\, f''(\mathbf{k}\cdot\mathbf{r}\pm\omega t)+k_z^2\, f''(\mathbf{k}\cdot\mathbf{r}\pm\omega t)$$
$$= (k_x^2+k_y^2+k_z^2)\, f''(\mathbf{k}\cdot\mathbf{r}\pm\omega t)$$
$$= k^2\, f''(\mathbf{k}\cdot\mathbf{r}\pm\omega t)$$

Here, f'' denotes the second derivative of the function f.

and

$$\frac{\partial^2}{\partial t^2} f(\mathbf{k}\cdot\mathbf{r}\pm\omega t) = (\pm\omega)^2 f''(\mathbf{k}\cdot\mathbf{r}\pm\omega t) = \omega^2 f''(\mathbf{k}\cdot\mathbf{r}\pm\omega t).$$

Thus the wave equation is fulfilled if $\mathbf{k}\cdot\mathbf{k}=k^2=\omega^2/c^2$, as given.

Solution 5.12

(a) We have

$$f(\mathbf{k}\cdot\mathbf{r}-\omega t) = f(k\mathbf{i}\cdot\mathbf{r}-kct) = f[k(x-ct)],$$

so f depends only on $x-ct$.

(b) Consider the function at position x at time $t=0$, which is $f(kx)$. At time $t>0$, we find the same value at position x', given by $k(x'-ct)=kx$, so $x'=x+ct$. So the wave pattern described by f has moved a distance ct along the $\mathbf{i}$ direction.

Solution 5.13

Using again the Fourier transform of the derivative, the wave equation becomes

$$\widetilde{u}_{xx}(x,\omega) = -\frac{\omega^2}{c^2}\widetilde{u}(x,\omega).$$

Solution 5.14

If we Fourier transform the wave equation with respect to time and position, both derivatives disappear, and we are left with

$$-k^2\widetilde{u}(k,\omega) = -\frac{\omega^2}{c^2}\widetilde{u}(k,\omega).$$

Hence

$$(\omega^2-k^2c^2)\,\widetilde{u}(k,\omega) = 0,$$

which implies the relation $\omega^2=k^2c^2$ for non-vanishing solutions $\widetilde{u}(k,\omega)$. However, we do not obtain an equation for $\widetilde{u}(k,\omega)$.

Solution 5.15

(a) The initial conditions are

$$\widetilde{u}(k,0) = \frac{1}{\sqrt{2\pi}}\int_{-\infty}^{\infty} dx\, u(x,0)\, e^{-ikx} = \frac{1}{\sqrt{2\pi}}\int_{-\infty}^{\infty} dx\, a(x)\, e^{-ikx} = \widetilde{a}(k)$$

and

$$\widetilde{u}_t(k,0) = \frac{1}{\sqrt{2\pi}}\int_{-\infty}^{\infty} dx\, u_t(x,0)\, e^{-ikx} = \frac{1}{\sqrt{2\pi}}\int_{-\infty}^{\infty} dx\, b(x)\, e^{-ikx} = 0.$$

(b) From equation (5.79), we find

$$\widetilde{u}(k,0) = A(k)+A^*(k) = 2\,\mathrm{Re}[A(k)]$$

and

$$\widetilde{u}_t(k,0) = i\omega[A(k)-A^*(k)] = -2\omega\,\mathrm{Im}[A(k)].$$

Thus $\mathrm{Im}[A(k)]=0$ and $A(k)$ is real, hence

$$A(k) = \tfrac{1}{2}\widetilde{a}(k).$$

(c) The Fourier transform of the solution is
$$\widetilde{u}(k,t) = \tfrac{1}{2}\widetilde{a}(k)(e^{i\omega t} + e^{-i\omega t}).$$
Using $\omega = kc$, the inverse transformation becomes
$$u(x,t) = \frac{1}{\sqrt{2\pi}} \int_{-\infty}^{\infty} dk \, \tfrac{1}{2}\widetilde{a}(k)(e^{ik(x+ct)} + e^{ik(x-ct)}).$$
We can evaluate this using the translation property of the Fourier transform discussed in Section 3.4, which gives
$$u(x,t) = \tfrac{1}{2}[a(x+ct) + a(x-ct)].$$
This corresponds to d'Alembert's solution for the initial conditions; compare equation (2.25).

Solution 5.16

(a) The initial conditions for $\widetilde{u}(\boldsymbol{k},t)$ are
$$\widetilde{u}(\boldsymbol{k},0) = \frac{1}{2\pi} \int_{\mathbb{R}^2} d\boldsymbol{r}\, a(\boldsymbol{r})\, e^{-i\boldsymbol{k}\cdot\boldsymbol{r}} = \widetilde{a}(\boldsymbol{k})$$
and
$$\widetilde{u}_t(\boldsymbol{k},0) = \frac{1}{2\pi} \int_{\mathbb{R}^2} d\boldsymbol{r}\, b(\boldsymbol{r})\, e^{-i\boldsymbol{k}\cdot\boldsymbol{r}} = \widetilde{b}(\boldsymbol{k}).$$

(b) Inserting the initial conditions into the general solution, and solving for the coefficients $A(\boldsymbol{k})$, we obtain
$$A(\boldsymbol{k}) = \frac{1}{2}\widetilde{a}(\boldsymbol{k}) + \frac{1}{2i\omega}\widetilde{b}(\boldsymbol{k}).$$

(c) The inverse Fourier transform gives
$$u(\boldsymbol{r},t) = \frac{1}{2\pi} \int_{\mathbb{R}^2} d\boldsymbol{k}\, \widetilde{u}(\boldsymbol{k},t)\, e^{i\boldsymbol{k}\cdot\boldsymbol{r}}$$
$$= \frac{1}{2\pi} \int_{\mathbb{R}^2} d\boldsymbol{k}\, \left(A(\boldsymbol{k})\, e^{i(\boldsymbol{k}\cdot\boldsymbol{r}+kct)} + A^*(\boldsymbol{k})\, e^{i(\boldsymbol{k}\cdot\boldsymbol{r}-kct)} \right).$$

Solution 5.17

(a) This follows from Exercise 5.16 and Example 5.2. The initial conditions for $\widetilde{u}(\boldsymbol{k},t)$ are given in Exercise 5.16 by the Fourier transforms $\widetilde{a}(\boldsymbol{k})$ and $\widetilde{b}(\boldsymbol{k})$, which according to Example 5.2 depend only on $k = |\boldsymbol{k}|$. The general solution for $\widetilde{u}(\boldsymbol{k},t)$ given in Exercise 5.16 involves functions $\exp(\pm ikct)$, which depend only on k, and coefficients $A(\boldsymbol{k})$, which in turn are expressed in terms of $\widetilde{a}$ and $\widetilde{b}$, so also depend only on k. Hence $\widetilde{u}(\boldsymbol{k},t)$ depends only on k.

(b) We have, with $\widetilde{u}(\boldsymbol{k},t) = \widetilde{u}(k,t)$,
$$u(\boldsymbol{r},t) = \frac{1}{2\pi} \int_{\mathbb{R}^2} d\boldsymbol{k}\, \widetilde{u}(k,t)\, e^{i\boldsymbol{k}\cdot\boldsymbol{r}}.$$
In terms of polar coordinates, defined as in Example 5.2, we have
$$\boldsymbol{k}\cdot\boldsymbol{r} = kr[\cos\psi\cos\phi + \sin\psi\sin\phi] = kr\cos(\psi-\phi),$$
thus
$$u(\boldsymbol{r},t) = \frac{1}{2\pi} \int_0^\infty dk \int_0^{2\pi} d\psi\, k\, \widetilde{u}(k,t)\, e^{ikr\cos(\psi-\phi)}.$$
Substituting $\theta = \psi - \phi$, and using once more the periodicity and symmetry of the cosine function, this becomes
$$u(\boldsymbol{r},t) = \frac{1}{\pi} \int_0^\infty dk \int_0^\pi d\theta\, k\, \widetilde{u}(k,t)\, e^{ikr\cos\theta}$$
$$= \int_0^\infty dk\, k\, \widetilde{u}(k,t)\, J_0(kr),$$

where J_0 is again the Bessel function; compare Example 5.2. With $\widetilde{u}(k,t)$ from Exercise 5.16, this becomes

$$u(\boldsymbol{r},t) = \int_0^\infty dk\, k\, [A(k)\, e^{ikct} + A^*(k)\, e^{-ikct}]\, J_0(kr),$$

with $A(k) = \widetilde{a}(k)/2 + \widetilde{b}(k)/(2ikc)$.

Solution 5.18

(a) For $f(r) = r^{-1}$, equation (5.96) becomes

$$\widetilde{f}(k) = \int_0^\infty dr\, r\, \frac{1}{r}\, J_0(kr) = \int_0^\infty dr\, J_0(kr) = \frac{1}{k} \int_0^\infty dz\, J_0(z) = \frac{1}{k},$$

where we have substituted $z = kr$ and used the normalisation (5.55) of the Bessel function J_0.

(b) From part (a), for $a(r) = 1/r$ we have $\widetilde{a}(k) = 1/k$. From Exercises 5.16 and 5.17, we obtain $A(k) = \widetilde{a}(k)/2 = 1/(2k)$, hence

$$\widetilde{u}(k,t) = \frac{1}{2k}[e^{ikct} + e^{-ikct}] = \frac{\cos(kct)}{k}.$$

The solution is

$$u(r,t) = \int_0^\infty dk\, k\, \frac{\cos(kct)}{k}\, J_0(kr) = \int_0^\infty dk\, \cos(kct)\, J_0(kr).$$

Solution 5.19

The Fourier transform in spherical coordinates is

$$\widetilde{f}(k) = \frac{1}{(2\pi)^{3/2}} \int_{\mathbb{R}^3} d\boldsymbol{r}\, f(r)\, e^{-i\boldsymbol{k}\cdot\boldsymbol{r}}$$

$$= \frac{1}{(2\pi)^{3/2}} \int_0^\infty dr \int_0^\pi d\theta \int_0^{2\pi} d\phi\, r^2 \sin\theta\, f(r)\, e^{-i\boldsymbol{k}\cdot\boldsymbol{r}}.$$

Again, we assume as instructed that we already know that the result depends only on $k = |\boldsymbol{k}|$, so we may choose $\boldsymbol{k}$ to be in the $\boldsymbol{k}$ direction. In spherical coordinates, we then have $\boldsymbol{k}\cdot\boldsymbol{r} = kr\cos\theta$. This gives

$$\widetilde{f}(k) = \frac{1}{(2\pi)^{3/2}} \int_0^\infty dr \int_0^\pi d\theta \int_0^{2\pi} d\phi\, r^2 \sin\theta\, f(r)\, e^{-ikr\cos\theta}.$$

Now the integration over ϕ just gives a factor 2π. Substituting $w = \cos\theta$ gives

$$\widetilde{f}(k) = \frac{1}{\sqrt{2\pi}} \int_0^\infty dr \int_{-1}^1 dw\, r^2\, f(r)\, e^{-ikrw}$$

$$= \sqrt{\frac{2}{\pi}} \int_0^\infty dr\, r^2\, f(r)\, \frac{\sin(kr)}{kr}$$

$$= \frac{\sqrt{2}}{\sqrt{\pi}\, k} \int_0^\infty dr\, r\, f(r)\, \sin(kr).$$

Solution 5.20

From the result of Exercise 5.19, the three-dimensional Fourier transform of $a(\boldsymbol{r})$ is

$$\widetilde{a}(\boldsymbol{k}) = \frac{\sqrt{2}}{\sqrt{\pi}\, k} \int_0^\infty dr\, r\, e^{-r^2/2}\, \sin(kr)$$

$$= \frac{1}{\sqrt{2\pi}\, k} \int_{-\infty}^\infty dr\, r\, e^{-r^2/2}\, \sin(kr),$$

using the fact that $r\sin(kr)\exp(-r^2/2)$ is an even function of r. This is a Fourier sine transform (compare Subsection 2.6.5), which we can write in terms of the imaginary part of the Fourier transform:

$$\widetilde{a}(\boldsymbol{k}) = -\frac{1}{\sqrt{2\pi}\, k}\, \mathrm{Im}\left[\int_{-\infty}^\infty dr\, r\, e^{-r^2/2}\, \exp(-ikr)\right].$$

The function $-r\exp(-r^2/2)$ is just the derivative of $\exp(-r^2/2)$, so the integral is the Fourier transform of the derivative of a Gaussian, thus

$$-\frac{1}{\sqrt{2\pi}}\int_{-\infty}^{\infty} dr\, r\, e^{-r^2/2} e^{-ikr} = ike^{-k^2/2},$$

because the Fourier transform of the derivative $f'(r)$ is $ik\widetilde{f}(k)$, and the one-dimensional Fourier transform of the Gaussian $\exp(-r^2/2)$ is just $\exp(-k^2/2)$. Thus

$$\widetilde{a}(\mathbf{k}) = \frac{1}{k}\operatorname{Im}\left(ike^{-k^2/2}\right) = e^{-k^2/2}.$$

Solution 5.21

(a) Using the result of the previous exercise for the Fourier transform $\widetilde{a}(\mathbf{k})$, equation (5.102) gives

$$A(\mathbf{k}) = \tfrac{1}{2}\widetilde{a}(\mathbf{k}) = \tfrac{1}{2}e^{-k^2/2},$$

which again depends only on $k = |\mathbf{k}|$. So the solution is, from equation (5.101),

$$\widetilde{u}(k,t) = \tfrac{1}{2}e^{-k^2/2}\left(e^{i\omega t} + e^{-i\omega t}\right) = e^{-k^2/2}\cos(\omega t),$$

with $\omega = kc$.

(b) Inserting the result of part (a) into equation (5.108) gives, with $\omega = kc$,

$$u(r,t) = \frac{\sqrt{2}}{\sqrt{\pi}\,r}\int_0^{\infty} dk\, k\, e^{-k^2/2}\cos(kct)\sin(kr)$$

$$= \frac{1}{\sqrt{2\pi}\,r}\int_{-\infty}^{\infty} dk\, k\, e^{-k^2/2}\cos(kct)\sin(kr),$$

since the integrand is an even function of k. Using the addition formula gives

$$u(r,t) = \frac{1}{2\sqrt{2\pi}\,r}\int_{-\infty}^{\infty} dk\, k\, e^{-k^2/2}\left(\sin[k(r-ct)] + \sin[k(r+ct)]\right)$$

$$= \frac{1}{2\sqrt{2\pi}\,r}\operatorname{Im}\left(\int_{-\infty}^{\infty} dk\, k\, e^{-k^2/2} e^{ik(r-ct)}\right)$$

$$+ \frac{1}{2\sqrt{2\pi}\,r}\operatorname{Im}\left(\int_{-\infty}^{\infty} dk\, k\, e^{-k^2/2} e^{ik(r+ct)}\right).$$

We can evaluate these terms as in Exercise 5.20; the only difference is that now $r \pm ct$ occurs in place of $-k$, while k takes the place of r. The result is

$$u(r,t) = \frac{r-ct}{2r}\exp\left(-\frac{(r-ct)^2}{2}\right) + \frac{r+ct}{2r}\exp\left(-\frac{(r+ct)^2}{2}\right).$$

It appears that $u(r,t)$ has a singularity at $r=0$; however, this is not the case. In fact,
$$u(0,t) = (1-c^2t^2)e^{-c^2t^2/2},$$
as can be seen by considering the limit as $r \to 0$.

Solution 5.22

(a) The eigenvalues are

$$k_{n,m}^2 = \pi^2\left(\frac{n^2}{L^2} + \frac{m^2}{M^2}\right) = \pi^2\left(\frac{n^2}{16} + m^2\right) = \frac{\pi^2}{16}(n^2 + 16m^2),$$

and the eigenfrequencies are given by $\omega_{n,m} = k_{n,m}c$, so

$$\omega_{n,m} = \frac{c\pi}{4}\sqrt{n^2 + 16m^2},$$

with $n, m = 1, 2, \ldots$.

(b) The corresponding eigenmodes are

$$f_{n,m}(\mathbf{r}) = \sin\left(\frac{\pi n x}{L}\right)\sin\left(\frac{\pi m y}{M}\right) = \sin\left(\frac{\pi n x}{4}\right)\sin(\pi m y),$$

with $n, m = 1, 2, \ldots$.

(c) The five lowest frequencies are

$$\omega_{1,1} = c\pi\sqrt{17}/4,$$
$$\omega_{2,1} = c\pi\sqrt{20}/4 = c\pi\sqrt{5}/2,$$
$$\omega_{3,1} = c\pi\sqrt{25}/4 = c5\pi/4,$$
$$\omega_{4,1} = c\pi\sqrt{32}/4 = c\pi\sqrt{2},$$
$$\omega_{5,1} = c\pi\sqrt{41}/4.$$

For all of these we have $m = 1$, so there are no nodal lines parallel to the **i** direction, and $f_{n,1}$ has $n - 1$ nodal lines parallel to the **j** direction.

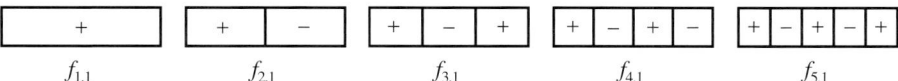

Figure 5.18 Nodal line patterns for the eigenmodes $f_{n,1}$ with $n = 1, 2, 3, 4, 5$

(d) The next seven lowest frequencies are

$$\omega_{6,1} = c\pi\sqrt{52}/4 = c\pi\sqrt{13}/2,$$
$$\omega_{7,1} = \omega_{1,2} = c\pi\sqrt{65}/4,$$
$$\omega_{2,2} = c\pi\sqrt{68}/4 = c\pi\sqrt{17}/2,$$
$$\omega_{3,2} = c\pi\sqrt{73}/4,$$
$$\omega_{8,1} = \omega_{4,2} = c\pi\sqrt{80}/4 = c\pi\sqrt{5}.$$

So there are two degeneracies.

(e) Systematic degeneracies occur for

$$n^2 + 16m^2 = n^2 + (4m)^2 = (n')^2 + (4m')^2.$$

We can choose $n' = 4m$ and $m' = n/4$ for the case that n is a multiple of 4. Thus $k_{n,m}^2 = k_{4m,n/4}^2$, which is a non-trivial degeneracy if n and $4m$ differ. As n in this formula must be a multiple of 4, we can equivalently write $k_{4n,m}^2 = k_{4m,n}^2$ for integer $n \ne m$. The first example is thus $k_{4,2}^2 = k_{8,1}^2$, which is in the list above. However, these are not all degeneracies, as the example $\omega_{7,1} = \omega_{1,2}$ shows.

(f) The differences of degenerate eigenmodes are

$$f_{n,m}^{(-)}(\boldsymbol{r}) = f_{4n,m}(\boldsymbol{r}) - f_{4m,n}(\boldsymbol{r}) = \sin(\pi nx)\sin(\pi my) - \sin(\pi mx)\sin(\pi ny).$$

The nodal lines are given by the condition $f_{n,m}^{(-)}(\boldsymbol{r}) = 0$, so

$$\sin(\pi nx)\sin(\pi my) = \sin(\pi mx)\sin(\pi ny).$$

This again holds for $x = y$; however, this is not the diagonal of the rectangle.

For $x = 4y$, this equation gives

$$\sin(4\pi ny)\sin(\pi my) = \sin(4\pi my)\sin(\pi ny),$$

or

$$\frac{\sin(4\pi ny)}{\sin(\pi ny)} = \frac{\sin(4\pi my)}{\sin(\pi my)}.$$

If $n = m$, this is satisfied, but this case is not interesting because $f_{n,n}^{(-)}(\boldsymbol{r}) = 0$. For different n and m, it is not possible to satisfy this equation for all $0 < y < 1$, because the left-hand side depends only on n, whereas the right-hand side involves only m. In order to be true for all values of y, both sides would need to be constant (for fixed y) if n or m changes, which is not the case. So we do not have nodal lines on the diagonal for the difference of normal modes, and thus cannot derive normal modes for the triangle in this way.

Solution 5.23

(a) As shown in Exercise 5.10, we obtain a set of eigenmodes

$$v_{0,m}(r,\theta) = J_0(k_{0,m}r),$$

for $n = 0$, and

$$v^{(1)}_{n,m}(r,\theta) = J_n(k_{n,m}r)\sin(n\theta), \quad v^{(2)}_{n,m}(r,\theta) = J_n(k_{n,m}r)\cos(n\theta),$$

for $n > 0$.

Here, m again labels the solutions $k_{n,m}$ of the equation $J_n(k_{n,m}l) = 0$.

(b) The modes $v_{0,m}(r,\theta)$ have no nodal lines at all, and these modes are non-degenerate. For $n > 0$, the modes $v^{(1)}_{n,m}(r,\theta)$ all have nodal lines for $\theta = 0$, whereas $v^{(2)}_{n,m}(r,0) = J_n(k_{n,m}r)$ does not vanish for all r. So the complete set of eigenmodes with a nodal line for $\theta = 0$ is given by the modes $v^{(1)}_{n,m}(r,\theta)$.

(c) The argument is the same as for the triangular membrane. Given an arbitrary eigenmode $v(r,\theta)$ for the semi-circular membrane, say $\theta \in [0,\pi]$, we can extend it to an eigenmode for the full circle by setting $v(r,\theta) = -v(r,2\pi-\theta)$. This then is an eigenmode for the circular membrane with the same eigenvalue, so it must be a linear combination of the eigenmodes of the circular membrane, with a nodal line for $\theta = 0$. This shows that we obtain all eigenmodes this way.

See page 195.

(d) For each n, we used one of the two degenerate modes of the circular membrane, so there is no degeneracy left in the spectrum. However, we do not have simple closed expressions for the zeros of the Bessel functions J_n, and it could in principle happen that, for a given radius l, there is an 'accidental' degeneracy because two zeros happen to coincide, i.e. $k_{n,m} = k_{n',m'}$. But there are no systematic degeneracies in the spectrum.

CHAPTER 6
Wave propagation

6.1 Introduction

In the preceding chapters, we emphasised the importance of normal modes for solutions of the wave equation on finite domains, such as the stretched string with fixed ends. The normal mode solutions correspond to *standing waves*, in the sense that the wave pattern as a whole does not move. However, in Chapter 2, we discussed how d'Alembert's solution can be interpreted as a *travelling wave* solution, in terms of a wave pulse moving along an infinite string, and how a standing wave solution can be obtained by repeated reflections of a travelling pulse at the boundaries of the finite system. This final chapter on waves considers some properties of wave motion, using the picture of travelling wave solutions. These properties are related to familiar phenomena of wave motion.

This topic was covered in Section 2.5, which is not assessed.

You may have noticed that the pitch of the sound of a train whistle or a car horn changes as it passes you. This phenomenon is called the Doppler effect and is discussed in Section 6.2. It is caused by the movement of the wave source – in the example the train or the car – relative to the observer.

We then return to the reflection of waves, which was discussed in Chapter 2 in connection with d'Alembert's solution. Here, we consider what happens if a wave encounters an interface, meaning a boundary between two different materials. For one-dimensional wave motion, an example might be the joining of two strings. Examples in higher dimensions are the partial reflection of light waves at a water surface and at a glass plate.

In Chapter 2, we found that a wave is reflected at a boundary, and you might expect that this also happens at interfaces. However, besides reflection there may also be *transmission*, which means that all or part of the wave passes through the interface, continuing to propagate in the other medium. After considering the one-dimensional case in some detail, we turn to the reflection and transmission of two-dimensional waves at a one-dimensional interface. Apart from a partial reflection at the surface, it transpires that the interface alters the direction of motion of a travelling wave pulse, a phenomenon known as *refraction*. This 'bending' of waves is the reason why a stick that is partially submerged in water appears to be bent, and is used, for instance, in lenses to focus light.

The linear wave equation derived in Chapter 1 involves a single parameter c, the wave speed. In systems that are described by this equation, all waves move with the same speed, irrespective of their frequency. This is true, for example, for electromagnetic waves, such as light, in a vacuum. However, for light travelling in other media, such as a crystal, this may not be true.

If the wave speed depends on the frequency of a wave, one speaks of *dispersion*, and such materials are called *dispersive media*. They can be characterised by the dependency of the angular frequency ω on the wave number k, called the *dispersion relation*, which replaces the equality $\omega = kc$ that holds for non-dispersive waves. Dispersive wave motion is discussed in the optional Section 6.4. In particular, we consider a discrete system of coupled oscillators, which is a crude model of the motion of atoms in a solid.

Compare equation (2.3).

6.2 The Doppler effect

You have probably experienced the effect where the sound of a moving object appears to change as it passes you, for instance that of a police car siren or a train whistle. This phenomenon is due to the wave nature of the sound and the relative movement between you, the observer, and the object that creates the wave. Let us discuss this for the example of water waves, which makes the effect easier to visualise.

Assume that you are standing next to a canal, watching waves moving along the canal with a certain speed c, which we assume to be constant and in the $\mathbf{i}$ direction. The wavelength, which is the distance between two wave crests, is denoted by λ. If the wave crests move towards you, the wave crests pass by at regular time intervals

This is easy to visualise, but may not be so easily observable unless the wind blows directly along the canal.

$$\tau = \frac{\lambda}{c}. \tag{6.1}$$

The corresponding frequency is $\nu = 1/\tau = c/\lambda$.

What happens if you now start walking along the canal? Clearly, the speed of the waves does not change, but now you are also moving, either in the direction of the wave motion or opposite to it. Let us suppose that you are walking against the wave motion, so your velocity is $-v\mathbf{i}$, with speed $v < c$. This situation is shown in Figure 6.1(a).

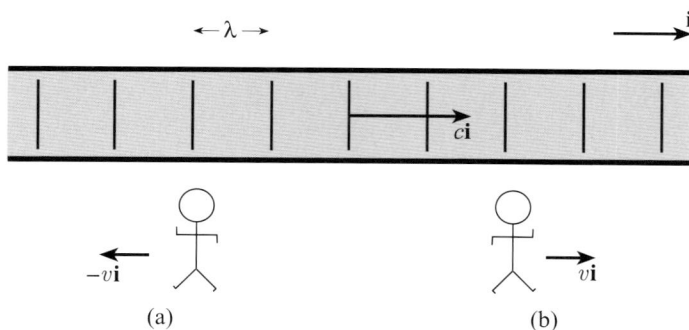

Figure 6.1 Two walkers moving along a canal, (a) moving opposite to, and (b) walking in the direction of, the wave motion

Obviously, the wave crests now arrive at your position at shorter time intervals. The time τ_+ between seeing two crests can be calculated as follows. The distance $v\tau_+$ that you have walked during the time interval τ_+, and the distance $c\tau_+$ that the wave has travelled, must add up to the wavelength λ, so

$$c\tau_+ + v\tau_+ = (c+v)\tau_+ = \lambda, \tag{6.2}$$

which gives

$$\tau_+ = \frac{\lambda}{c+v}. \tag{6.3}$$

6.2 The Doppler effect

This means that you now experience a wave of frequency

$$\nu_+ = \frac{1}{\tau_+} = \frac{c+v}{\lambda} = \left(1 + \frac{v}{c}\right)\nu, \tag{6.4}$$

which is *higher* than the frequency of the wave you saw when you did not move.

Let us now consider what happens in the situation of Figure 6.1(b). Now, you walk in the direction of wave motion rather than opposite to it. It now takes longer between two wave crests arriving, because the wave crests have to travel not just one wavelength to reach you, but also the distance that you have walked ahead.

Let τ_- denote the time between seeing two crests. Now, the wave travels a distance $c\tau_-$ which is larger than λ. It is larger by the distance $v\tau_-$ you have walked during the time interval τ_-. So we now obtain the equation

$$c\tau_- - v\tau_- = (c-v)\tau_- = \lambda, \tag{6.5}$$

which gives

$$\tau_- = \frac{\lambda}{c-v}, \tag{6.6}$$

which makes sense only for $v \neq c$. (For $v = c$, the period τ_- becomes infinite, as you walk at the same speed as the moving wave – so from your perspective you see a stationary wave.) So the frequency

$$\nu_- = \frac{1}{\tau_-} = \frac{c-v}{\lambda} = \left(1 - \frac{v}{c}\right)\nu \tag{6.7}$$

is now *lower* than the frequency when you did not move. The shift in the frequency is determined by the ratio of the observer's speed v and the wave speed c. As v approaches c, the frequency goes to zero – it takes longer and longer for the wave to catch up with you, and if you walk at exactly the same speed, no wave crest will pass you at all. The case where the source is moving can be analysed analogously; you are asked to derive the corresponding result in Exercise 6.1.

In conclusion, the observed frequency of a wave depends on the *relative motion* between you, the observer, and the wave source. If you move in the direction of the wave motion, you experience a lower frequency than at rest; if you move opposite to that direction, the frequency appears to be higher. This phenomenon is known as the *Doppler effect*, named after the Austrian physicist Christian Andreas Doppler (1803–1853). The relative motion can, as in our example above, be the result of the observer's motion with respect to the wave. But equally well it could originate from a moving wave source, while you are standing still. When you hear a car honking its horn, it appears that the pitch of the sound changes while it passes you, whereas people in the car will not hear any difference. As the car approaches you, the frequency is higher than that you would hear as a passenger of the car. When it moves away from you, it is lower, giving the impression that the pitch of the sound has changed.

The Doppler effect has various important scientific and technical applications. It makes it possible to measure the relative speed of some moving object if one can detect a frequency difference in some kind of wave that is emitted by the object. For instance, by observing the frequency shift of light emitted by distant stars or galaxies (which can be done for light originating from particular atomic transitions for which the frequency is known), one can determine the relative speed between that distant object and the Earth. These observations provide evidence for an expansion of our universe.

Doppler radar measures the shift in the frequency of radiation reflected by moving objects. It is widely used for weather prediction, as the intensity of the reflected radiation contains information about rainfall, and the frequency shift allows us to measure wind speed in the atmosphere.

Exercise 6.1

Consider the situation where you are standing still while a wave-emitting object moves towards you with speed $v < c$, where c is the wave speed. This situation can be analysed as follows. Imagine that the source emits some characteristic signals (corresponding to the wave crests, say) at regular time intervals τ, which travel with the wave speed c.

(a) Suppose that the source emits a signal at time $t = 0$ when it is at a (sufficiently large) distance x_0 away from you. At what time t_0 does the signal reach you?

(b) What is the distance x_1 of the source when it emits the next signal?

(c) When does this second signal arrive?

(d) What time τ_+ has elapsed between the arrival of the two signals? What is the frequency ν_+ for you, compared to the frequency $\nu = 1/\tau$ of the source at rest?

Consider the case of a moving wave source as discussed in Exercise 6.1. If the speed of the source relative to the observer exceeds the wave speed, other interesting phenomena may occur. An aeroplane travelling at supersonic speed emits sound waves only into a part of space – it cannot emit them into the direction it is moving, because it moves *faster* than the sound waves. On the ground, you hear the sound well after the aeroplane has already passed overhead, starting with a 'sonic boom' which stems from a density shock wave that is created at the boundary of the region in which the aeroplane transmits waves.

For the same reason, because gun bullets usually travel faster than sound, you are safe from a bullet once you have heard the report of the gun shot, whereas you will not hear a bullet before it hits you. So whenever you see an action movie hero jumping aside after hearing a gun being fired and managing to evade the bullet, you should take it for what it is – pure fiction.

6.3 Reflection and transmission

In Chapter 2, we considered the reflection of a wave pulse at a boundary when deriving d'Alembert's solution for the finite string by repeated reflections at the ends of the string. Similarly, we can study reflection at other types of boundaries, such as the sprung or damped ends introduced in Chapter 4.

This was done in the non-assessed Section 2.5.

More interesting is what happens when a travelling wave pulse encounters an *interface* between two regions. As an example, you may think of light that is partially reflected at a glass window or the surface of water. Then what usually happens is that some of the light is reflected at the interface, and some is transmitted through the interface – there are light waves travelling in the glass and in the water.

6.3 Reflection and transmission

The motion in air, water or glass is described by the wave equation. However, the speed of light in glass or water differs from that in air. Therefore we need to solve the wave equation separately in different regions, then combine the solutions, ensuring that they match properly along an interface.

An important application of reflection and transmission of waves is in geological surveys. For instance, to locate possible oil deposits underground, one needs detailed information about the various layers of different materials in the ground. This can be obtained by studying the partial reflection of seismic waves, set off by either small explosions or a vibrating device as depicted in Figure 6.2, at geological fault lines. The reflected waves are detected by an array of geophones (i.e. sound detectors), and from the time delays and intensities of the reflections, one can extract information about the type and depth of geological layers.

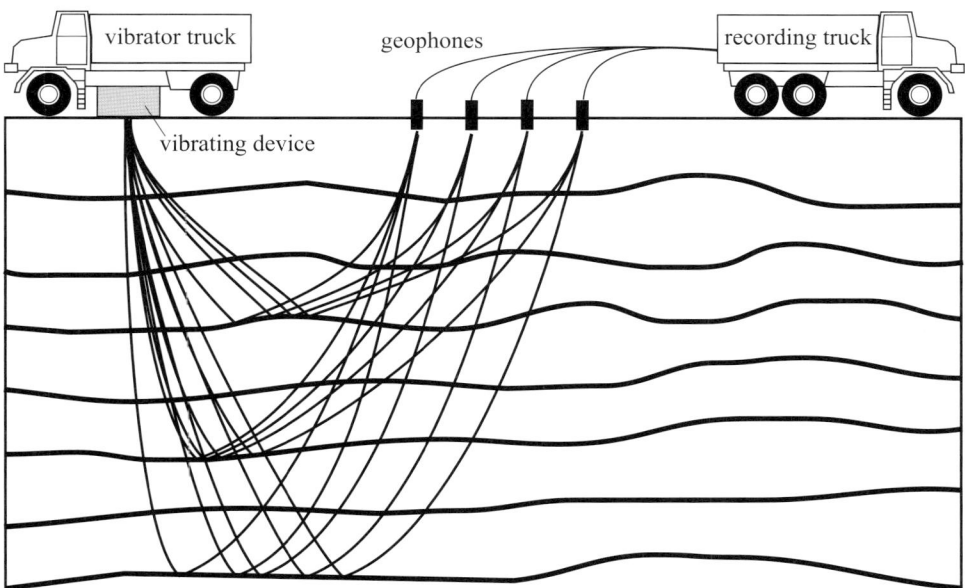

Figure 6.2 Schematic representation of a seismic survey using reflection of seismic waves at interfaces between different layers

We start by analysing the one-dimensional situation, using the same approach as for the reflection at a boundary studied previously. We then briefly discuss how we can analyse reflection and transmission in terms of plane harmonic waves, and apply this approach to the case of a one-dimensional interface encountered by two-dimensional waves. We show how the laws of *reflection* and *refraction*, which describe the reflected and transmitted wave, are obtained. We note that these laws were known – long before the wave description of light was introduced – within the framework of *ray optics*, which views light as line geometry (see Block III). The law of reflection is much older still; it goes back to Euclid (around 300 BC).

The law of refraction is also known as *Snell's law*, named after Willebrord van Roijen Snell (1580–1626).

6.3.1 Reflection and transmission at interfaces

Reflection occurs not only at boundaries, but also at *interfaces* between different materials. As an example, think of two different strings, joined together to form a single, non-uniform string. We now address the question of what happens when a transverse wave pulse travelling on one string reaches the point where the two are joined – will it continue to move across, or be reflected, or both?

We consider two semi-infinite taut strings which are joined at $x = 0$. The tension T is constant throughout the composite string. The individual strings have different wave speeds, say c_1 for the semi-infinite string occupying the region $x < 0$, and c_2 for the second string, over $x > 0$. The different wave speeds can be realised by different linear densities of the strings, so you may think of a thinner string to one side of $x = 0$ and a thicker string to the other. A sketch of the partial reflection and transmission of an incident wave pulse is shown in Figure 6.3.

Compare Exercise 1.18, which shows this for a composite spring.

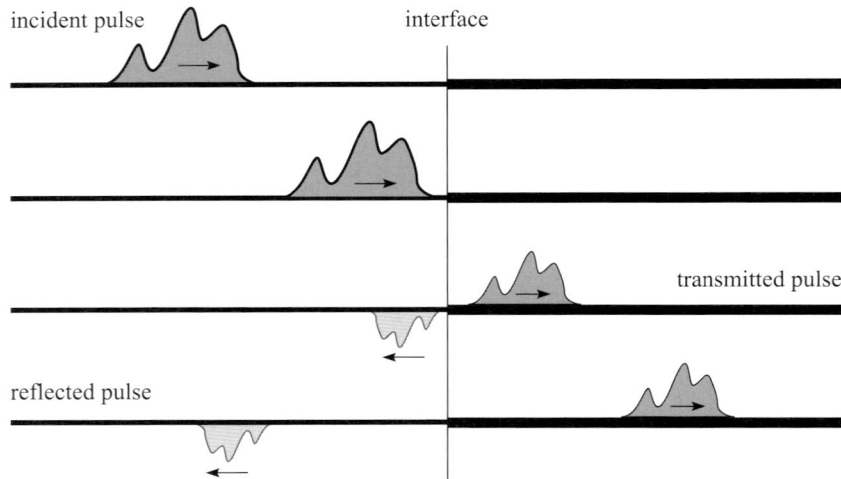

Figure 6.3 Sketch of the reflection and transmission of a wave pulse at an interface

Using d'Alembert's solution, we can write the two parts of the solution

$$u(x,t) = \begin{cases} u_\text{L}(x,t), & x < 0, \\ u_\text{R}(x,t), & x > 0, \end{cases} \tag{6.8}$$

in the form

$$u_\text{L}(x,t) = F_1(x - c_1 t) + G_1(x + c_1 t), \quad x < 0, \tag{6.9}$$
$$u_\text{R}(x,t) = F_2(x - c_2 t) + G_2(x + c_2 t), \quad x > 0. \tag{6.10}$$

Here, F_1, G_1, F_2 and G_2 are at least twice differentiable functions. We require that the functions u_L and u_R match properly at the interface at $x = 0$, where there are two boundary conditions.

The first boundary condition is that the function $u(x,t)$ is *continuous* at $x = 0$, because the displacement of the string at the interface must not jump. This means

$$u(0,t) = \lim_{x \to 0_-} u_\text{L}(x,t) = \lim_{x \to 0_+} u_\text{R}(x,t), \tag{6.11}$$

where $x \to 0_-$ means that $x = 0$ is approached from $x < 0$, and $x \to 0_+$ means approach from $x > 0$. Inserting the expressions of equations (6.9) and (6.10) yields

$$F_1(-c_1 t) + G_1(c_1 t) = F_2(-c_2 t) + G_2(c_2 t). \tag{6.12}$$

The second boundary condition is that the slope of the string changes continuously across the interface. This is required because a discontinuity in the slope corresponds to a discontinuity of the transverse force component along the string, which would mean that the string moves with an infinite velocity near $x = 0$. Therefore we require that the derivative $u_x(x,t)$ is continuous at $x = 0$:

Compare equation (1.18).

$$u_x(0,t) = \lim_{x \to 0_-} u_x(x,t) = \lim_{x \to 0_+} u_x(x,t). \tag{6.13}$$

6.3 Reflection and transmission

This boundary condition gives
$$F_1'(-c_1t) + G_1'(c_1t) = F_2'(-c_2t) + G_2'(c_2t), \tag{6.14}$$
where $F'(z)$ denotes the derivative of F at argument z.

Here, we are interested in what happens to an incident wave pulse when it encounters the interface, as depicted in Figure 6.3. Let us start from a situation at $t = 0$ where we have a right-moving pulse on the left part, $x < 0$, and the right part of the string, $x > 0$, is undisturbed. Upon encountering the interface, part of the pulse may be reflected, and part may be transmitted. We assume that there is no left-moving pulse for $x > 0$, so $G_2 = 0$. The right-moving incident pulse corresponds to a term $F_1(x - c_1t)$ in the solution $u(x,t)$. We require that $F_1(z) = 0$ for $z \geq 0$, because the pulse starts entirely on the left part of the string. The other two terms in the solution $u(x,t)$ are $G_1(x + c_1t)$, which corresponds to the left-moving *reflected* pulse, and $F_2(x - c_2t)$, which describes the *transmitted* pulse. At $t = 0$, our assumptions mean that $G_1(z) = 0$ for $z \leq 0$ (no reflected pulse at $t = 0$) and $F_2(z) = 0$ for $z \geq 0$ (no transmitted pulse at $t = 0$). Thus we have

$$\begin{cases} F_1(z) = F_2(z) = 0 & \text{for } z \geq 0, \\ G_1(z) = 0 & \text{for } z \leq 0, \\ G_2(z) = 0 & \text{for all } z. \end{cases} \tag{6.15}$$

The boundary conditions become
$$F_1(-c_1t) + G_1(c_1t) = F_2(-c_2t), \tag{6.16}$$
$$F_1'(-c_1t) + G_1'(c_1t) = F_2'(-c_2t). \tag{6.17}$$

Integrating equation (6.17) with respect to t gives
$$-\frac{1}{c_1}F_1(-c_1t) + \frac{1}{c_1}G_1(c_1t) = -\frac{1}{c_2}F_2(-c_2t) + C, \tag{6.18}$$
where C is a constant. At $t = 0$, all terms vanish according to our conditions (6.15) on the functions F_1, F_2 and G_1, thus $C = 0$. Multiplying equation (6.18) by $-c_1$, we then have

$$F_1(-c_1t) - G_1(c_1t) = \frac{c_1}{c_2}F_2(-c_2t). \tag{6.19}$$

From equations (6.16) and (6.19), we can now express the reflected wave G_1 and the transmitted wave F_2 in terms of the incident wave F_1:

$$G_1(z) = -\frac{c_1 - c_2}{c_1 + c_2}F_1(-z) = \alpha F_1(-z), \tag{6.20}$$

where $\alpha = -(c_1 - c_2)/(c_1 + c_2)$, and

$$F_2(z) = \frac{2c_2}{c_1 + c_2}F_1\left(\frac{c_1}{c_2}z\right) = \beta F_1\left(\frac{c_1}{c_2}z\right), \tag{6.21}$$

where $\beta = 1 + \alpha$. The full solution is thus

$$u(x,t) = \begin{cases} F_1(x - c_1t) + \alpha F_1(-x - c_1t), & x < 0, \\ \beta F_1\left(\frac{c_1}{c_2}x - c_1t\right), & x > 0. \end{cases} \tag{6.22}$$

Let us consider two special cases.

If $c_1 = c_2 = c$, then we effectively have no interface. From equation (6.20), we see that the reflected wave G_1 vanishes in this case, and equation (6.21) reduces to $F_2(z) = F_1(z)$, so the pulse passes through the hypothetical 'interface' without being affected at all.

If $c_2 = 0$, then we effectively have the situation of a reflecting fixed end. In this case, there is no transmitted wave, as F_2 vanishes, and $G_1(x + c_1t) = -F_1(-x - c_1t)$ is the inverted reflected wave that we met in Chapter 2.

In the general situation, both reflected and transmitted waves will be present.

To quantify how much of the incident wave is reflected and how much is transmitted, we introduce the *reflectivity* $\mathcal{R}$ and the *transmissivity* $\mathcal{T}$ of the interface. For this, one considers how much energy the reflected and transmitted parts of the wave carry in comparison to that of the incident pulse: the reflectivity and transmissivity are defined as the fractions of the incident wave energy that are reflected and transmitted at the interface, respectively. In this course, we do not discuss the expression for the wave energy, which in our example corresponds to the sum of the kinetic and elastic energy of the moving string, but it may appear rather natural that the energy is proportional to the square of the amplitude of the wave, and that the wave speed enters because it matters how fast the wave is moving.

If turns out that the reflectivity $\mathcal{R}$ is

$$\mathcal{R} = \alpha^2, \qquad (6.23)$$

and the transmissivity $\mathcal{T}$ is

$$\mathcal{T} = \frac{c_1}{c_2}\beta^2. \qquad (6.24)$$

Note that while $\mathcal{R}$ is just the squared ratio of the coefficients of the incoming and reflected waves, $\mathcal{T}$ is given by the squared ratio of the coefficients of the incoming and transmitted waves multiplied by the ratio of the two wave speeds c_1 and c_2. This is consistent with the conservation of energy because $\mathcal{R} + \mathcal{T} = 1$, as you are asked to verify in Exercise 6.2.

Exercise 6.2

Using equations (6.20) and (6.21), and the definitions (6.23) and (6.24) of $\mathcal{R}$ and $\mathcal{T}$, show that $\mathcal{R} + \mathcal{T} = 1$.

Exercise 6.3

Consider the ratio $\alpha = -(c_1 - c_2)/(c_1 + c_2)$ that relates the amplitude of the reflected pulse to that of the incident pulse.

(a) Rewrite α in terms of the ratio $\gamma = c_2/c_1$.

(b) Show that α is monotonically increasing with $\gamma > 0$, by computing $d\alpha/d\gamma$.

(c) What is the range of values that α can assume?

(d) What follows for the reflectivity $\mathcal{R}$?

6.3.2 Reflection and transmission in two dimensions

We now move on to consider reflection and transmission at an interface in two dimensions. By analogy with our string, you may think of the motion of an infinite membrane consisting of two semi-infinite parts that have been sewn together along a straight line. A second example is surface waves on shallow water, where an interface can be realised by a sudden change in water depth. Most familiar will be the reflection and transmission of light, for instance at a surface of water or at a glass window.

6.3 Reflection and transmission

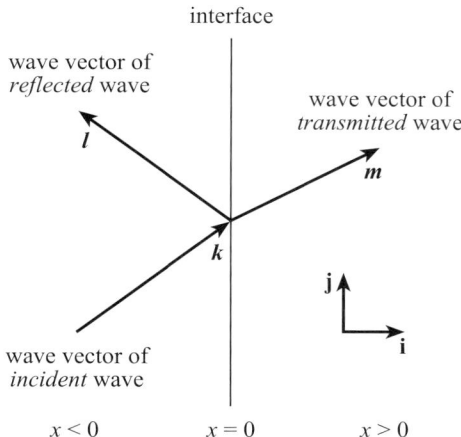

Figure 6.4 Reflection and transmission at an interface in two dimensions

As a simplification, we restrict ourselves to the case where the interface is straight. Let the interface be along the y-axis in the (x,y)-plane. It divides the plane into two regions, a region with $x<0$, where we have the incident and reflected waves, and a region with $x>0$, where we find the transmitted wave; see Figure 6.4. Wave motion on both sides is supposed to be described by the linear wave equation, with different wave speeds c_1 for $x<0$ and c_2 for $x>0$. In principle, we can proceed as in the one-dimensional case, starting from general d'Alembert-type solutions for both regions, and matching these at the interface. However, it turns out to be more convenient to exploit the superposition principle and consider the reflection and transmission of plane harmonic waves only. This is sufficient, because we can express any solution of the wave equation as a linear combination of such solutions, as discussed in Chapter 2.

We consider an incident plane wave of wave vector $\boldsymbol{k} = k_x\mathbf{i} + k_y\mathbf{j}$ and angular frequency $\omega = kc_1$, with $k = |\boldsymbol{k}| = \sqrt{k_x^2 + k_y^2}$ and $k_x > 0$, so the wave vector has a positive component along the $\mathbf{i}$ direction. A plane wave moving in the direction of $\boldsymbol{k}$ is described by the real part of the complex function $A\exp[i(\boldsymbol{k}\cdot\boldsymbol{r} - \omega t)]$. Because of the interface, we expect that there will be a reflected wave $\boldsymbol{l}$ and a transmitted wave $\boldsymbol{m}$; we assume that these will again be plane harmonic waves (and not superpositions of those). Therefore we are looking for a solution $u(\boldsymbol{r},t)$, with $\boldsymbol{r} = x\mathbf{i} + y\mathbf{j}$, of the two-dimensional wave equation which in the region $x<0$ consists of two terms, i.e.

This was discussed in Chapter 5.

This assumption will be justified by finding a solution. One can show that the solution is unique.

$$u(\boldsymbol{r},t) = A_1 e^{i(\boldsymbol{k}\cdot\boldsymbol{r} - kc_1 t)} + B_1 e^{i(\boldsymbol{l}\cdot\boldsymbol{r} - lc_1 t)}, \quad x<0, \qquad (6.25)$$

corresponding to the incident and reflected waves, with angular frequencies kc_1 and lc_1, respectively. In the region $x>0$, the solution $u(\boldsymbol{r},t)$ consists of the transmitted wave only, so it has the form

$$u(\boldsymbol{r},t) = A_2 e^{i(\boldsymbol{m}\cdot\boldsymbol{r} - mc_2 t)}, \quad x>0. \qquad (6.26)$$

So the full solution $u(\boldsymbol{r},t)$ contains three (in general complex) coefficients A_1, B_1, A_2, and three wave vectors $\boldsymbol{k}$, $\boldsymbol{l}$, $\boldsymbol{m}$, of respective lengths

$$\begin{aligned} k &= |\boldsymbol{k}| = \sqrt{k_x^2 + k_y^2}, \\ l &= |\boldsymbol{l}| = \sqrt{l_x^2 + l_y^2}, \\ m &= |\boldsymbol{m}| = \sqrt{m_x^2 + m_y^2}. \end{aligned} \qquad (6.27)$$

At the interface, i.e. at $x=0$, we demand that the solution $u(\boldsymbol{r},t)$ is continuous and that its partial derivative $u_x(\boldsymbol{r},t)$ is continuous. The first condition means

$$\lim_{x\to 0_-} u(x\mathbf{i} + y\mathbf{j}, t) = \lim_{x\to 0_+} u(x\mathbf{i} + y\mathbf{j}, t), \qquad (6.28)$$

which yields
$$A_1 e^{i(k_y y - k c_1 t)} + B_1 e^{i(l_y y - l c_1 t)} = A_2 e^{i(m_y y - m c_2 t)}. \tag{6.29}$$

Then continuity of the partial derivative u_x at $x = 0$ gives
$$i k_x A_1 e^{i(k_y y - k c_1 t)} + i l_x B_1 e^{i(l_y y - l c_1 t)} = i m_x A_2 e^{i(m_y y - m c_2 t)}. \tag{6.30}$$

The two relations (6.29) and (6.30) must hold for *all* values of y and t.

First, we consider the case $y = t = 0$. Equation (6.29) gives $A_1 + B_1 = A_2$, and from (6.30) we find
$$k_x A_1 + l_x B_1 = m_x A_2 = m_x (A_1 + B_1), \tag{6.31}$$

hence
$$B_1 = -\frac{k_x - m_x}{l_x - m_x} A_1, \quad A_2 = \frac{l_x - k_x}{l_x - m_x} A_1. \tag{6.32}$$

Equations (6.32) express the coefficients B_1 and A_2 of the reflected and transmitted waves in terms of the coefficient A_1 of the incident wave and the **i** components of the wave vectors. To derive relations between the wave vectors, we use the results of the following exercise.

Exercise 6.4

(a) Assume that the equation
$$A e^{iax} + B e^{ibx} = 0$$
holds for all x, with some constants $A \neq 0$, $B \neq 0$, a and b. Show that $a = b$. [Hint: Divide by e^{ibx}, and take the derivative with respect to x.]

(b) Assume that the equation
$$A e^{iax} + B e^{ibx} + C = 0$$
holds for all x, with some constants $A \neq 0$, $B \neq 0$, C, a and b. Show that $a = b$. [Hint: Take the derivative with respect to x.]

(c) Assume that the equation
$$A e^{iax} + B e^{ibx} + C e^{icx} = 0$$
holds for all x, with some constants $A \neq 0$, $B \neq 0$, $C \neq 0$, a, b and c. Show that $a = b = c$. [Hint: Divide by e^{icx}, and use the previous result to show that $a = b$. Repeat this argument with division by e^{iax} or e^{ibx}.]

In this exercise, we concentrate on showing that the coefficients in the exponents are the same. In fact, this implies relations for the other coefficients; in part (a) it follows that $A = -B$.

For $t = 0$, the continuity relation (6.29) becomes
$$A_1 e^{i k_y y} + B_1 e^{i l_y y} = A_2 e^{i m_y y}, \tag{6.33}$$

which, by Exercise 6.4, implies
$$k_y = l_y = m_y. \tag{6.34}$$

For $y = 0$, equation (6.29) becomes
$$A_1 e^{-i k c_1 t} + B_1 e^{-i l c_1 t} = A_2 e^{-i m c_2 t}, \tag{6.35}$$

which, by Exercise 6.4, implies
$$k = l = \frac{c_2}{c_1} m. \tag{6.36}$$

Equations (6.36) and (6.34) give $l^2 = l_x^2 + l_y^2 = k^2 = k_x^2 + k_y^2$ and $k_y = l_y$, so $l_x^2 = k_x^2$ and thus $l_x = \pm k_x$. For the reflected wave we have $l_x < 0$, so
$$l_x = -k_x \tag{6.37}$$

is the appropriate solution.

6.3 Reflection and transmission

We also have $m^2 = m_x^2 + m_y^2 = n^2 k^2 = n^2 k_x^2 + n^2 k_y^2$, where $n = c_1/c_2$ denotes the ratio of the two wave speeds. With $m_y = k_y$, this gives

$$m_x^2 = n^2 k_x^2 + (n^2 - 1)k_y^2 = n^2 k_x^2 + (n^2 - 1)(k^2 - k_x^2)$$
$$= k_x^2 + (n^2 - 1)k^2. \qquad (6.38)$$

Provided that this expression is positive, we have

$$m_x = \sqrt{k_x^2 + (n^2 - 1)k^2}, \qquad (6.39)$$

with $n = c_1/c_2$. Here, we choose the positive square root, because for the transmitted wave we need $m_x > 0$.

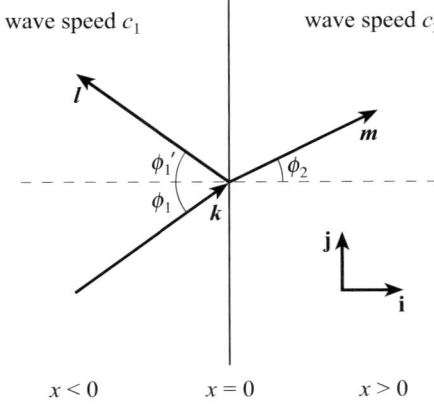

Figure 6.5 Reflection and transmission at an interface

Geometrically, these results mean the following. If ϕ_1 denotes the angle of the incident wave vector $\boldsymbol{k}$ with respect to the $\boldsymbol{i}$ direction (see Figure 6.5), then the angle ϕ_1' between the reflected wave vector $\boldsymbol{l}$ and the $-\boldsymbol{i}$ direction is

Remember that $\boldsymbol{k}$ denotes the wave vector, not the Cartesian unit vector $\mathbf{k}$.

Law of reflection
$$\phi_1' = \phi_1. \qquad (6.40)$$

This is the *law of reflection*, which states that the angle of incidence equals the angle of reflection.

Exercise 6.5

Show that the law of reflection (6.40) follows from equations (6.34)–(6.37).

The transmitted wave does not, in general, move in the same direction as the incident wave. This phenomenon is known as *refraction*. Denoting the angle between the refracted wave vector $\boldsymbol{m}$ and the $\boldsymbol{i}$ direction, the *angle of refraction*, by ϕ_2, equation (6.34) implies $k \sin \phi_1 = m \sin \phi_2$. Using equation (6.36), this is $c_1^{-1} \sin \phi_1 = c_2^{-1} \sin \phi_2$, which yields the relation

Snell's law of refraction
$$n_1 \sin \phi_1 = n_2 \sin \phi_2 \qquad (6.41)$$

between the incident and refracted waves, where n_1 and n_2 are characteristic properties called the *refractive indices* of the materials occupying the half-spaces $x < 0$ and $x > 0$, respectively, which are defined as $n_i = c/c_i$, $i = 1, 2$, where c is a reference speed (such as the speed of light in a vacuum for light waves). The ratio of the two refractive indices satisfies

$$\frac{n_2}{n_1} = \frac{m}{k} = \frac{c_1}{c_2} = n, \qquad (6.42)$$

which follows from equation (6.36).

The refractive index of a medium is defined by the ratio of the wave speed in a reference system to the wave speed in the medium. In the case of light waves, the refractive index is the ratio of the speed of light in a vacuum to the speed of light in the medium. By definition, the refractive index of a vacuum is then $n_{\text{vacuum}} = 1$. The refractive index of air, $n_{\text{air}} = 1.0003$, differs only very slightly from that of a vacuum. For comparison, the refractive index of water is $n_{\text{water}} = 1.33$, for glass we find refractive indices n_{glass} in the range 1.5–1.6, and diamond has a refractive index $n_{\text{diamond}} = 2.42$. The higher the ratio n in equation (6.42), i.e. the ratio of the two refractive indices at an interface, the more the angles ϕ_1 and ϕ_2 differ. The refraction of light at curved surfaces is used in lenses and optical instruments that contain lenses, such as cameras or telescopes.

To visualise how the interface leads to refraction, consider the following analogy that goes back to Newton. Imagine troops of soldiers marching in file – you may think of the lines as representing the wave crests of a plane wave all moving in the same direction. The soldiers walk on dry ground at a steady pace, say with velocity v_1, but then encounter a swamp where they are forced to slow down to velocity v_2. As the first soldiers enter the swamp and slow down, others in the same line may still walk on dry ground; the soldiers fighting the mud fall behind. This results not only in a lower walking speed, but also in a change in the direction of the lines, which appear to bend at the boundary between the two regions, and hence in the direction of the motion; see Figure 6.6.

See Figure 5.16.

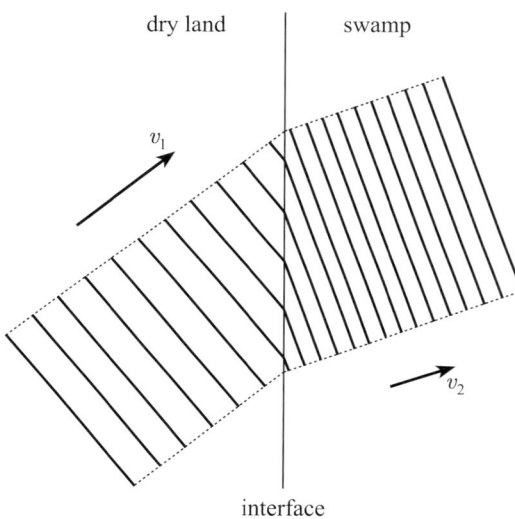

Figure 6.6 Different walking velocities of soldiers marching in file on dry ground and swamp result in a bending of their lines

We now return our attention to the coefficients. Because of linearity, only the ratios B_1/A_1 and A_2/A_1 are important; they determine the reflectivity

Reflectivity
$$\mathcal{R} = \left(\frac{B_1}{A_1}\right)^2 \tag{6.43}$$

and the transmissivity

Transmissivity
$$\mathcal{T} = \frac{c_1}{c_2}\frac{\cos\phi_2}{\cos\phi_1}\left(\frac{A_2}{A_1}\right)^2 \tag{6.44}$$

of the interface at this particular incident angle; compare equations (6.23) and (6.24), respectively. In contrast to the one-dimensional situation, the energy transmission through the interface depends not only on the ratio of the wave speeds c_1/c_2, but also on the angles of the incident and refracted

6.3 Reflection and transmission

waves, via a factor $\cos\phi_2/\cos\phi_1$. You are asked to show in Exercise 6.6 that these definitions are consistent with $\mathcal{R} + \mathcal{T} = 1$ (i.e. conservation of energy).

Inserting the results for the wave vectors in equations (6.32) yields rather lengthy expressions. You are not supposed to memorise these, but you should know how they can be obtained. Expressing the ratios B_1/A_1 and A_2/A_1 in terms of the incident wave vector $\boldsymbol{k}$ gives

$$\frac{B_1}{A_1} = \frac{k_x - \sqrt{k_x^2 + (n^2-1)k^2}}{k_x + \sqrt{k_x^2 + (n^2-1)k^2}} = \frac{1 - \sqrt{1 + (n^2-1)k^2/k_x^2}}{1 + \sqrt{1 + (n^2-1)k^2/k_x^2}} \quad (6.45)$$

and

$$\frac{A_2}{A_1} = \frac{2k_x}{k_x + \sqrt{k_x^2 + (n^2-1)k^2}} = \frac{2}{1 + \sqrt{1 + (n^2-1)k^2/k_x^2}}. \quad (6.46)$$

Alternatively, we can express the ratios in terms of the incident angle ϕ_1, using $\cos\phi_1 = k_x/k$. This gives

$$\frac{B_1}{A_1} = \frac{\cos\phi_1 - \sqrt{\cos^2\phi_1 + (n^2-1)}}{\cos\phi_1 + \sqrt{\cos^2\phi_1 + (n^2-1)}} = \frac{\cos\phi_1 - \sqrt{n^2 - \sin^2\phi_1}}{\cos\phi_1 + \sqrt{n^2 - \sin^2\phi_1}} \quad (6.47)$$

and

$$\frac{A_2}{A_1} = \frac{2\cos\phi_1}{\cos\phi_1 + \sqrt{\cos^2\phi_1 + (n^2-1)}} = \frac{2\cos\phi_1}{\cos\phi_1 + \sqrt{n^2 - \sin^2\phi_1}}. \quad (6.48)$$

Exercise 6.6

(a) Show that equation (6.44) can be written as

$$\mathcal{T} = \frac{m_x}{k_x}\left(\frac{A_2}{A_1}\right)^2.$$

[Hint: Express the cosines in terms of the vector components, and use equation (6.36).]

(b) Hence deduce that the reflectivity $\mathcal{R}$ and the transmissivity $\mathcal{T}$ defined in equations (6.43) and (6.44) satisfy $\mathcal{R} + \mathcal{T} = 1$. [Hint: Insert the expressions of equations (6.32) for the ratios B_1/A_1 and A_2/A_1, and use the relations between the wave vectors $\boldsymbol{k}$, $\boldsymbol{l}$ and $\boldsymbol{m}$ derived in the text.]

Exercise 6.7

(a) A light wave is refracted at an interface from air (assume $n_1 = 1$) to glass ($n_2 = 1.5$) with an incident angle $\phi_1 = 30°$. Calculate numerical values for the angle of refraction ϕ_2, the reflectivity $\mathcal{R}$, and the transmissivity $\mathcal{T}$, and check that the result is consistent with $\mathcal{R} + \mathcal{T} = 1$.

(b) A light wave is refracted at an interface from glass ($n_1 = 1.5$) to air ($n_2 = 1$) with an incident angle $\phi_1 = 30°$. Calculate numerical values for the angle of refraction ϕ_2, the reflectivity $\mathcal{R}$, and the transmissivity $\mathcal{T}$, and check that the result is consistent with $\mathcal{R} + \mathcal{T} = 1$.

Exercise 6.8

Consider the case where the incident wave is normal to the interface, which means $\phi_1 = 0$.

(a) Why is it sufficient to consider the $\boldsymbol{i}$ direction in this case?

(b) Show that equations (6.47) and (6.48), obtained for the two-dimensional case, are consistent with equations (6.20) and (6.21), which describe reflection and transmission in the one-dimensional case.

6.3.3 Total reflection

We now address the question of what happens if the argument of the square root in equation (6.39) becomes negative, which is the case for $(k_x/k)^2 < 1 - n^2$. In terms of the angle of incidence ϕ_1, we have $(k_x/k)^2 = \cos^2 \phi_1 = 1 - \sin^2 \phi_1$, so the condition becomes

$$\sin \phi_1 > n = \frac{n_2}{n_1} = \frac{c_1}{c_2}. \tag{6.49}$$

If this holds with $n_1 > n_2$, then equation (6.41) for the refracted angle ϕ_2 becomes $\sin \phi_2 = \sin \phi_1 / n > 1$, for which there is no real solution. So for an incident angle ϕ_1 such that inequality (6.49) is satisfied, there is no real vector $\boldsymbol{m}$. Hence there is *no* refracted wave, which means that there is *no transmission* at all. This is the case of *total reflection*; for $n < 1$, the angle ϕ_{tot} with

Angle of total reflection
$$\sin \phi_{\text{tot}} = n \tag{6.50}$$

is called the *angle of total reflection*. For angles of incidence ϕ_1 such that $\phi_{\text{tot}} < \phi_1 < 90°$, there will be total reflection. Clearly, if $n > 1$, there is no solution to equation (6.50) for ϕ_{tot}, so any incident angle $\phi_1 < 90°$ will lead to a refracted wave. For $n < 1$, the smaller n, the smaller the angle of total reflection.

Total reflection occurs only for $n_1 > n_2$, which for light means that it can take place at an interface only if the light is incident from an optically denser medium (higher refractive index). So while total reflection can occur at an interface from glass to air, it cannot happen at an interface from air to glass.

Total reflection is used, for instance, in binoculars where glass prisms rather than mirrors reflect the light. Another important application is where optical fibres act as wave-guides for the transmission of light. By repeated internal reflection inside an optical fibre, light can effectively be bent in the direction of the fibre.

Exercise 6.9

What is the minimum incident angle for total reflection at each of the following interfaces?
(a) From diamond ($n_1 = 2.42$) to air (assume $n_2 = 1$)
(b) From glass ($n_1 = 1.5$) to air (assume $n_2 = 1$)
(c) From water ($n_1 = 1.33$) to air (assume $n_2 = 1$)
(d) From glass ($n_1 = 1.5$) to water ($n_2 = 1.33$)

For $\sin \phi_1 > n$, we still need to make sense of the complex wave vector $\boldsymbol{m}$ and the imaginary angle ϕ_2 which we obtain formally from equations (6.39) and (6.41), respectively. For the wave vector $\boldsymbol{m} = m_x \mathbf{i} + m_y \mathbf{j}$ of the refracted wave we obtain, from equation (6.39),

$$\begin{aligned} m_x &= \sqrt{k_x^2 + (n^2 - 1)k^2} \\ &= k\sqrt{\cos^2 \phi_1 + n^2 - 1} \\ &= k\sqrt{-\sin^2 \phi_1 + n^2} \\ &= \pm ik\sqrt{\sin^2 \phi_1 - n^2}, \end{aligned} \tag{6.51}$$

and from equation (6.34) we have $m_y = k_y = k \sin \phi_1$. The solution (6.26) in the region $x > 0$, also known as the *evanescent wave*, becomes

$$u(\boldsymbol{r}, t) = A_2 e^{\pm kx\sqrt{\sin^2 \phi_1 - n^2} + ik(\sin \phi_1\, y - c_1 t)}, \tag{6.52}$$

where we have used $mc_2 = kc_1$ from equation (6.36).

6.4 Dispersion

The positive sign in the exponent leads to a solution that increases exponentially with x and hence corresponds to a case with infinite transmitted energy; it does not apply to our situation of an incident wave of finite energy, and thus is discarded.

Inserting the coefficient A_2 from equation (6.48), we obtain

$$u(\boldsymbol{r},t) = \frac{2A_1 \cos\phi_1 e^{-kx\sqrt{\sin^2\phi_1 - n^2}}}{\cos\phi_1 + i\sqrt{\sin^2\phi_1 - n^2}} e^{ik(\sin\phi_1\, y - c_1 t)}. \tag{6.53}$$

This solution decays exponentially with x, and in practice will be very hard to see because the distance over which it decays is of the same order as the wavelength, so after a few wavelengths no wave motion will be detectable. So, even though there is some motion in the region $x > 0$, very close to the interface, no propagating wave enters this region, which is why the term *total reflection* is justified. What happens is that the energy corresponding to the motion in the region $x > 0$ is eventually carried back in the reflected wave, as within our set-up energy is conserved and there is no dissipation of energy.

From equations (6.25) and (6.47), the reflected wave is

$$B_1 e^{i(\boldsymbol{l}\cdot\boldsymbol{r} - kc_1 t)} = A_1 \frac{\cos\phi_1 - i\sqrt{\sin^2\phi_1 - n^2}}{\cos\phi_1 + i\sqrt{\sin^2\phi_1 - n^2}} e^{i(\boldsymbol{l}\cdot\boldsymbol{r} - kc_1 t)}, \tag{6.54}$$

with $|\boldsymbol{l}| = l = k$. The ratio B_1/A_1 of equation (6.47) is now complex, and its modulus is given by

$$\left|\frac{B_1}{A_1}\right|^2 = \left|\frac{\cos\phi_1 - i\sqrt{\sin^2\phi_1 - n^2}}{\cos\phi_1 + i\sqrt{\sin^2\phi_1 - n^2}}\right|^2 = 1, \tag{6.55}$$

because B_1/A_1 has the form z/z^*. Because the modulus of the ratio is unity, the ratio has the form $\exp(i\psi)$, with a real argument ψ that depends on ϕ_1 and n. The reflected plane wave is thus of the form

See Block 0, Subsection 1.2.6.

$$A_1 e^{i(\boldsymbol{l}\cdot\boldsymbol{r} - kc_1 t + \psi)}, \tag{6.56}$$

so it is just a phase-shifted plane wave, with the same amplitude as the incident plane wave.

So for $\sin\phi_1 < n$, we have a real ratio B_1/A_1 with $|B_1/A_1| < 1$, which means that only part of the incident wave is reflected. For $\sin\phi_1 > n$, we have a complex ratio with $|B_1/A_1| = 1$, which corresponds to a phase-shifted total reflection of the incident plane wave. In the complex case, the reflectivity (6.43) and transmissivity (6.44) are each defined by the square of the *modulus* of the coefficient ratio, hence $\mathcal{R} = |B_1/A_1|^2 = 1$ and $\mathcal{T} = 0$, consistent with $\mathcal{R} + \mathcal{T} = 1$.

6.4 Dispersion (optional)

Refraction of waves explains, for instance, why a stick that is partially submerged in water appears to be bent. However, there is another effect that can be observed when light passes through water or glass – it can be split into colours, as in a prism, or in raindrops which then may lead to a beautiful rainbow. The colours that are obtained in this way are called *spectral colours*. The different colours correspond to different frequencies, and hence different wavelengths, of the light. The angle of refraction of light at an interface with water or glass is dependent on the wavelength of the light, and so on its colour. It is this property that explains the splitting of light into its spectral colours. This means that for light waves in glass, the refractive index (and hence the wave speed) depends on the frequency of the wave.

More generally, a medium which can transmit waves, and where the wave velocity depends on the frequency of the wave, is called a *dispersive medium*; the corresponding phenomenon is known as *dispersion*. In the wave equation, the constant parameter c turned out to be the wave velocity; in this case all wave motion described by the wave equation moves with the very same speed c. In reality, however, this is often not true – the behaviour of light waves is just one example. The wave equation with constant wave velocity c applies for light waves moving in a vacuum, and accurately describes light waves moving in air. However, in a dense medium like water or glass, the interaction of the light with the atoms of the medium influences the wave motion, and this leads not only to a reduced light velocity, but also to a frequency-dependence of the velocity of light, and of the refractive index of the medium. So it is the presence of the medium rather than the underlying fundamental physics of light waves that is responsible for dispersion.

6.4.1 Dispersion relation

In deriving the wave equation, we made a number of simplifying approximations, which is why we ended up with a rather simple partial differential equation. Depending on what effects are included, one obtains various wave equations, often involving higher derivatives, which apply to particular physical problems such as, for instance, water waves or stiff strings. Provided that the resulting equation is linear and has constant coefficients, it has plane wave solutions of the form $A\exp[i(kx \pm \omega t)]$, because taking derivatives reproduces the exponential function up to multiplicative factors. One is left with an equation involving k, ω, c (where now $\omega \neq kc$) and any other constant coefficients that enter in the modified wave equation. This equation is called the *dispersion relation*.

We restrict consideration to the one-dimensional case for simplicity.

As an example, consider the equation
$$u_{tt}(x,t) - c^2 u_{xx}(x,t) + \kappa^2 u(x,t) = 0, \tag{6.57}$$
which describes a string that is anchored to its equilibrium position by an elastic force that is proportional to its displacement – similar to the sprung end considered in Chapter 4, but now a force per unit length acts along the whole string. You might think of a string on a rubber sheet as an example. Inserting $u(x,t) = A\exp[i(kx \pm \omega t)]$ in equation (6.57) gives
$$\left(-\omega^2 + c^2 k^2 + \kappa^2\right) A e^{i(kx \pm \omega t)} = 0, \tag{6.58}$$
which shows that $u(x,t) = A\exp[i(kx \pm \omega t)]$ is a solution provided that $\omega^2 = c^2 k^2 + \kappa^2$, which yields the dispersion relation
$$\omega(k) = \sqrt{c^2 k^2 + \kappa^2}, \tag{6.59}$$
where we choose the positive square root because we have $\pm \omega t$ in the exponent of the plane wave solution, and we prefer to work with positive angular frequencies. For $\kappa = 0$, we recover the usual wave equation, and the corresponding dispersion relation $\omega^2 = c^2 k^2$, so ω depends linearly on k.

This equation is derived in Block III. It also arises in quantum mechanics, where it is known as the Klein–Gordon equation.

Note that the dispersion relation (6.59), for real k, implies $\omega^2 \geq \kappa^2$. Real values of k correspond to travelling wave solutions, which means that such solutions exist only for angular frequencies $\omega > \kappa$. (The case $\omega = \kappa$ corresponds to $k = 0$.) For $0 \leq \omega < \kappa$, equation (6.59) can be satisfied only if k is imaginary, which leads to solutions which decay exponentially; compare the discussion about total reflection in Subsection 6.3.3. This means that waves of angular frequencies $0 \leq \omega < \kappa$ cannot be transmitted by the string. So the elastically anchored string obeying equation (6.57) acts like a *filter*: waves of sufficiently high frequencies are allowed to pass, while those with low frequencies are not.

6.4.2 Phase and group velocity

From the form of the solution $A\exp[i(kx \pm \omega t)]$, we find that the velocity of a wave is still given by the ratio ω/k. This velocity, which is called the *phase velocity* $v_{\text{phase}}(k)$, now depends on k. In our string example, using equation (6.59) gives

$$v_{\text{phase}} = \frac{\omega}{k} = \sqrt{\frac{c^2k^2 + \kappa^2}{k^2}} = c\sqrt{1 + \frac{\kappa^2}{c^2k^2}}. \tag{6.60}$$

See the discussion of plane waves in Chapter 5.

This implies that if we consider a wave pulse comprised of a linear combination of many frequencies, the form of the pulse will change during the motion, because the components of different frequencies move at different speeds. Hence it becomes meaningless to define the speed of the wave pulse.

In many applications, we are interested in the velocity of a signal superimposed on a wave, such as a radio or television programme transmitted by electromagnetic waves. This speed can be quite different from the wave speed, as will become clear in what follows. A commonly used method to transmit a signal is by *modulating* a *carrier wave*, and we base our discussion on this example. *Modulation* means that the signal is imposed on a wave by changing its amplitude slowly in comparison to the frequency of the carrier wave – which is known as *amplitude modulation* – or by changing its frequency – which is called *frequency modulation*. On a radio set, you find these abbreviated as 'AM' and 'FM', respectively.

As a simple example, consider the linear combination of two sinusoidal waves

$$u(x,t) = \sin(k_1 x - \omega_1 t) + \sin(k_2 x - \omega_2 t), \tag{6.61}$$

where ω_1, k_1 and ω_2, k_2 both satisfy the dispersion relation (6.59). Equation (6.61) can be written as

$$u(x,t) = 2\cos\left(\frac{k_1 - k_2}{2}x - \frac{\omega_1 - \omega_2}{2}t\right)$$
$$\times \sin\left(\frac{k_1 + k_2}{2}x - \frac{\omega_1 + \omega_2}{2}t\right), \tag{6.62}$$

which we may interpret as a harmonic carrier wave moving with velocity $(\omega_1 + \omega_2)/(k_1 + k_2)$ whose amplitude is modulated by a harmonic wave which has velocity $(\omega_1 - \omega_2)/(k_1 - k_2)$. An example is shown in Figure 6.7; this shows *beats* such as may be observed in coupled pendulums, with which you might be familiar from a previous course. The carrier wave has the higher angular frequency $\omega_1 + \omega_2$.

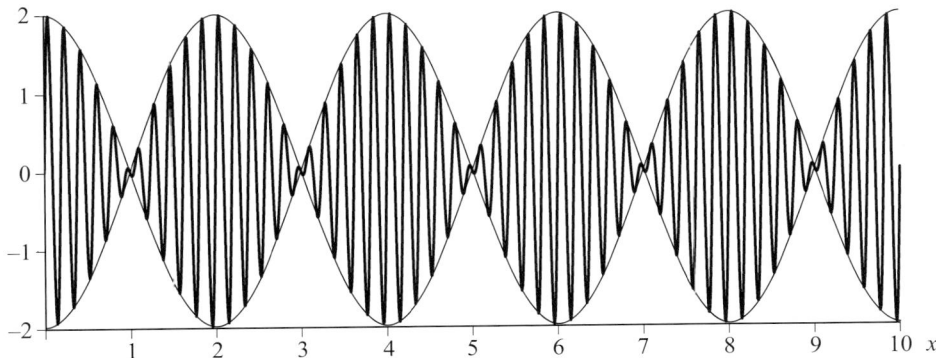

Figure 6.7 A linear combination of two sinusoidal waves. The rapidly oscillating thick line is the graph of $\sin(11\pi x) + \sin(10\pi x)$, and the two thin lines are the graphs of $\pm 2\cos(\pi x/2)$, describing the amplitude modulation; compare equations (6.61) and (6.62), with $k_1 = 11\pi$, $k_2 = 10\pi$ and $t = 0$.

We consider the case where k_1 and k_2 are close to each other. If k_1 and k_2 approach k, so that $\omega_1 = \omega(k_1)$ and $\omega_2 = \omega(k_2)$ approach $\omega(k)$, then the carrier wave moves at the phase velocity $v_{\text{phase}} = \omega(k)/k$, while the modulation moves with the *group velocity*

$$v_{\text{group}}(k) = \lim_{k_1, k_2 \to k} \frac{\omega(k_1) - \omega(k_2)}{k_1 - k_2} = \frac{d\omega}{dk}. \tag{6.63}$$

In many applications, it is this velocity that can be regarded as an effective signal velocity. It determines the speed of slowly varying modulations of a harmonic wave. In our example, equation (6.59) gives

$$v_{\text{group}}(k) = \frac{d\omega}{dk} = \frac{c}{\sqrt{1 + \frac{\kappa^2}{c^2 k^2}}} \tag{6.64}$$

for the group velocity.

6.4.3 A discrete system of coupled oscillators

An interesting example for a wave motion with dispersion can be found in a discrete system consisting of a chain (or lattice) of masses coupled by springs; see Figure 6.8.

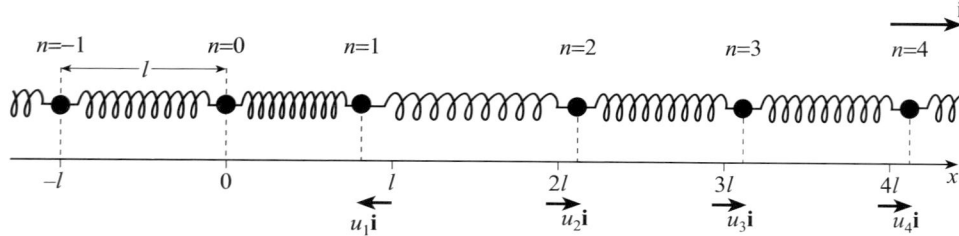

Figure 6.8 A chain of coupled oscillators

The system consists of equal masses m, which at equilibrium are uniformly spaced with lattice spacing l, so they are located at positions $x_n = nl$, $n = \ldots, -3, -2, -1, 0, 1, 2, 3, \ldots$, along the **i** direction. Each pair of neighbouring masses is coupled by a spring. The springs are considered to be identical massless model springs, so they have no mass, natural length l, and elastic modulus K. Systems of this type can be used as simple models for the investigation of elastic waves in crystals.

We denote the displacement of the nth mass in the **i** direction, from an equilibrium arrangement with position $x_n = nl$, by $u_n(t)$. Considering the nth mass, the two attached springs provide linear forces according to Hooke's law. The forces on the nth mass are $\boldsymbol{F}_1 = -\mathbf{i}K(u_n - u_{n-1})/l$ from the spring connecting it to the $(n-1)$th mass, and $\boldsymbol{F}_2 = \mathbf{i}K(u_{n+1} - u_n)/l$ from the spring connecting it to the $(n+1)$th mass. By Newton's second law, the equation of motion for the nth mass is

See Subsection 1.2.2.

See equation (1.8).

$$m \frac{d^2 u_n}{dt^2} \mathbf{i} = \frac{K}{l}(u_{n+1} - u_n)\mathbf{i} + \frac{K}{l}(u_n - u_{n-1})(-\mathbf{i}). \tag{6.65}$$

Resolving in the **i** direction, this becomes

$$m \frac{d^2 u_n}{dt^2} = \frac{K}{l}(u_{n+1} - 2u_n + u_{n-1}). \tag{6.66}$$

Now, because we are dealing with a system that is discrete in space, we do not obtain a partial differential equation. However, you may recognise that the bracketed term on the right-hand side of equation (6.66), in the limit that the lattice spacing l tends to zero, approaches l^2 times a second

6.4 Dispersion

derivative of a function $u(x)$ with respect to $x = nl$. More precisely, define a function $u(x)$ such that $u(nl) = u_n$. Then we have

$$\lim_{l \to 0,\, n \to \infty} \frac{u(nl+l) - 2u(nl) + u(nl-l)}{l^2}$$

$$= \lim_{l \to 0,\, n \to \infty} \frac{\frac{u(nl+l) - u(nl)}{l} - \frac{u(nl) - u(nl-l)}{l}}{l}$$

$$= \lim_{l \to 0,\, n \to \infty} \frac{u_x(nl+l) - u_x(nl)}{l}$$

$$= \lim_{l \to 0,\, n \to \infty} u_{xx}(nl), \tag{6.67}$$

provided that the limit exists. In this case, equation (6.66) can be written as

$$u_{tt} = c^2 u_{xx}, \tag{6.68}$$

i.e. we recover the wave equation with wave speed $c = \lim_{l \to 0} \sqrt{Kl/m}$, provided that this limit exists. This means that in order to arrive at the wave equation with a finite wave speed, the masses m should go to zero as l vanishes, in such a way that the linear density $\rho = m/l$ remains constant. In this sense, you can think of the bracketed term in equation (6.66) as a discrete analogue of the second-order derivative, and equation (6.66) is thus a discrete version of the wave equation.

The analogue of harmonic wave motion in this discrete system is

$$u_n = A e^{i(kx_n - \omega t)} = A e^{i(knl - \omega t)}. \tag{6.69}$$

Inserting this into equation (6.66) yields

$$-m\omega^2 u_n = \frac{K}{l}(u_{n+1} - 2u_n + u_{n-1})$$

$$= \frac{K}{l}\left(e^{ikl} - 2 + e^{-ikl}\right) u_n. \tag{6.70}$$

Using $e^{ix} + e^{-ix} = 2\cos x$ and dividing by u_n gives

$$-m\omega^2 = \frac{2K}{l}[\cos(kl) - 1] = -\frac{4K}{l}\sin^2\frac{kl}{2}, \tag{6.71}$$

hence

$$\omega = \sqrt{\frac{4K}{ml}} \sin\frac{kl}{2} = \omega_{\max}\sin\frac{kl}{2}, \tag{6.72}$$

where $\omega_{\max} = \sqrt{\frac{4K}{ml}}$ is the maximum possible value of ω. This dispersion relation $\omega(k)$ is shown in Figure 6.9.

Expanding the sine function for small arguments, we have

See Block 0, Subsection 1.2.3.

$$\omega = \sqrt{\frac{Kl}{m}} k \left(1 - \frac{(kl)^2}{24} + \cdots\right). \tag{6.73}$$

So, for small kl, we find a phase velocity $v_{\text{phase}} \simeq \sqrt{Kl/m}$, which decreases with increasing k. Because $|\sin\frac{kl}{2}| \leq 1$, the possible angular frequencies are $0 \leq \omega \leq \omega_{\max} = \sqrt{4K/ml}$; higher frequencies cannot be transmitted by this system. So we have the opposite effect from that observed for the elastically anchored string, in that there is a maximum frequency that the system can transmit. Note that the smaller the lattice spacing l, the larger this cut-off frequency becomes. In the limit as l becomes zero, which is when we recover a continuous system, all frequencies can again be transmitted. This is in accordance with our observation that equation (6.66) becomes the wave equation in this limit.

Let us compare the phase and group velocities for this example. The phase velocity

$$v_{\text{phase}} = \frac{\omega}{k} = \sqrt{\frac{Kl}{m}} \frac{\sin(kl/2)}{kl/2} \qquad (6.74)$$

and the group velocity

$$v_{\text{group}} = \frac{d\omega}{dk} = \sqrt{\frac{Kl}{m}} \cos\frac{kl}{2} \qquad (6.75)$$

agree in the limit as $k \to 0$, because $\lim_{x \to 0}(\sin x/x) = 1$. For $k > 0$, the group velocity is smaller than the phase velocity, and vanishes for $k = \pi/l$, which corresponds to the maximum angular frequency $\omega = \omega_{\max} = \sqrt{4K/ml}$.

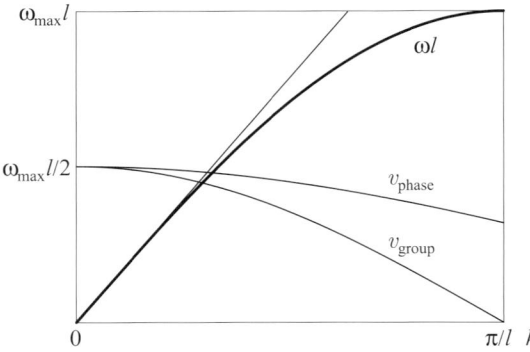

Figure 6.9 Dispersion relation $\omega(k)$ (multiplied by the constant l) and corresponding phase and group velocities for waves in a one-dimensional lattice of coupled oscillators. The thin straight line shows the linear approximation of equation (6.73).

Essentially, what happens is the following. For small k, the wavelength $\lambda = 2\pi/k$ is large compared with the lattice spacing, so the wave motion is not much affected by the discrete nature of the system, and the wave is propagated as in a continuous string. The maximum angular frequency $\omega_{\max} = \sqrt{4K/ml}$ occurs when $k = \pi/l$, i.e. when the wavelength is $\lambda = 2l$. So the shortest wavelength that can be transmitted by the system is such that the lattice spacing corresponds to half the wavelength, which makes sense because it is hard to see how a discrete system with lattice spacing l could possibly sustain a periodic wave motion of a shorter wavelength. The shortest wavelength corresponds, for example, to a motion where all the masses oscillate with equal amplitude, but neighbouring masses oscillate with opposite phases; see Exercise 6.10.

Exercise 6.10

(a) Show that $u_n = Ae^{i(knl-\omega t)}$ becomes $u_n = (-1)^n Ae^{-i\omega t}$ for $k = \pi/l$, which describes a motion where neighbouring masses oscillate with equal amplitudes and opposite phases. Compare equation (6.69).

(b) Hence show that this is a solution of equation (6.66) for $\omega = \pm\omega_{\max}$.

6.5 Summary and Outcomes

In this chapter, we considered some phenomena associated with wave propagation. The first concerned the Doppler effect, which can be explained as a consequence of a wave motion perceived by an observer who moves with respect to the source of the wave. Then we considered the reflection and transmission of waves at interfaces. One of the main results is Snell's law, which describes the refraction of waves at an interface. After that, we briefly discussed the phenomenon of dispersion, which occurs in many situations found in nature.

After working through this chapter, you should:
- understand the origin of the Doppler effect;
- know how to calculate the Doppler shift in frequency;
- be able to calculate reflection and transmission of pulses at interfaces;
- know why waves are refracted at interfaces;
- understand Snell's law of refraction;
- (optional) know what dispersive wave motion means.

There are various wave phenomena that we could also discuss at this stage; however, we stop here to move on to the next block, which introduces another important partial differential equation of mathematical physics, the diffusion equation.

6.6 Further Exercises

The three final exercises are difficult and rather long; you may find them rather challenging. They have been broken into parts to guide you through the argument. The first two exercises consider reflection and transmission at a double interface, the second showing the interesting phenomenon that total reflection is overcome if there is another interface close by, so the evanescent wave can 'tunnel' through the small gap between the interfaces.

> The term 'tunnelling' originates from quantum mechanics, where the analogous effect is very important, for instance to explain radioactive decay of atoms. In optics, this effect is called *frustrated total internal reflection*. Experiments concerning this effect received a lot of attention because the light seemingly passes the gap at superluminal speed; see, for instance, G. Nimtz, 'On superluminal tunneling', *Progress in Quantum Electronics* **27** (2003) pp. 417–50, for details.

The final exercise discusses a discrete system of masses and springs as in Subsection 6.4.3, but now for the case where two different masses occupy alternating positions along the chain.

Exercise 6.11 *This exercise is optional.*

Consider reflection and transmission on a string consisting of three parts, so there are two interfaces. The string has the same properties for $x < 0$ and for $x > 1$, where the wave speed is c. In the region $0 \leq x \leq 1$, the wave speed is smaller, given by c/n where n denotes the refractive index.

Consider a solution of the wave equation of the form

$$u(x,t) = \begin{cases} A_1 e^{ik_1(x-ct)} + B_1 e^{-ik_1(x+ct)}, & x < 0, \\ A_2 e^{ik_2(x-ct/n)} + B_2 e^{-ik_2(x+ct/n)}, & 0 \leq x \leq 1, \\ A_3 e^{ik_3(x-ct)}, & 1 < x, \end{cases}$$

which can be interpreted as the reflection and transmission of an incident wave moving in the positive x direction. The coefficients A_1, B_1, A_2, B_2 and A_3 can, in general, be complex. The meaning of the five amplitudes is shown schematically in Figure 6.10.

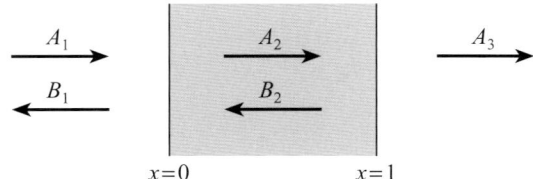

Figure 6.10 Schematic representation of the two-interface arrangement and the amplitudes that occur in the solution

(a) Show that the continuity conditions (6.11) at the interfaces imply $k_1 = k_2/n = k_3$ and $A_1 + B_1 = A_2 + B_2$, $e^{ik} A_3 = e^{ink} A_2 + e^{-ink} B_2$, where we write $k = k_1 = k_3 = k_2/n$ for simplicity. [Hint: Use Exercise 6.4.]

(b) Show that continuity of derivatives with respect to x at the interfaces (see equation (6.13)) implies $A_1 - B_1 = n(A_2 - B_2)$ and $e^{ik} A_3 = n(e^{ink} A_2 - e^{-ink} B_2)$.

(c) Show that this implies $B_2 = e^{2ink} A_2 (n-1)/(n+1)$.

(d) We now have four independent equations for the five unknowns A_1, A_2, A_3, B_1 and B_2. Show that we can express these coefficients in terms of A_1 as

$$A_2 = \frac{2(n+1)}{(n+1)^2 - (n-1)^2 e^{2ink}} A_1,$$

$$B_2 = \frac{2(n-1)e^{2ink}}{(n+1)^2 - (n-1)^2 e^{2ink}} A_1,$$

$$B_1 = \frac{(n^2-1)(e^{2ink}-1)}{(n+1)^2 - (n-1)^2 e^{2ink}} A_1,$$

$$A_3 = \frac{4n e^{i(n-1)k}}{(n+1)^2 - (n-1)^2 e^{2ink}} A_1.$$

(e) Calculate the reflectivity $\mathcal{R} = |B_1/A_1|^2$ and the transmissivity $\mathcal{T} = |A_3/A_1|^2$, and check that $\mathcal{R} + \mathcal{T} = 1$.

(f) For which values of k does the reflected wave amplitude B_1, and hence the reflectivity, vanish? Do you understand what happens in this case? What are the wavelengths of the incident wave and the wave in the region $0 \leq x \leq 1$, and what are the values of A_2, B_2 and A_3 in this situation? Can you interpret the results?

Exercise 6.12 *This exercise is optional.*

With the method described in this chapter, you can analyse what happens when a wave passes through a region of material which lowers its wave speed, such as light being refracted by a glass plate. In this exercise, we look at the opposite situation, where the wave speed c in the plate is higher than outside. You may think of light waves travelling in two glass blocks, of refractive index n_{glass}, which are brought close together such that there is a small gap of width a, filled with air of refractive index 1, as shown in Figure 6.11.

6.6 Further Exercises

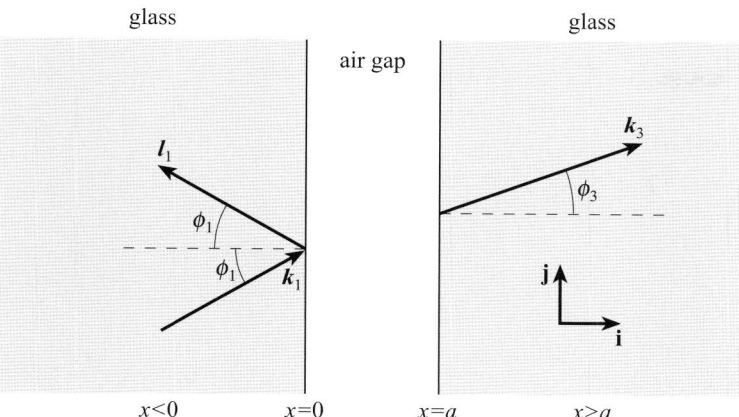

Figure 6.11 Refraction at a gap

We are mainly interested in what happens in the situation where the incident angle of the light wave is larger than the angle of total reflection, to answer the question about whether or not any light is transmitted. From our discussion above, particularly equation (6.53), you might expect that some light will penetrate the air gap, with an amplitude that decreases exponentially with the width a of the gap, and that the presence of the second interface might mean that some, albeit small, part of the wave motion is transmitted.

We consider a solution of the wave equation in the form

$$u(\mathbf{r}, t) = \begin{cases} A_1 e^{i(\mathbf{k}_1 \cdot \mathbf{r} - \omega_1 t)} + B_1 e^{i(\mathbf{l}_1 \cdot \mathbf{r} - \omega_1 t)}, & x < 0, \\ A_2 e^{i(\mathbf{k}_2 \cdot \mathbf{r} - \omega_2 t)} + B_2 e^{i(\mathbf{l}_2 \cdot \mathbf{r} - \omega_2 t)}, & 0 \leq x \leq a, \\ A_3 e^{i(\mathbf{k}_3 \cdot \mathbf{r} - \omega_3 t)}, & a < x, \end{cases}$$

with $\mathbf{k}_p = k_{p,x}\mathbf{i} + k_{p,y}\mathbf{j}$, $\mathbf{l}_p = l_{p,x}\mathbf{i} + l_{p,y}\mathbf{j}$, $p = 1, 2, 3$, and $\omega_1 = k_1 c/n_{\text{glass}} = l_1 c/n_{\text{glass}}$, $\omega_2 = k_2 c = l_2 c$ and $\omega_3 = k_3 c/n_{\text{glass}}$.

(a) What is the condition on the incident angle ϕ_1 for total reflection?

(b) Write down the continuity relations for u and its partial derivative with respect to x at the interfaces $x = 0$ and $x = a$.

(c) As in Exercise 6.11, use Exercise 6.4 and the continuity relations to express the quantities ω_p, $\mathbf{k}_p$, $\mathbf{l}_p$, $p = 1, 2, 3$, in terms of the incident wave data $k = k_1 = |\mathbf{k}_1|$ and $\phi = \phi_1$ as

$$\mathbf{k}_1 = \mathbf{k}_3 = k(\alpha \mathbf{i} + \sin\phi \mathbf{j}),$$
$$\mathbf{l}_1 = k(-\alpha \mathbf{i} + \sin\phi \mathbf{j}),$$
$$\mathbf{k}_2 = k(\beta \mathbf{i} + \sin\phi \mathbf{j}),$$
$$\mathbf{l}_2 = k(-\beta \mathbf{i} + \sin\phi \mathbf{j}),$$

where $\alpha = \cos\phi$ and $\beta = \sqrt{n_{\text{glass}}^{-2} - \sin^2\phi}$.

(d) Use the results of part (c) to simplify the equations for the case $y = t = 0$, and derive an expression for the coefficient A_3 in terms of A_1.

[Hint: Use $\alpha = \cos\phi$ and $\beta = \sqrt{n_{\text{glass}}^{-2} - \sin^2\phi}$ as abbreviations. Proceed by first using the two equations involving A_3, expressing A_3 and B_2 in terms of A_2 only. Then use the two equations involving A_1 and B_1 to express A_1 in terms of A_2 and B_2, which then allows you to obtain an expression for A_2 involving only A_1.]

(e) In the case of total reflection, $\beta = ib$ is imaginary, where we have written $b = \sqrt{\sin^2\phi - n_{\text{glass}}^{-2}}$. Calculate the transmissivity $\mathcal{T} = |A_3/A_1|^2$ of the air gap.

(f) Show that the result of part (e) implies that for large gap width a, $\mathcal{T}$ decays exponentially with a.

(g) Show that in the limit as the gap width a tends to 0, we recover $\mathcal{T} = 1$.

Exercise 6.13 *This exercise is optional.*

In this exercise, we consider the dispersion relation for a chain like the one shown in Figure 6.8, but now consisting of two different particles, of masses $m_1 = \frac{1}{5}$ and $m_2 = \frac{1}{10}$, arranged in an alternating fashion, which are coupled by equal springs of natural length $l = \frac{1}{5}$ and elastic modulus $K = 400$. Again, we denote by u_n the deviation of the position of the nth mass from an equilibrium situation where the masses are at positions $x_n = nl$ with $n = \ldots, -2, -1, 0, 1, 2, \ldots$. We choose n such that even values $n = 2j$ correspond to masses m_1, and odd values $n = 2j+1$ correspond to masses m_2, where $j \in \mathbb{Z}$.

(a) Using Newton's second law, determine the equation of motion for particle $n = 2j$ of mass m_1, and for particle $n = 2j+1$ of mass m_2.

(b) Insert the trial solution given by

$$u_{2j} = A\exp[i(2jkl - \omega t)] \quad \text{and} \quad u_{2j+1} = B\exp[i((2j+1)kl - \omega t)]$$

into these equations. From this, derive two equations for A, B, k and ω.

(c) Eliminate A and B by solving one of the equations for A and inserting the result into the other.

(d) Derive a quadratic equation for ω^2.

(e) Calculate the two positive solutions ω of this equation.

[For each k, there are two such solutions: we speak of two *branches* of the dispersion relation. Traditionally, from the modelling of crystal vibrations, these branches are called the *acoustical branch* (lower frequency branch) and the *optical branch* (higher frequency branch).]

(f) Calculate the values of ω for $k = 0$ and $kl = \pi/2$, and draw a rough sketch of the dispersion relations for $0 \leq k \leq \pi/(2l)$.

(g) Consider the ratio of coefficients A/B. What is the sign of this ratio for each branch? Calculate the values of A/B for $k = 0$. How does the motion compare to that of the uniform system considered in Subsection 6.4.3?

Solutions to Exercises in Chapter 6

Solution 6.1

(a) The signal is emitted at distance x_0 with a wave speed c. The signal thus travels a distance x_0 at speed $c + v$, and arrives at time $t_0 = x_0/c$.

(b) The source emits the next signal at time $t = \tau$, i.e. after a time interval τ. During that time interval, it has moved a distance $v\tau$ towards you. Thus it emits the next signal at a distance $x_1 = x_0 - v\tau$ from you.

(c) The signal emitted at time $t = \tau$ travels the distance x_1 at speed c. It arrives at time
$$t_1 = \tau + \frac{x_1}{c} = \tau + \frac{x_0 - v\tau}{c} = \frac{x_0}{c} + \left(1 - \frac{v}{c}\right)\tau = t_0 + \left(1 - \frac{v}{c}\right)\tau.$$

(d) The time between the arrivals of the signals is
$$\tau_+ = t_1 - t_0 = \left(\frac{c-v}{c}\right)\tau.$$
The frequency ν_+ for you is thus
$$\nu_+ = 1/\tau_+ = \left(\frac{c}{c-v}\right)\nu = \frac{\nu}{1 - v/c}.$$

Solution 6.2

We obtain
$$\mathcal{R} + \mathcal{T} = \alpha^2 + \frac{c_1}{c_2}\beta^2$$
$$= \left(\frac{c_1 - c_2}{c_1 + c_2}\right)^2 + \frac{c_1}{c_2}\left(\frac{2c_2}{c_1 + c_2}\right)^2$$
$$= \frac{(c_1 - c_2)^2 + 4c_1 c_2}{(c_1 + c_2)^2} = \frac{(c_1 + c_2)^2}{(c_1 + c_2)^2}$$
$$= 1.$$

Solution 6.3

(a) We have $\alpha = -(c_1 - c_2)/(c_1 + c_2) = -(1 - \gamma)/(1 + \gamma)$.

(b) The derivative is
$$\frac{d\alpha}{d\gamma} = \frac{1 - \gamma}{(1 + \gamma)^2} + \frac{1}{1 + \gamma} = \frac{2}{(1 + \gamma)^2} > 0,$$
hence α increases monotonically with γ.

(c) Because the wave speeds c_1 and c_2 cannot be negative, we must have $\gamma = c_2/c_1 \geq 0$. Thus $\alpha \geq -1$, the value for $\gamma = 0$, and $\alpha \leq 1$, where α approaches 1 in the limit as $\gamma \to \infty$.

(d) It follows that $0 \leq \mathcal{R} \leq 1$, with $\mathcal{R} = 0$ if $c_1 = c_2$, and $\mathcal{R} = 1$ if one of the wave speeds is zero or infinite.

Solution 6.4

(a) Dividing by e^{ibx} gives $Ae^{i(a-b)x} + B = 0$. Taking the derivative with respect to x yields
$$i(a - b)Ae^{i(a-b)x} = 0,$$
which implies $a - b = 0$, i.e. $a = b$ (since $A \neq 0$).

(b) Differentiating with respect to x gives
$$iaAe^{iax} + ibBe^{ibx} = 0.$$
It follows from the previous result that again $a = b$, unless a or b is zero. However, if $a = 0$, necessarily $b = 0$ as well, and vice versa, so we still have $a = b$.

(c) Dividing by e^{icx} gives
$$Ae^{i(a-c)x} + Be^{i(b-c)x} + C = 0.$$
From the previous result, we find $a - c = b - c$, hence $a = b$.

Dividing by e^{ibx} instead gives
$$Ae^{i(a-b)x} + Ce^{i(c-b)x} + B = 0,$$
which implies $a - b = c - b$, hence $a = c$.

It follows that $a = b = c$.

Solution 6.5

From Figure 6.5, the angles ϕ_1 and ϕ_1' satisfy $\cos\phi_1 = k_x/k$ and $\cos\phi_1' = -l_x/l$. Equations (6.36) and (6.37) yield $\cos\phi_1' = k_x/k = \cos\phi_1$. Similarly, using equation (6.34) one can show that $\sin\phi_1' = l_y/l = k_y/k = \sin\phi_1$, so the angles agree.

Solution 6.6

(a) We have $\cos\phi_1 = k_x/k$ and $\cos\phi_2 = m_x/m$, so equation (6.44) gives
$$\mathcal{T} = \frac{c_1}{c_2}\frac{\cos\phi_2}{\cos\phi_1}\left(\frac{A_2}{A_1}\right)^2 = \frac{c_1}{c_2}\frac{m_x k}{k_x m}\left(\frac{A_2}{A_1}\right)^2 = \frac{m_x}{k_x}\left(\frac{A_2}{A_1}\right)^2,$$
because $k/m = c_2/c_1$ by equation (6.36).

(b) Inserting in equations (6.43) and (6.44) the coefficients from equations (6.32), and using equation (6.37) and part (a), gives
$$\mathcal{R} = \left(\frac{k_x - m_x}{l_x - m_x}\right)^2 = \left(\frac{k_x - m_x}{-k_x - m_x}\right)^2 = \left(\frac{k_x - m_x}{k_x + m_x}\right)^2$$
and
$$\mathcal{T} = \frac{m_x}{k_x}\left(\frac{l_x - k_x}{l_x - m_x}\right)^2 = \frac{m_x}{k_x}\left(\frac{2k_x}{-k_x - m_x}\right)^2$$
$$= \frac{m_x}{k_x}\frac{4k_x^2}{(k_x + m_x)^2} = \frac{4k_x m_x}{(k_x + m_x)^2}.$$

Thus
$$\mathcal{R} + \mathcal{T} = \frac{(k_x - m_x)^2}{(k_x + m_x)^2} + \frac{4k_x m_x}{(k_x + m_x)^2}$$
$$= \frac{k_x^2 - 2k_x m_x + m_x^2 + 4k_x m_x}{(k_x + m_x)^2}$$
$$= \frac{(k_x + m_x)^2}{(k_x + m_x)^2} = 1,$$
as required.

Solution 6.7

(a) Snell's law (6.41) with $n_1 = 1$ and $n_2 = 1.5$ gives
$$\sin\phi_2 = \frac{n_1}{n_2}\sin\phi_1 = \tfrac{2}{3}\sin 30° = \tfrac{1}{3}. \qquad \sin 30° = 1/2$$

Thus $\phi_2 = \arcsin(\tfrac{1}{3})$, which is approximately $19.47°$.

Equations (6.47) and (6.48) with $n = n_2/n_1 = \tfrac{3}{2}$ give
$$\frac{B_1}{A_1} = \frac{\sqrt{3} - 2\sqrt{2}}{\sqrt{3} + 2\sqrt{2}} \simeq -0.2404, \qquad \frac{A_2}{A_1} = \frac{2\sqrt{3}}{\sqrt{3} + 2\sqrt{2}} \simeq 0.7596. \qquad \cos 30° = \sqrt{3}/2$$

Solutions to Exercises in Chapter 6

So, with $\cos\phi_2 = \sqrt{1 - \sin^2\phi_2} = \sqrt{1 - \tfrac{1}{9}} = 2\sqrt{2}/3$, and $c_1/c_2 = n$, we obtain

$$\mathcal{R} = \left(\frac{B_1}{A_1}\right)^2 \simeq 0.0578, \quad \mathcal{T} = \frac{3}{2} \times \frac{4\sqrt{2}}{3\sqrt{3}}\left(\frac{A_2}{A_1}\right)^2 \simeq 0.9422,$$

in agreement with $\mathcal{R} + \mathcal{T} = 1$.

(b) Snell's law (6.41) with $n_1 = 1.5$ and $n_2 = 1$ gives

$$\sin\phi_2 = \frac{n_1}{n_2}\sin\phi_1 = \tfrac{3}{2}\sin 30° = \tfrac{3}{4}.$$

Thus $\phi_2 = \arcsin(\tfrac{3}{4})$, which is approximately $48.59°$.

Equations (6.47) and (6.48) with $n = n_2/n_1 = \tfrac{2}{3}$ give

$$\frac{B_1}{A_1} = \frac{3\sqrt{3} - \sqrt{7}}{3\sqrt{3} + \sqrt{7}} \simeq 0.3252, \quad \frac{A_2}{A_1} = \frac{6\sqrt{3}}{3\sqrt{3} + \sqrt{7}} \simeq 1.3252.$$

So, with $\cos\phi_2 = \sqrt{1 - \sin^2\phi_2} = \sqrt{1 - \tfrac{9}{16}} = \sqrt{7}/4$, we obtain

$$\mathcal{R} = \left(\frac{B_1}{A_1}\right)^2 \simeq 0.1058, \quad \mathcal{T} = \frac{2}{3} \times \frac{\sqrt{7}}{2\sqrt{3}}\left(\frac{A_2}{A_1}\right)^2 \simeq 0.8942,$$

in agreement with $\mathcal{R} + \mathcal{T} = 1$.

Solution 6.8

(a) As $k_y = 0$, equation (6.34) implies $k_y = l_y = m_y = 0$. Hence all wave vectors are in the **i** direction, so it suffices to consider this direction.

(b) With $\phi_1 = 0$, equations (6.47) and (6.48) become

$$\frac{B_1}{A_1} = \frac{1-n}{1+n} = \frac{1 - c_1/c_2}{1 + c_1/c_2} = \frac{c_2 - c_1}{c_2 + c_1}$$

and

$$\frac{A_2}{A_1} = \frac{2}{1+n} = \frac{2}{1 + c_1/c_2} = \frac{2c_2}{c_2 + c_1}.$$

These are the coefficients α and β of equations (6.20) and (6.21), respectively.

Solution 6.9

The incident angle ϕ_1 must exceed ϕ_{tot}, so ϕ_1 must be larger than $\arcsin(n) = \arcsin(n_2/n_1)$.

(a) $\phi_1 > \arcsin(1/2.42)$, which gives $\phi_1 \gtrsim 24.4°$.

(b) $\phi_1 > \arcsin(1/1.5)$, which gives $\phi_1 \gtrsim 41.8°$.

(c) $\phi_1 > \arcsin(1/1.33)$, which gives $\phi_1 \gtrsim 48.8°$.

(d) $\phi_1 > \arcsin(1.33/1.5)$, which gives $\phi_1 \gtrsim 62.5°$.

Solution 6.10

(a) For $k = \pi/l$, we obtain

$$u_n = Ae^{i(knl - \omega t)} = Ae^{i(\pi n - \omega t)} = Ae^{i\pi n}e^{-i\omega t} = (-1)^n A e^{-i\omega t}.$$

(b) Inserting this expression into equation (6.66) yields

$$m(-1)^n A(-i\omega)^2 e^{-i\omega t} = \frac{K}{l}[-(-1)^n - 2(-1)^n - (-1)^n]Ae^{-i\omega t}$$

$$= -\frac{4K}{l}(-1)^n A e^{-i\omega t}.$$

Dividing both sides by u_n gives

$$-m\omega^2 = -\frac{4K}{l},$$

so u_n is a solution of equation (6.66) if

$$\omega^2 = \frac{4K}{ml},$$

which means $\omega = \pm\sqrt{4K/ml} = \pm\omega_{\max}$.

Solution 6.11

(a) At $x=0$, continuity of u requires
$$(A_1 + B_1)e^{-ik_1 ct} = (A_2 + B_2)e^{-ik_2 ct/n}$$
to hold for all t, so $k_2 = nk_1$ by Exercise 6.4, and $A_1 + B_1 = A_2 + B_2$ by setting $t = 0$.

At $x = 1$, we find
$$(e^{ik_2}A_2 + e^{-ik_2}B_2)e^{-ik_2 ct/n} = e^{ik_3}A_3 e^{-ik_3 ct},$$
which implies, by Exercise 6.4, $k_3 = k_2/n$, thus $k_1 = k_3 = k$ and $k_2 = nk$. For $t = 0$, the equation reduces to $e^{ink}A_2 + e^{-ink}B_2 = e^{ik}A_3$.

(b) Continuity of the derivatives at $x = 0$ and $x = 1$, evaluated at $t = 0$, yields $ik_1(A_1 - B_1) = ik_2(A_2 - B_2)$ and $ik_2(e^{ik_2}A_2 - e^{-ik_2}B_2) = ik_3 e^{ik_3}A_3$. The required relations follow by inserting $k_1 = k_2/n = k_3$.

(c) Equating the two expressions for $e^{ik}A_3$ gives
$$e^{ink}A_2 + e^{-ink}B_2 = n(e^{ink}A_2 - e^{-ink}B_2).$$
Solving for B_2 gives the desired result.

(d) Inserting the expression for B_2 into the two equations that involve A_1 and B_1 gives
$$A_1 + B_1 = \left(1 + \frac{n-1}{n+1}e^{2ink}\right)A_2,$$
$$A_1 - B_1 = \left(1 - \frac{n-1}{n+1}e^{2ink}\right)nA_2.$$
Taking the sum and solving for A_2 yields the desired expression for A_2. The result for B_2 then follows from that. Next, B_1 is obtained by taking the difference of the two equations above, solving for B_1, and expressing A_2 in terms of A_1. Finally, for A_3 we use one of the equations that relates A_3 to A_2 and B_2, and express the latter two variables in terms of A_1.

(e) The reflectivity is
$$\mathcal{R} = \left|\frac{B_1}{A_1}\right|^2$$
$$= \left|\frac{(n^2-1)(e^{2ink}-1)}{(n+1)^2 - (n-1)^2 e^{2ink}}\right|^2$$
$$= \frac{(n^2-1)^2(e^{2ink}-1)(e^{-2ink}-1)}{[(n+1)^2 - (n-1)^2 e^{2ink}][(n+1)^2 - (n-1)^2 e^{-2ink}]}$$
$$= \frac{(n^2-1)^2(1-\cos(2nk))}{n^4 + 6n^2 + 1 - (n^2-1)^2 \cos(2nk)}$$
$$= \frac{(n^2-1)^2 \sin^2(nk)}{4n^2 + (n^2-1)^2 \sin^2(nk)},$$

For complex z, we use the result $|z|^2 = z\,z^*$.

where the last step follows by using the trigonometric identity $\cos 2x = \cos^2 x - \sin^2 x = 1 - 2\sin^2 x$ in the numerator and in the denominator.

The transmissivity is
$$\mathcal{T} = \left|\frac{A_3}{A_1}\right|^2$$
$$= \left|\frac{4ne^{i(n-1)k}}{(n+1)^2 - (n-1)^2 e^{2ink}}\right|^2$$
$$= \frac{16n^2 e^{i(n-1)k} e^{-i(n-1)k}}{[(n+1)^2 - (n-1)^2 e^{2ink}][(n+1)^2 - (n-1)^2 e^{-2ink}]}$$
$$= \frac{4n^2}{4n^2 + (n^2-1)^2 \sin^2(nk)},$$
which satisfies $\mathcal{R} + \mathcal{T} = 1$.

(f) Of course, the reflected amplitude B_1 vanishes for $n = 1$ when there is no interface, but we are not interested in this case. For $n \neq 1$, B_1 vanishes for those wave numbers k that satisfy $e^{2ink} = 1$, i.e. for $k = j\pi/n$ with integer j. The wavelengths in the regions are $\lambda_1 = \lambda_3 = 2\pi/k = 2n/j$ and $\lambda_2 = 2\pi/k_2 = 2\pi/(nk) = \lambda_1/n = 2/j$. So the length of the interval $0 \leq x \leq 1$ is a half-integer multiple of the wavelength λ_2. The reflectionless transmission is a resonance phenomenon; it happens whenever the wave 'fits' perfectly in the interval $0 \leq x \leq 1$, which means when we have a standing wave solution at the particular value of the wave number k. For the coefficients, we find $B_1 = 0$, $A_2 = (n+1)/(2n)$, $B_2 = (n-1)/(2n)$ and $A_3 = e^{i(n-1)k} = e^{i\pi j(n-1)/n}$, which is a pure phase factor, i.e. $|A_3| = 1$. So there is no reflection, and complete transmission apart from a change in the phase of the wave motion, which depends on n and on j.

Solution 6.12

(a) From equation (6.49), with $n_1 = n_{\text{glass}}$ and $n_2 = 1$, we find the condition $\sin \phi_1 > 1/n_{\text{glass}}$ for total reflection.

(b) The four continuity relations are

$$A_1 e^{i(k_{1,y}y - \omega_1 t)} + B_1 e^{i(l_{1,y}y - \omega_1 t)}$$
$$= A_2 e^{i(k_{2,y}y - \omega_2 t)} + B_2 e^{i(l_{2,y}y - \omega_2 t)},$$

$$ik_{1,x} A_1 e^{i(k_{1,y}y - \omega_1 t)} + il_{1,x} B_1 e^{i(l_{1,y}y - \omega_1 t)}$$
$$= ik_{2,x} A_2 e^{i(k_{2,y}y - \omega_2 t)} + il_{2,x} B_2 e^{i(l_{2,y}y - \omega_2 t)},$$

$$A_2 e^{i(k_{2,x}a + k_{2,y}y - \omega_2 t)} + B_2 e^{i(l_{2,x}a + l_{2,y}y - \omega_2 t)}$$
$$= A_3 e^{i(k_{3,x}a + k_{3,y}y - \omega_3 t)},$$

$$ik_{2,x} A_2 e^{i(k_{2,x}a + k_{2,y}y - \omega_2 t)} + il_{2,x} B_2 e^{i(l_{2,x}a + l_{2,y}y - \omega_2 t)}$$
$$= ik_{3,x} A_3 e^{i(k_{3,x}a + k_{3,y}y - \omega_3 t)}.$$

(c) Using Exercise 6.4, we can immediately deduce $\omega_1 = \omega_2 = \omega_3$ and $k_{1,y} = l_{1,y} = k_{2,y} = l_{2,y} = k_{3,y}$. Equality of the angular frequencies also implies $k_1 = l_1 = k_3 = n_{\text{glass}} k_2 = n_{\text{glass}} l_2$ for the lengths of the vectors $\boldsymbol{k}_p$ and $\boldsymbol{l}_p$. Hence $\boldsymbol{k}_1 = \boldsymbol{k}_3$, so unless $A_3 = 0$, we have transmission with the same wave vector as the incident wave. The missing components of the wave vectors are determined by the conditions $k_{p,x} > 0$ and $l_{p,x} < 0$, which yield $l_{1,x} = -k_{1,x}$ and $l_{2,x} = -k_{2,x}$. Finally, using $k_1^2 = k_{1,x}^2 + k_{1,y}^2$ and $k_1^2/n_{\text{glass}}^2 = k_2^2 = k_{2,x}^2 + k_{2,y}^2 = k_{2,x}^2 + k_{1,y}^2$, we can calculate

$$k_{2,x} = \sqrt{\frac{1 - n_{\text{glass}}^2}{n_{\text{glass}}^2} k_1^2 + k_{1,x}^2} = \sqrt{\frac{1 - n_{\text{glass}}^2}{n_{\text{glass}}^2} k_1^2 + \cos^2 \phi_1 \, k_1^2}$$
$$= \sqrt{n_{\text{glass}}^{-2} - \sin^2 \phi_1} \, k_1.$$

Using $k = k_1$, $\phi = \phi_1$, $\alpha = \cos \phi$ and $\beta = \sqrt{n_{\text{glass}}^{-2} - \sin^2 \phi}$, this gives the desired expressions.

(d) Inserting the results of part (c) into the continuity relations for $y = t = 0$ gives

$$A_1 + B_1 = A_2 + B_2,$$
$$\alpha(A_1 - B_1) = \beta(A_2 - B_2),$$
$$A_3 e^{i\alpha ka} = A_2 e^{i\beta ka} + B_2 e^{-i\beta ka},$$
$$\alpha A_3 e^{i\alpha ka} = \beta(A_2 e^{i\beta ka} - B_2 e^{-i\beta ka}).$$

The last two equations, upon eliminating A_3, give

$$\alpha(A_2 e^{i\beta ka} + B_2 e^{-i\beta ka}) = \beta(A_2 e^{i\beta ka} - B_2 e^{-i\beta ka}),$$

which yields

$$B_2 = \frac{\beta - \alpha}{\beta + \alpha} e^{2i\beta ka} A_2.$$

Adding the same two equations after multiplication by β gives

$$(\beta + \alpha)A_3 e^{i\alpha ka} = 2\beta A_2 e^{i\beta ka},$$

thus

$$A_3 = \frac{2\beta}{\beta + \alpha} e^{i(\beta - \alpha)ka} A_2.$$

Combining the first two equations that involve A_1 and B_1, we find

$$2\alpha A_1 = (\beta + \alpha)A_2 - (\beta - \alpha)B_2 = \left(\beta + \alpha - \frac{(\beta - \alpha)^2}{\beta + \alpha} e^{2i\beta ka}\right) A_2,$$

which we can solve to give

$$A_2 = \frac{2\alpha(\beta + \alpha)}{(\beta + \alpha)^2 - (\beta - \alpha)^2 e^{2i\beta ka}} A_1.$$

This finally gives

$$A_3 = \frac{4\alpha\beta e^{-i\alpha ka}}{(\beta + \alpha)^2 e^{-i\beta ka} - (\beta - \alpha)^2 e^{i\beta ka}} A_1.$$

(e) With $\beta = ib$ and $\phi_3 = \phi_1$, the transmissivity is

$$\mathcal{T} = \frac{4ib\alpha e^{-i\alpha ka}}{(ib + \alpha)^2 e^{bka} - (ib - \alpha)^2 e^{-bka}} \times \frac{-4ib\alpha e^{i\alpha ka}}{(-ib + \alpha)^2 e^{bka} - (-ib - \alpha)^2 e^{-bka}}$$

$$= \frac{8b^2\alpha^2}{(\alpha^2 + b^2)^2 \cosh(2bka) - b^4 + 6b^2\alpha^2 - \alpha^4}.$$

(f) For large a, we have $2\cosh(2bka) = \exp(2bka) + \exp(-2bka) \simeq \exp(2bka)$, and the exponential term dominates, to give

$$\mathcal{T} \simeq \frac{4b^2\alpha^2}{(b^2 + \alpha^2)^2} e^{-2bka}.$$

This shows that there is transmission of light through the system even for incident angles which are larger than the angle of total reflection, and that the transmissivity decays exponentially with the gap width a.

(g) For $a = 0$, we obtain

$$\mathcal{T} = \frac{8b^2\alpha^2}{(\alpha^2 + b^2)^2 - b^4 + 6b^2\alpha^2 - \alpha^4} = 1.$$

Solution 6.13

(a) The equation of motion follows in complete analogy to the case of a uniform chain. There is no difference between the forces acting on the two different masses, because the force depends only on the extension of the springs, which we assumed to be identical. Thus the equations of motion, resolved in the **i** direction, are

$$\frac{d^2 u_{2j}}{dt^2} = \frac{K}{lm_1}(u_{2j+1} - 2u_{2j} + u_{2j-1}) = 10\,000\,(u_{2j+1} - 2u_{2j} + u_{2j-1}),$$

$$\frac{d^2 u_{2j+1}}{dt^2} = \frac{K}{lm_2}(u_{2j+2} - 2u_{2j+1} + u_{2j}) = 20\,000\,(u_{2j+2} - 2u_{2j+1} + u_{2j});$$

compare equation (6.66).

(b) Inserting the given trial solutions gives

$$0 = \exp\left(i\frac{2jk - 5\omega t}{5}\right)\left(A(20\,000 - \omega^2) - 10\,000\,B\left[\exp\left(\frac{ik}{5}\right) + \exp\left(-\frac{ik}{5}\right)\right]\right),$$

$$0 = \exp\left(i\frac{2(j+1)k - 5\omega t}{5}\right)\left(B(40\,000 - \omega^2) - 20\,000\,A\left[\exp\left(\frac{ik}{5}\right) + \exp\left(-\frac{ik}{5}\right)\right]\right).$$

This yields the two equations

$$A(20\,000 - \omega^2) - 20\,000\,B\cos(k/5) = 0,$$
$$B(40\,000 - \omega^2) - 40\,000\,A\cos(k/5) = 0.$$

(c) Solving the first equation for A gives
$$A = \frac{20\,000\, B \cos(k/5)}{20\,000 - \omega^2}.$$
Using this result, the second equation becomes, upon dividing by B,
$$40\,000 - \omega^2 - \frac{8 \times 10^8 \cos^2(k/5)}{20\,000 - \omega^2} = 0.$$

(d) Multiplying through by $20\,000 - \omega^2$ gives
$$\omega^4 - 60\,000\, \omega^2 + 8 \times 10^8 [1 - \cos^2(k/5)]$$
$$= \omega^4 - 60\,000\, \omega^2 + 8 \times 10^8 \sin^2(k/5) = 0.$$

(e) The solutions of the quadratic equation are
$$\omega^2 = 10\,000 \left(3 \pm \sqrt{9 - 8\sin^2(k/5)}\right) = 10\,000 \left(3 \pm \sqrt{5 + 4\cos(2k/5)}\right).$$

In the last step, we used the trigonometric identity $\cos 2x = \cos^2 x - \sin^2 x = 1 - 2\sin^2 x$.

The positive solutions are
$$\omega = 100 \sqrt{3 \pm \sqrt{5 + 4\cos(2k/5)}}.$$

(f) For $k = 0$, we obtain $\omega = 100\sqrt{3 \pm 3}$, so the solutions are
$$\omega = 0 \quad \text{and} \quad \omega = 100\sqrt{6} \simeq 244.95.$$
For $k = \pi/(2l) = 5\pi/2$, we obtain $\omega = 100\sqrt{3 \pm 1}$, which gives
$$\omega = 100\sqrt{2} \simeq 141.42 \quad \text{and} \quad \omega = 200.$$
The resulting dispersion relations are shown in Figure 6.12.

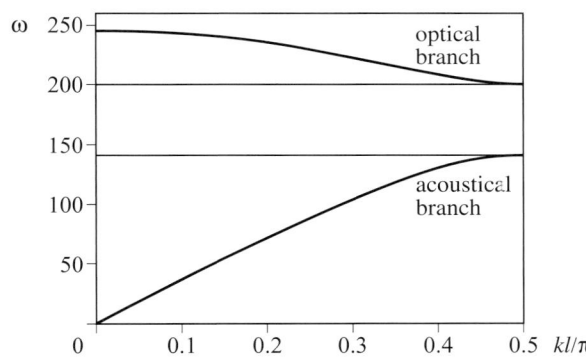

Figure 6.12 The two branches of the dispersion relation

(g) The ratio is given by
$$\frac{A}{B} = \frac{20\,000 \cos(k/5)}{20\,000 - \omega^2}.$$
As $1 \geq \cos(k/5) \geq 0$ for $0 \leq k \leq 5\pi/2$, the denominator determines the sign of this ratio. For the acoustical branch we have $0 \leq \omega^2 \leq 20\,000$, so the sign is positive; for the optical branch we have $40\,000 \leq \omega^2 \leq 60\,000$, so the sign is negative.

For $k = 0$, we have $A = B$ for the acoustical branch, and $A = -B/2$ for the optical branch. Essentially, this means that the motion in the acoustical branch is similar to that of the uniform system, the masses moving in phase with the same amplitude. In contrast, the optical branch corresponds to a motion where the two types of masses are moving with opposite phases, m_1 with half the amplitude of m_2.

INDEX

acceleration 20, 21, 35
    due to gravity 14
    of spring element 30
acoustical branch 254
aeroplane, supersonic 234
air 41, 157, 244
    frequencies in 44
    refractive index 242
    speed of sound in 9, 31, 44, 58, 63
air gap (refraction) 253
aluminium 14
AM 247
ambient pressure 40
amplitude 62, 112
    complex 113
amplitude modulation (AM) 247
angle of incidence 241
angle of reflection 241
angle of refraction 241
angle of total reflection 244
angular eigenfrequency 188
angular frequency 61
argon 41
atmospheric pressure 40, 44

beats 247
Bernoulli, Daniel 36
Bernoulli, Johann 36
Bessel, Friedrich Wilhelm 198
Bessel differential equation 198
Bessel functions 198
    and the Fourier transform 214
    behaviour for large arguments 201
    behaviour for small arguments 200
    of the first kind 199, 201, 203
    of the second kind 199
    power series 220
    zeros of 202, 203
blurred image 122, 123, 126
boundary condition 22, 59, 68–71, 79
    at an interface 236, 237
    for a damped end 150, 151
    for a fixed end 144
    for a free end 149, 150
    for a sprung end 151
    inhomogeneous 157
boundary-value problem 59, 75, 80

carrier wave 247
catenary 39
Cauchy, Augustin Louis 60
Cauchy problem 60
chain of coupled oscillators 248, 250, 254
characteristic function 114, 115, 127
circular membrane 196
    eigenmodes 198, 204
    nodal lines 198, 204
    normal modes 203
coefficients of Fourier series 82, 83
    complex exponential 106

colour 8, 245
commensurate eigenmodes 191
complementary function 158
complex amplitude 113
complex conjugate 24, 107
complex exponential Fourier series 106
complex solution 23–24, 85
complex wave vector 244
compression wave 28, 42
concrete 14
conjugate variable 113, 128
continuity at an interface 236
contour plots
    rectangular membrane 188
    triangular membrane 194
convolution 122, 123, 132
    discrete 123
convolution theorem 124, 132
Cotes, Roger 106
coupled oscillators 81, 248, 250, 254
critical damping 162
crystal
    elastic waves in 248
    vibration in 254
cylinder functions 198
cylindrical harmonics 198

d'Alembert, Jean le Rond 36
d'Alembert's solution 64, 65, 67, 83, 207
damped wave equation 161, 163
    inhomogeneous 165
damping 150, 160
    critical 162
    strong 162
    weak 162, 163
damping constant 151
definite integral 116
degeneracy 190
    degree of 198
degenerate eigenvalue 190
degree of degeneracy 198
density 18
    linear 18
    of a gas 42
    of a membrane 186
    of a rod 18
    of a spring 18, 26, 27, 29
    of a string 18, 165
    planar 186
    volume 14, 18, 42
density wave 42, 234
diamond 244
    refractive index 242
differential equation 36, 77, 103, 105
    Fourier transform of 128
Dirichlet, Johann Peter Gustav Lejeune 59
Dirichlet condition 59, 186
discrete convolution 123
dispersion 245, 246

interface   234
inverse Fourier transform   128
irrational number   189
isosceles triangular membrane   194
isospectral drums   206
isotropic system   186, 206, 207
isotropic wave equation   206
    three-dimensional   207, 215, 216
    two-dimensional   186, 213, 215

Kac, Mark   205
kinetic energy   9
Klein–Gordon equation   246
Kronecker delta symbol   82, 107

Lagrange, Joseph-Louis   36
Laplace, Pierre-Simon   36
Laplacian operator   207
lattice of coupled oscillators   248, 250
law of reflection   241
law of refraction   241
left-moving pulse   69, 73
lens   242
light wave   7
    in a vacuum   231, 246
    in glass   245
    refraction   242
    speed of   9, 63, 242
linear combination   10
linear density   18, 26–28
linear equation   10, 41
linear regime   14
linear wave   11
longitudinal wave   28–34
Lorentz, Hendrik Antoon   114
Lorentzian function   114, 127

mass   14, 18
    of a spring   15, 30
    of a string   21
matrices   80
Maxwell's equations   35
mechanical wave   7, 11
membrane   185
    circular   196, 198, 203, 204
    rectangular   186, 192, 193
    semi-circular   219
    square   190, 191
    triangular   194
Mersenne, Marin   36
method of images   71
Mexican wave   9
microwave oven   63
mirror pulse   70, 73
mode   61, 62, 81
model spring   15
modulation   247
modulus of elasticity   14
multi-dimensional Fourier transform   131
multi-dimensional wave motion   211
multiple reflections   73, 75
music   36, 62, 157

natural length   15
negative tension   17
neon   41
Neumann, Carl Gottfried   150
Neumann condition   150
Neumann functions   199
Newton's laws of mechanics   11
Newton's second law   20, 21, 30, 35, 43
nitrogen   41
nodal lines   188
    circular membrane   198, 204
    rectangular membrane   188, 192
    square membrane   190
    triangular membrane   194
node   23, 81
non-periodic function   105
non-uniform spring   28
non-uniformly deformed spring   24–28
normal mode   77, 81, 152
    circular membrane   203
    finite string   61, 80
    rectangular membrane   188, 189
    string with one free end   153
    string with two free ends   154
    superposition   81

odd periodic extension   146, 180
optical branch   254
optical fibre   244
organ pipe   63
orthogonality   82
orthogonality relation   155
oxygen   41

partial derivative notation   8
partial differential equation   8, 36, 60
    inhomogeneous   165
    separation of variables   77, 78
partial sum of a Fourier series   104
particle   11
particular integral   158
particular solution   158
period   61
periodic extension   105, 115
    even   156, 180
    odd   146, 180
periodic function   104, 196
periodic motion   61, 62, 76
phase   112
phase velocity   247, 248
pipe organ   63
piston   40–42
pitch   62, 63, 162
    Doppler effect   231, 233
plane wave   208
plastic deformation   13
plucked string   19, 62, 144, 162
polystyrene   14
positive tension   17
power series   105
    for Bessel functions   220
pre-tensioned spring   16

pressure   40–41
pressure wave   42
proportional limit   13
pulse   66, 69, 71
Pythagoras   62
Pythagorean comma   62

quantum mechanics   11

radio wave   7, 63
real part of a complex solution   23, 24
rectangular membrane   186, 192, 193, 219
    contour plots   188
    eigenmodes   188, 189, 192
    initial-value problem   192
    nodal lines   188, 192
    normal modes   188, 189
reflected wave   237
reflection   11, 68, 70, 71
    angle   241
    at an interface   235, 236, 238, 241
    law of   241
    total   244
reflectivity   238, 242
refraction   11, 241
    angle   241
    at an air gap   253
    law of   241
refractive index   242
relative motion   233
resonance   160, 169
right-moving pulse   71, 75
ring on a string   149, 150
rod   13–14, 16, 31, 35
    density of   18
    force on   13
    wave speed in   30
rotationally symmetric function   211
    three-dimensional   217
    two-dimensional   214
rotationally symmetric initial condition   215

scalar wave   185
seismic wave   235
semi-circular membrane   219
semi-infinite string   68, 69, 72, 84
separation of variables   76, 77, 161
sign function   109
simple harmonic motion   79
sinc function   111, 216
single-frequency driving   167
sinusoidal wave   61
Snell, Willebrord van Roijen   235
Snell's law of refraction   241
solution of the wave equation   22–24, 59, 72
sonic boom   234
sound detector   235
sound wave   7, 10, 36, 62
    Doppler effect   232
    from a drum   189
    from a string   19, 162
    from an aeroplane   234
    in a gas   40, 42, 43

    in a rod   31
    in a tube   40–44
    in a tunnel   10
    in a vacuum   11
    in air   9, 58, 63
    in an organ pipe   63
    in helium   58
spectral colours   245
spectrum of eigenvalues   80, 81, 189, 205
    degeneracies in   192
    for isospectral drums   206
speed of sound   9
    in a gas   43
    in a rod   31
    in air   44, 63
    in helium   44, 58
spherical coordinates   216
spherically symmetric initial condition   216
spring   15–16, 24–33, 38, 151, 152
    constant   152
    cut   17
    density of   18, 26, 29
    double   16, 37
    force in   15, 30
    longitudinal motion   28
    longitudinal wave in   30–32, 38
    mass of   30
    model   15
    non-uniform   28
    non-uniformly deformed   24, 25, 27, 49, 50
    parameters   15
    suspended   24, 53, 54
    tension in   16, 17, 26, 27, 29
    uniform   15
sprung end   151
square membrane   190
    eigenmodes   190, 191
    nodal lines   190
square wave function   109, 110, 112, 116
square-shaped pulse   70, 171
    odd periodic extension   180
standing wave   23
steel   14, 31
stiffness   15
strain   13, 14
stress   13, 14
string   19–24, 34, 39
    density of   18
    in three parts   251
    interface   235
    plucked   144, 162
    with a damped end   151
    with moving ends   149
    with one fixed and one free end   152
    with two free ends   154
string of beads   36
strong damping   162
superposition
    of modes   62
    of normal modes   81
    of waves   10
surface tension   186

Index

Taylor expansion   20, 35, 43
Taylor series   105
tension   16–17
    in a spring   16, 17, 26, 27, 29
    in a string   17, 20, 34, 36
three-dimensional isotropic wave equation   207, 215, 216
three-dimensional rotationally symmetric function   217
three-dimensional wave equation   207, 215, 216
time-delay process   123
top-hat function   114
total reflection   244, 245
    angle   244
transient   167
transmission at an interface   235, 236, 238, 241
transmissivity   238, 242
transmitted wave   237
transverse wave   19–24, 34
travelling wave   23
triangular membrane   194, 219
    contour plots   194
    eigenmodes   194
    nodal lines   194
triangular wave function   111, 112, 144
    even periodic extension   156, 180
    odd periodic extension   147
trigonometric Fourier series   36, 81, 104, 105, 108
trigonometric identities   83
trigonometric series   105
trivial solution   22
two-dimensional isotropic wave equation   186, 213, 215
two-dimensional rotationally symmetric function   214
two-dimensional wave equation   186, 213, 215
two-interface system   252

uneven mark-space ratio   110
uniform spring   15
unit vector   15

vacuum
    light wave in   231, 242, 246

refractive index   242
    sound wave in   11
vibrating string   19, 36
visible light   63
volume   40–41
volume density   14, 18, 41, 42
volume element   211, 216
    in spherical coordinates   216

water   235
    dispersion in   246
    refraction in   11, 245
    refractive index   242
water wave   7, 9, 10, 35, 232
wave energy   238
wave equation   35–36
    for a rod   30
    for a spring   30, 32
    for a string   40
    for sound in a gas   43
    isotropic   206
    one-dimensional linear   8, 22, 59
    three-dimensional isotropic   207, 215, 216
    two-dimensional isotropic   186, 213
wave node   23
wave number   62, 113, 152
wave speed   9, 44
    in a gas   43
    in a rod   30
    in a spring   32, 33
    in a string   34
    light   9
    sound   9
    transverse   22
wave vector   207, 208
    complex   244
wavelength   44, 62, 245
weak damping   162, 163
Werckmeister, Andreas   63
wood   14

yield point   13, 14
Young's modulus   14